CW00692554

Atmospheric Modelling and Millimetre Wave Propagation

Atmospheric Modelling and Millimetre Wave Propagation

G. Brussaard, Eindhoven University of Technology, The Netherlands
and
P.A. Watson, University of York, UK

with major contributions from
L. Leyten and A. D. Papatsoris

CHAPMAN & HALL
London · Glasgow · Weinheim · New York · Tokyo · Melbourne · Madras

Published by Chapman & Hall, 2–6 Boundary Row, London SE1 8HN, UK

Chapman & Hall, 2–6 Boundary Row, London SE1 8HN, UK

Blackie Academic & Professional, Wester Cleddens Road, Bishopbriggs, Glasgow G64 2NZ, UK

Chapman & Hall GmbH, Pappelallee 3, 69469 Weinheim, Germany

Chapman & Hall USA, One Penn Plaza, 41st Floor, New York NY 10119, USA

Chapman & Hall Japan, ITP-Japan, Kyowa Building, 3F, 2-2-1 Hirakawacho, Chiyoda-ku, Tokyo 102, Japan

Chapman & Hall Australia, Thomas Nelson Australia, 102 Dodds Street, South Melbourne, Victoria 3205, Australia

Chapman & Hall India, R. Seshadri, 32 Second Main Road, CIT East, Madras 600 035, India

---

First edition 1995

Typeset in 10/12pt Times by Keytec Typesetting Ltd, Bridport, Dorset
Printed in Great Britain by St Edmundsbury Press Ltd, Bury St Edmunds, Suffolk

ISBN 0 412 56230 8

A catalogue record for this book is available from the British Library
Library of Congress Catalog Card Number: 94-72637

∞ Printed on acid-free text paper, manufactured in accordance with ANSI/NISO Z39.48-1992 (Permanence of Paper).

Contents

Preface

As a result of a shift in the focus of research studies for satellite communications and remote sensing from the microwave to the millimetre wave region, a need was identified by the authors to re-examine the modelling assumptions frequently made for the prediction of radiowave interactions with the earth's atmosphere.

For telecommunications applications, interest in the millimetre wave region is driven by the continuing demand for radio spectrum for the provision of new services, especially mobile services. In the context of atmospheric remote sensing, there is growing interest in the use of the millimetre wavelengths for radiometry, limb sounding and radar, especially in the context of climate modelling for the understanding of global warming mechanisms.

For the past two decades, extensive measurements and modelling have taken place for the 10–20 GHz frequency range, in the context of both terrestrial communication systems and earth–space paths. The authors have been actively involved in these studies and have helped to develop some of the prediction 'models' that are now an accepted part of the literature.

As interest shifts to higher frequencies, it is clear that many of the assumptions made for 10–20 GHz are no longer on firm ground. A simple example relates to attenuation in rainfall, where we can no longer assume that only convective or showery rain determines system outages or neglect the contribution from associated rain clouds. Interest in clouds and light rainfall then leads on to a need to reassess assumptions on raindrop size distributions and the modelling of extinction in ice particle populations.

The essentially different modelling assumptions that must be made for the prediction of attenuation in rainfall for the millimetre wave region illustrate but one instance of the change of behaviour of the atmosphere in moving from 10 GHz to 50 GHz. Such a change is perhaps not too surprising when we note the change in wavelength involved (3 cm to 0.6 cm) and recall the sizes of scatterers in the atmosphere. Other and more fundamental problems arise when we consider gaseous spectroscopy at millimetre wavelengths, where we find considerable uncertainty in the underlying physics, especially in relation to the water vapour continuum.

The research study that led to our undertaking this work was funded by the European Space Agency (ESA), and we are most grateful to our ESA colleagues Bertram Arbesser-Rastburg, Pedro Poiares Baptista and Maurice Borgeaud for discussions in the context of that study. We should especially like to acknowledge the work of our research assistants Lukas Leyten and Tasos Papatsoris, who made contributions to several of the chapters in the context of our ESA research study. We are indebted to Dr M.H.A.J. Herben (EUT) for contributions to Chapter 3 on tropospheric scintillation and we would especially acknowledge the contribution from Max van de Kamp (EUT) in going through the text, unifying the use of symbols and providing such an excellent list of symbols. Finally we would wish to thank our secretaries, Mrs Doret Pellegrino and Mrs Freda Hewitson, for coping so patiently and well with such a highly mathematical text.

One motivation for turning our research study into a book was the diversity of specialized classical texts and papers that we have drawn on. Many of these are not readily available and several use notations or definitions that are not familiar to the radio engineer. It is always educational for a researcher to re-examine seminal texts and we would not wish to discourage our readers from doing just that; nevertheless, we believe that our treatment will help in the understanding of those texts.

In this book we have re-examined the physical assumptions that apply to radiowave propagation in the earth's atmosphere for the millimetre wavebands. We have uncovered many more problems than we have been able to provide solutions for. There is still plenty of work to do!

List of symbols

Local variables which have no global meaning are not included in this list, such as variables which are only used to ease the representation of large formulae.

CHAPTER 2

a_e	radius of the earth (6370 km)
$a_{1-5,i}$	coefficients of oxygen absorption lines
A	attenuation (dB)
$b_{1-3,i}$	coefficients for water vapour absorption lines
B	correction term for M_w (m^3)
B	broadening coefficient C_s/C_1
c	velocity of light in vacuum (3×10^8 m s^{-1})
$C_{1,2}$	scale factors
$C_{s,i}$	self- and foreign broadening coefficient of absorbing gas
d	width parameter for the Debye spectrum (GHz)
D	electric flux density (C m^{-2})
e	constant (2.71828 . . .)
e	partial pressure of water vapour (mbar)
$\mathbf{E}$	electric field (V m^{-1})
f	frequency (GHz)
Δf_{pdz}	line width, due to pressure, Doppler and Zeeman broadening (GHz)
F	function
F_0	non-resonant contribution to the absorption coefficient α (ns)
$F_{K^\pm}$	Lorentzian shape factor (ns)
F	broadening coefficient C_2/C_1
g	ratio
h	altitude (km)
$h_{o,w}$	oxygen and water vapour equivalent height (km)
h_{w0}	water vapour equivalent height in the window regions (km)
$h'_{o,w}$	corrected oxygen and water vapour equivalent height (km)

H	magnetic field strength (T)
i	index
j	index
j	imaginary unit
J	total angular momentum number
$\boldsymbol{J}$	total angular momentum
k	Boltzmann's constant ($1.380\,62 \times 10^{-23}$ J K^{-1})
K	orbital momentum number
$\boldsymbol{K}$	orbital momentum
L	path length (km)
m	molecular mass (u)
M	number density (m^{-3})
$M_{\mathrm{da,w}}$	dry air and water vapour molecule density (m^{-3})
M_{d}	dimer density
n	refractive index
N	complex refractivity (ppm)
$N'^{,}{''}$	real and imaginary parts of complex refractivity N (ppm)
N_0	frequency-independent term of complex refractivity N (ppm)
$N_{\mathrm{d,w}}$	dry and wet complex refractivity (ppm)
p	atmospheric (partial) pressure or dry air (mbar)
$\boldsymbol{p}$	elementary dipole moment (C m)
$\boldsymbol{p}'$	dipole moment density (C m m^{-3})
P	total barometric pressure (mbar)
$\boldsymbol{P}$	electric polarization (C m^{-2})
R	gas constant (8.3143 J mol^{-1} K^{-1})
s	correction or extension factor
S	strength of a resonance (F m^2 s^{-2})
S	molecular spin number
$\boldsymbol{S}$	molecular spin
t	relative inverse temperature
T	temperature (K)
V	volume (m^3)
w	liquid water content (cm)
x	variable
y	variable
y_K	parameters (mbar^{-1})
z	variable
α	absorption coefficient (dB km^{-1})
$\alpha_{\mathrm{o,w}}$	oxygen and water vapour absorption coefficient (dB km^{-1})
α_{cont}	continuum absorption of water vapour (dB km^{-1})
$\alpha_{\mathrm{l,w}}$	water vapour attenuation coefficient due to local lines and far wings (dB km^{-1})
$\alpha_{\mathrm{a,d}}$	liquid water attenuation coefficient due to aerosols and dimers (dB km^{-1})

α_x	correction term for the water vapour attenuation coefficient α (dB km^{-1})
β	phase dispersion (rad km^{-1})
γ_t	molecular polarizability (F m^2)
$\gamma_{e,i,d}$	electronic, ionic, dipole part of molecular polarizability (F m^2)
Γ	damping constant (rad s^{-1})
$\Delta()$	small element or deviation of argument
ε	complex permittivity (F m^{-1})
$\varepsilon^{\prime,\prime\prime}$	real and imaginary parts of complex permittivity ε (F m^{-1})
ε_0	permittivity of vacuum ($8.854\,19 \times 10^{-12}$ F m^{-1})
ε_r	complex relative permittivity
$\varepsilon_r^{\prime,\prime\prime}$	real and imaginary parts of complex relative permittivity ε_r
ϵ	elevation angle (deg)
κ	temperature factor for M_d (m^3)
Λ	electronic axial quantum number
$\mu_{K^\pm}^2$	transition matrix element
$\mu_{K_0}^2$	non-resonant transition matrix element
π	constant (3.314 159. . .)
ρ	water vapour density (g m^{-3})
ρ_s	saturation water vapour density (g m^{-3})
τ	relaxation time (s)
χ_e	electric susceptibility
ω	angular frequency (rad s^{-1})

CHAPTER 3

a_e	effective earth radius (8.5×10^6 m)
A	long-term cumulative distribution function of fading due to scintillation (dB)
C_n^2	structure constant of refractive index (m$^{-2/3}$)
C_{n0}^2	value of C_n^2 at ground level (m$^{-2/3}$)
D	antenna diameter (m)
D_{eff}	effective antenna diameter (m)
E	amplitude of electric field (V m^{-1})
f	frequency (GHz)
$\mathcal{F}_v$	structure function of the velocity variations (m^2 s^{-2})
$\mathcal{F}_n$	refractive index structure function
$\mathcal{F}_p$	phase structure function (rad^2)
F_n	three-dimensional spectrum of the refractive index fluctuations (m^3)
g	log-amplitude variations
h	height above the earth (m)

h_0	reference height (m)
k	wavenumber (mm^{-1})
$l_{1,2,\text{t}}$	outer, inner and integral scale of turbulence (m)
L	length of the turbulent part of the path (km)
n	refractive index
N_{wet}	wet term of ground refractivity (ppm)
p	probability density
P	probability
r	distance
s	ratio $\sigma_{\text{g}}^2/\sigma_{\text{m}}^2$
t	time (s)
v	air velocity (m s^{-1})
v_{t}	velocity component transverse to the propagation path (m s^{-1})
V	volume (m^3)
$W_g^{0,\infty}$	asymptotes of spectral density function
x	$0.0584kD^2/L$
$\Delta()$	fluctuation of argument
ϵ	elevation angle (deg)
η	antenna efficiency
κ	spatial wavenumber (m^{-1})
λ	wavelength (mm)
σ_g^2	variance of the log-amplitude variations g
$\ln \sigma_m^2$	mean of $\ln \sigma_g^2$
σ_{gg}	standard deviation of $\ln \sigma_g^2$
σ_{ref}	estimation of the intensity of fluctuations
σ_{pre}	predicted standard deviation of the amplitude variation
ω	angular frequency (rad s^{-1})
ω_{t}	angular frequency belonging to v_{t} (rad s^{-1})
ω_{s}	angular frequency above which aperture smoothing effects become dominant (rad s^{-1})
$\overline{}$	mean of argument

CHAPTER 4

a	radius of sphere (m)
a_{mn}	expansion coefficient (V m^{-1})
A	amplitude
A_0	reference amplitude
b_{mn}	expansion coefficient (V m^{-1})
c	velocity of light in a vacuum ($3 \times 10^8\ \text{m s}^{-1}$)
c_{mn}	expansion coefficient (V m^{-1})
$\mathbf{C}$	Fourier coefficient for electric field (V)

d_{mn}	expansion coefficient ($\mathrm{V\,m^{-1}}$)
D	diameter of a particle (m)
$\boldsymbol{e}_m$	Fourier coefficient for electric field ($\mathrm{V\,m^{-1}}$)
$E_{\mathrm{i,r,s,t}}$	incident, coherent, scattered and total electric field ($\mathrm{V\,m^{-1}}$)
$E_{\varphi,\theta}$	φ and θ components of electric field ($\mathrm{V\,m^{-1}}$)
$E_{+,-}$	right- and left-hand circularly polarized component of electric field ($\mathrm{V\,m^{-1}}$)
$\boldsymbol{E}$	electric field ($\mathrm{V\,m^{-1}}$)
$\mathcal{E}$	electromagnetic field in dyadic notation ($\mathrm{V\,m^{-1}}$)
$\mathcal{E}_j$	discrete Fourier coefficient for electric field ($\mathrm{V\,m^{-1}}$)
f	frequency (Hz)
$\boldsymbol{f}$	scattering amplitude (m)
$\boldsymbol{F}_{\mathrm{t,r}}$	angular representation of a plane wave component for the transmitting and receive antenna
$\mathbf{f}$	scattering amplitude in dyadic notation (m)
$\mathbf{G}$	Green's function in dyadic notation ($\mathrm{m^{-1}}$)
$\boldsymbol{h}_m$	Fourier coefficient for magnetic field ($\mathrm{A\,m^{-1}}$)
$H_n^{(1)}$	Hankel function of the first kind and order n
$\boldsymbol{H}_{\mathrm{i,s,t}}$	incident, scattered and total magnetic field ($\mathrm{A\,m^{-1}}$)
$I_{\mathrm{c,i}}$	coherent and incoherent intensity ($\mathrm{W\,m^{-2}}$)
$\mathbf{I}$	identity matrix
J_n	Bessel function of the first kind and order n
J	source function ($\mathrm{W\,m^{-2}}$)
$\boldsymbol{J}_{\mathrm{cc,i}}$	coherent and incoherent part of Stokes vector
$\mathbf{J}_i$	function in dyadic notation ($\mathrm{V\,m^{-1}}$)
k	wavenumber ($\mathrm{m^{-1}}$)
k_0	free space wavenumber ($\mathrm{m^{-1}}$)
k_{s}	wavenumber inside a scatterer or a slab ($\mathrm{m^{-1}}$)
$\boldsymbol{k}_{\mathrm{i,s}}$	wave vector of incident and scattered field ($\mathrm{m^{-1}}$)
$\mathbf{K}$	function in dyadic notation ($\mathrm{m^{-2}}$)
l	linear dimension of an object (m)
M	number density ($\mathrm{m^{-3}}$)
M	number
$\boldsymbol{M}_{mn}^{(3)}$	outgoing vector spherical harmonic
n	refractive index in the particle
$\mathbf{N}$	matrix representation of refractive index
$N_{mn}^{(3)}$	outgoing vector spherical harmonic
P_{a}	absorbed power (W)
$P_{\mathrm{i,s,d}}$	power in incident and scattered field, and cross-product (W)
P_n^1	associated Legendre function of the first kind and order n
r	distance to the observation point (m)
$\boldsymbol{r}$	place vector to the observation point (m)
$r_{1,2}$	distance from a point to the transmitter and to the receiver (m)
$\boldsymbol{r}^{\mathrm{a}}$	vector from origin to particle a (m)

S	surface (m^2)
$S_{i,r}$	power density of incident and received plane wave ($W\,m^{-2}$)
S_{mn}	complex scattering matrix element (m)
$\mathbf{S}_{i,s,d}$	Poynting vector of incident and scattered field, and cross-product ($W\,m^{-2}$)
t	time (s)
T	transmission coefficient
$T_{s,d}$	transmission coefficient of scattered and unscattered wave
T_{mn}	transmission matrix element
u_b^a	scattering characteristic of the particle located at $\boldsymbol{r}^b$ as observed at $\boldsymbol{r}^a$
$\boldsymbol{u}_i$	unit vector in the propagation direction of the incident field
$\boldsymbol{u}_e$	unit vector in the polarization direction of the wave
$\boldsymbol{u}_s$	unit vector in the direction from the object to the observation point
$\boldsymbol{u}_n$	unit vector normal to surface S and directed outwards
$\boldsymbol{u}_x$	unit vector in the positive x direction
$\boldsymbol{u}_z$	unit vector in the positive z direction
$\boldsymbol{u}_c$	unit vector in propagation direction of coherent radiation
U	complex electric or complex magnetic field
U_f	fluctuating electric or fluctuating magnetic field
U_0	reference electric or reference magnetic field
V	volume (m^3)
V	electric or magnetic field
x	coordinate (m)
z	coordinate (m)
Z_0	matched load (Ω)
γ	propagation coefficient of the medium (m^{-1})
$\Gamma_{c,f}$	correlation function of the coherent and the fluctuating field
δ	Dirac function (for vectors)
ε	permittivity ($F\,m^{-1}$)
ε_0	permittivity of vacuum ($F\,m^{-1}$)
ε_r	relative permittivity
ζ	optical distance
θ	coordinate (rad)
ϑ	antenna beamwidth (deg)
λ	wavelength (m)
μ_0	permeability of vacuum ($H\,m^{-1}$)
σ	cross-section per unit volume (m^{-1})
$\sigma_{a,s,t}$	absorption, scattering and total cross-section (m^2)
Σ_c	matrix, representing scattering cross-section per unit solid angle and per unit volume ($m^2\,m^{-3}\,sr^{-1}$)
Υ	power radiation per unit volume per unit solid angle ($W\,m^{-3}$)
ϕ	phase (rad)
ϕ_0	reference phase (rad)
φ	complex phase

$\varphi^{\prime,\prime\prime}$	real and imaginary parts of complex phase
Φ	phase function
Φ_0	scattering albedo
ψ	azimuth coordinate
Ψ_{a}^{b}	scalar field from particle a, measured at $\boldsymbol{r}^b$
$\Psi_{\mathrm{i}}^{\mathrm{a}}$	incident scalar field at $\boldsymbol{r}^{\mathrm{a}}$
$\Omega_{\mathrm{i,s}}$	solid angle related to incident and scattered flux density (sr)
$\overline{}$	time average of the argument
$\langle\ \rangle$	ensemble average of the argument
Re( )	real part of the argument
( )*	complex conjugate of the argument
( )′	derivative of function
$\mathbf{O}(\ )$	mathematical order of argument
$\mid\ \mid$	magnitude of a vector
$(\ )_{\perp,\parallel}$	normal and parallel component of a vector
$\boldsymbol{\nabla}$	nabla operator

CHAPTER 5

B	brightness (per unit bandwidth) ($\mathrm{W\,m^{-2}\,sr^{-1}\,(Hz^{-1})}$)
c	velocity of light ($2.997\,924\,6 \times 10^8\ \mathrm{m\,s^{-1}}$)
E	emission coefficient ($\mathrm{W\,m^{-3}}$)
f	frequency (Hz)
h	Planck's constant ($6.6262 \times 10^{-34}\ \mathrm{J\,Hz^{-1}}$)
$h_{\mathrm{a,s}}$	height of the atmosphere and the station (m or km)
I	intensity ($\mathrm{W\,m^{-2}}$)
J	source function ($\mathrm{W\,m^{-2}}$)
k	Boltzmann's constant ($1.380\,62 \times 10^{-23}\ \mathrm{J\,K^{-1}}$)
M	number density ($\mathrm{m^{-3}}$)
P	power (W)
$P_{\mathrm{i,r}}$	incident power and power received by an antenna (W)
T	temperature (K)
T_{b}	brightness temperature (K)
S	complex scattering function (m)
$\boldsymbol{u}_{\mathrm{i,s}}$	unit vector in the incident and the scattering direction
W	energy density (per unit bandwidth) ($\mathrm{J\,m^{-3}\,(Hz^{-1})}$)
z	distance (m or km)
z_{e}	optical distance
ϵ	elevation angle (deg)
λ	wavelength (m)
$\sigma_{\mathrm{t,s,a}}$	extinction, scattering and absorption cross-section ($\mathrm{m^2}$)
σ_{bi}	bistatic cross-section ($\mathrm{m^2}$)

Φ	phase function
Φ_0	scattering albedo
Ω	solid angle (sr)
Ω_s	solid angle in the scattering direction (sr)

CHAPTER 6

$a_{1,2,3}$	semi-axes of ellipsoid (mm)
$b_{1,2,3}$	variables related to ellipsoid shape (F m^2)
$C_{1,2,3}$	parameters related to axes of ellipsoid
$D_{1,2,3}$	parameters related to axes of ellipsoid
$E_{01,2,3}$	static electric field in direction of ellipsoid axis (V m^{-1})
$\boldsymbol{E}$	electric field (V m^{-1})
h	height (m)
L_e	effective path length (m)
$\boldsymbol{p}$	dipole moment (C m)
q_m	mass fraction
q_v	volume fraction
$u_{1,2,3}$	shape factors related to axes of ellipsoid
$\boldsymbol{u}_{x,y,z}$	unit vectors in the directions of coordinate axes
V	volume (m^3)
z_v	visibility (km)
$\alpha_{1,2,3}$	specific attenuation along axes of ellipsoid (dB km^{-1})
$\beta_{1,2,3}$	specific phase shift along axes of ellipsoid (deg km^{-1})
δ	polarization tilt angle from vertical (deg)
$\varepsilon_r^{\prime,\prime\prime}$	real and imaginary parts of the relative permittivity
ε_{rh}	relative permittivity of homogeneous particle
λ	wavelength (cm)
ρ	mass density (km m^{-3})
ρ_{2m}	density at a reference height of 2 m (kg m^{-3})

CHAPTER 7

a	radius of droplet (μm)
a_c	critical radius of droplet (μm)
a_e	effective earth radius (8497 km)
A	attenuation (dB)
$A_{d,w}$	dry and wet term attenuation (dB)
$A_{z,cl}$	zenith and cloud attenuation (dB)
b	parameter
c	parameter

f	frequency (GHz)
h_e	effective height (km)
$M(a)$	drop size distribution function ($m^{-3}\ \mu m^{-1}$)
M_0	parameter of drop size distribution (m^{-3})
n	complex refractive index
N	complex refractivity (ppm)
N_0	frequency-independent part of refractivity (ppm)
$N^{\prime,\prime\prime}$	real and imaginary parts of refractivity (ppm)
$N_{w,i}$	refractivities of water and ice (ppm)
R	rain rate ($mm\,h^{-1}$)
s	correction factor
t	relative inverse temperature
T	temperature (°C or K)
u	shape factor
Z	reflectivity (dBZ)
α	(power) attenuation coefficient ($dB\,km^{-1}$)
$\alpha_{m,r}$	normalized attenuations for Mie and Rayleigh scattering ($dB\,km^{-1}\,m^3\,g^{-1}$)
β	phase shift coefficient ($deg\,km^{-1}$)
ε_r	complex relative permittivity
$\varepsilon_r^{\prime,\prime\prime}$	real and imaginary parts of relative permittivity
$\varepsilon_{rw,i}$	relative permittivities of liquid water and ice
ϵ	elevation angle (deg)
λ	wavelength (cm)
ρ	liquid water content or absolute humidity ($g\,m^{-3}$)
$\rho_{w,i}$	specific masses of water ($1.000\ Mg\,m^{-3}$) and ice ($0.916\ Mg\,m^{-3}$)
σ	standard deviation
τ	propagation delay coefficient ($ps\,km^{-1}$)
$\overline{\ }$	mean of argument

CHAPTER 8

a	variable
A	attenuation (dB)
$A_{0.01}$	attenuation exceeded for 0.01% of time (dB)
$A(P)$	cumulative distribution of attenuation (dB)
b	exponent
$C_{1,2,3,4,5}$	constants
d	rate of decay of a horizontal profile
h	height (m)
h_r	effective rain height (m)
$h_{fy,m}$	yearly and monthly mean freezing heights (m)

h_0 station altitude (m)
$l_{c,d}$ linear dimensions of a rain cell and a piece of debris (km)
L path length through rain (km)
L_e effective path length (km)
L_c path length through a rain cell (km)
$p(R)$ probability density of point rain intensity (mm^{-1} h)
P cumulative probability
$P_{c,d}$ cumulative probability in cells and in debris
r distance from cell centre (km)
R rain rate ('rainfall intensity') ($mm\,h^{-1}$)
$R_{c,d}$ rain rate in a rain cell and a piece of debris ($mm\,h^{-1}$)
R_m maximum rain rate ($mm\,h^{-1}$)
R_p point rain rate ($mm\,h^{-1}$)
$R_p(P)$ cumulative distribution of point rain rate ($mm\,h^{-1}$)
$R_{0.01}$ rain rate exceeded for 0.01% of time ($mm\,h^{-1}$)
s reduction or correction factor
S horizontal area (km^2)
u shape parameter
z distance along the path (km)
α attenuation coefficient ($dB\,km^{-1}$)
$\alpha_{0.01}$ specific attenuation exceeded for 0.01% of time ($dB\,km^{-1}$)
δ polarization angle with respect to the horizontal (deg)
$\Delta(\ )$ small change of argument
ϵ elevation angle (deg)
ρ rain cell radius (km)
σ standard deviation
φ latitude (deg)
$\overline{\ \ }$ mean of argument
$\ldots^{(3)}$ third-order derivative

CHAPTER 9

A copolar attenuation (dB)
ΔA differential attenuation (dB)
C constant (dB)
$E_{x,y}$ components of electric field ($V\,m^{-1}$)
E_{ij} electric field at the receiver, transmitted in polarization state i and received in orthogonal state j ($i, j = x, y$) ($V\,m^{-1}$)
$E_{i,s}$ incident and scattered electric fields ($V\,m^{-1}$)
$\mathbf{E}$ complex electric field vector ($V\,m^{-1}$)
f frequency (GHz)
L path length through the scattering region (km)

R	rain rate (mm h^{-1})
$\boldsymbol{R}$	rotation matrix
$\boldsymbol{R}^{-1}$	inverse rotation matrix
S	constant (dB)
t	time (s)
T	temperature (°C)
$T_{1,2}$	transmission coefficients
T'_{xx}	component of $\boldsymbol{T}'$
$\boldsymbol{T}$	transmission matrix
$\boldsymbol{T}'$	rotated transmission matrix
$\boldsymbol{u}_{x,y}$	unit vectors in the x and y direction
U	constant (dB)
V	constant (dB)
XPD	cross-polarization discrimination (dB)
XPD_v	XPD for vertical polarization (dB)
XPI	cross-polarization isolation (dB)
α	attenuation coefficient (dB km^{-1})
β	phase shift coefficient (deg km^{-1})
ϵ	elevation angle (deg)
θ	canting angle (deg)
ξ	angle of incidence (deg)
ϕ	phase difference (rad)
$\Delta\phi$	differential phase (deg)
ω	angular frequency (rad s^{-1})
$\lvert\ \rvert$	amplitude of argument

CHAPTER 10

$a_{\parallel,\perp}$	anisotropy terms
k_0	free space wavenumber (m^{-1})
$k_{\parallel,\perp}$	wavenumber of the field polarized parallel and perpendicular to the symmetry axis of the particles (m^{-1})
n	refractive index
q	fractional volume (%)
$T_{12,21}$	diagonal terms in transmission matrix
XPD	cross-polarization discrimination (dB)
Z_{dr}	differential reflectivity (dB)
Z	reflectivity (dB)
δ	polarization angle (deg)
ϵ	elevation angle (deg)
λ	wavelength (m)
ψ	aximuth orientation (deg)
$\bar{\ }$	mean of argument

CHAPTER 11

C	constant (m^{-2})
f	scatter amplitude (m)
k	wavenumber (m^{-1})
l	thickness (km)
$M\sigma_s$	average scattering cross-section per unit volume ($m^2\,m^{-3}$)
$P_{r,t}$	received and transmitted powers (W)
$\boldsymbol{u}_i$	unit vector in the direction of incidence
V	scattering volume (m^3)
α	specific attenuation (km^{-1})
σ_t	extinction cross-section (m^2)
σ_{bi}	bistatic cross-section (m^2)
σ_d	differential cross-section (m^2)
Σ	cross-section of receiving antenna beam (m^2)
Φ_0	scattering albedo
Im ( )	imaginary part of argument
( )*	complex conjugate of argument

CHAPTER 12

a	radius of curvature of the ray (m)
a_e	radius of the earth (km)
A	attenuation (dB)
$\mathcal{A}$	amplitude
b	constant (km^{-1})
e	water vapour pressure (mbar)
h	height (km)
h_d	duct thickness (m)
j	imaginary unit
k	effective earth radius factor
k_0	free space wavenumber (m^{-1})
L_b	basic transmission loss (dB)
L_{bf}	free space loss (dB)
M	modified refractivity (ppm)
n	refractive index
n_m	modified refractive index
N	refractivity (ppm)
$N_{d,w}$	dry and wet component of refractivity (ppm)
N_0	constant (ppm)
P	atmospheric pressure (mbar)
r	radial coordinate (m)
T	temperature (K)

z	distance (m)
Δz	range delay (m)
$\Delta z_{\mathrm{d,w}}$	range delays due to dry air and due to moisture (m)
ε_0	permittivity of vacuum ($\mathrm{F\,m^{-1}}$)
ϵ	elevation angle (deg)
θ	angle (deg)
θ	coordinate (rad)
λ	wavelength (m)
λ_{c}	critical wavelength (m)
ξ	grazing angle of incidence (deg)
ξ_{c}	critical grazing angle of incidence (deg)
ρ	water vapour density ($\mathrm{g\,m^{-3}}$)
φ	angle (deg)
ϕ	coordinate (rad)
Ψ	field component
$\nabla(\ )$	nabla operator working on argument
$\mathrm{d}(\ )/\mathrm{d}(\ )$	derivative of argument
$\partial(\ )/\partial(\ )$	partial derivative of argument
$\Delta(\ )$	small change of argument

CHAPTER 13

a	radius (mm)
D	electric flux density ($\mathrm{C\,m^{-2}}$)
D	diameter (mm)
E	electric field ($\mathrm{V\,m^{-1}}$)
h	height (m)
l	length (mm)
m	mass (mg)
$M(a)$	drop size distribution function ($\mathrm{m^{-3}\,mm^{-1}}$)
M_0	parameter for size distribution ($\mathrm{m^{-3}}$)
n	complex refractive index
$n^{\prime,\prime\prime}$	real and imaginary parts of refractive index
q	fraction
$q_{\mathrm{i,a,w}}$	fraction of volume occupied by ice, air and water
R	rain rate ($\mathrm{mm\,h^{-1}}$)
s	spread parameter
t	relative inverse temperature
u	form number
v	velocity ($\mathrm{m\,s^{-1}}$)
v_{d}	Doppler velocity ($\mathrm{m\,s^{-1}}$)
Z	reflectivity (dBZ)
α_{j}	parameter for Ray model

$\beta_{i,ij}$	parameters for Ray model
$\gamma_{i,ij}$	parameters for Ray model
Δ_i	parameter for Ray model
ε	permittivity ($F\,m^{-1}$)
$\varepsilon_{s,i,a,w}$	permittivity of snow, ice, air and water ($F\,m^{-1}$)
ε_0	permittivity of vacuum ($F\,m^{-1}$)
ε_r	complex relative permittivity
$\varepsilon_r^{\prime,\prime\prime}$	real and imaginary parts of relative permittivity
$\varepsilon_r^{0,\infty}$	low- and high-frequency value of relative permittivity
θ	temperature (K)
λ	wavelength (cm or μm)
λ_{0i}	centre of a band in the Ray model (μm)
Λ	parameter for size distribution (mm^{-1})
$\rho_{s,a}$	density of snow and air ($g\,cm^{-3}$)
σ	conductivity ($\Omega^{-1}\,m^{-1}$)
τ	relaxation time (s)
ω	angular frequency ($rad\,s^{-1}$)
ω_{0ij}	parameter for Ray model

Introduction

Our knowledge of mechanisms of radiowave propagation in the troposphere is subject to continuous improvement especially in quantitative terms. While, with a few important exceptions, the dominant mechanisms have been well understood for some time, the needs for accurate performance prediction for telecommunications systems or improved accuracy of inversion for remote sensing purposes are still leading to significant advances in detailed knowledge and improvements in tools and techniques for prediction. Also, while individual propagation mechanisms may be identified, understood, modelled and characterized in isolation, the problem of predictions based on climatic inputs, especially in the millimetre wave range where a multitude of mechanisms coexist, remains a significant challenge.

In this text we examine in some depth the physical models that describe the interaction of millimetre waves with the medium. These include models for the medium (i.e. the atmosphere), mathematical models for the interaction processes with electromagnetic waves and eventually a prediction method or 'model' for use by the radio system designer.

Often, for the frequency range 10–20 GHz, prediction methods for the system designer have not been based on detailed physical models of the atmosphere, but on empirical or at best semi-empirical formulations adjusted for a limited set of climatological circumstances or range of frequencies. Evidently such approaches are unlikely to be satisfactory for extrapolation into the millimetre wave region.

It is quite commonly the case that our knowledge of the mathematical modelling of individual interaction processes (for example scattering by hydrometeors or scintillation by turbulence) is highly developed, but problems arise when we come to develop prediction methods (or 'models') relying on meteorological inputs with limited data samples taken on ground-based instruments. Furthermore, individual radiowave interaction processes cannot be measured in isolation: rain attenuation, cloud attenuation, gaseous absorption, scintillation and refraction will all occur simultaneously on a satellite–earth link and those developing models are faced with the task of firstly attempting to identify each process within a set of experimental data, then developing an approach to prediction which models individual processes and then, finally, combining contributions from all such processes in a manner appropriate to the climate under consideration.

We thus have at first to analyse each interaction process and then attempt to develop prediction models based on consideration of meteorology and climate. In the chapters of this book we attempt to summarize our knowledge of the interaction processes and to show in each case how this may be used for prediction. Chapter 1 provides a succinct guideline for the reader in this respect.

The chapter on gaseous absorption (Chapter 2) covers only essential ground. This is a particularly important area for system prediction, where there are still fundamental uncertainties in the physical models (in explaining the origins of the continuum absorption) and practical difficulties in the magnitude of databases and complexity of the models needed to describe atmospheric composition and structure.

In Chapter 3 we cover the classical treatment of scintillation but at the same time note the significant gap between such treatments and the semi-empirical formulae used for prediction.

Chapter 4 allows us to apply some rigor to particle scattering and we hope that the reader will be able to use this as a short-cut to some of the classical texts in the field as well as a resumé of the most recent work.

Thermal radiation from hydrometeors and atmospheric gases (Chapter 5) is discussed in order to outline the basic modelling underlying the interpretation of measurements of brightness temperature by radiometers. Thus it has applications both to the estimation of attenuation by radiometry and to passive remote sensing of the physical properties of the atmosphere.

Scattering and absorption from sand and dust (Chapter 6) are not of major systems importance, but the material is included for completeness and more especially the treatment of particle alignment for cross-polarization (based on Dr N.J. McEwan's work) is identical to that needed for ice particle alignment.

Chapter 7 covers an area of particular importance to millimetre wave propagation (attenuation by clouds and the melting layer). Much work is currently in progress here in the context of both telecommunications and millimetre wave cloud radar studies. Our coverage gives only an introduction to this rapidly developing field.

For attenuation by rain (Chapter 8), we emphasize a physical structure approach to prediction and avoid empirical treatments with only limited applicability to the millimetre wavebands.

Depolarization by rain and ice particles (Chapters 9 and 10) is also given rigorous introductory physical treatment, although, in the case of rain, for prediction purposes semi-empirical relations are developed between cross-polarization discrimination and attenuation. In the case of ice depolarization, even empirical relations are conspicuously absent and our treatment is only in general physical terms, with a number of questions yet to be answered.

The chapters on interference (Chapter 11) and reflection and refraction

(Chapter 12) are of a general and introductory nature only, being applicable to both microwave and millimetre regions of the spectrum.

The overriding importance of hydrometeors in the millimetre wavebands justifies the inclusion of a special chapter on their physical properties (Chapter 13) in which we consider their composition, dielectric properties, shape, size, orientation and fall velocities.

1
Outline of interaction processes

Atmospheric models are used in conjunction with electromagnetic theory to describe the effects of the atmosphere on radiowave propagation, both in telecommunications and in remote sensing applications. In this study we concentrate initially on the processes of radiowave interaction and hence give a comprehensive description of the models of the physical processes used to describe radiowave propagation. Furthermore, we look at how the models of physical processes are currently taken, simplified or extended to make predictions for system design. Finally, we identify and examine critically experiments that have been set up to test the validity of the various 'process-oriented' models and examine potential weaknesses in the models.

1.1 OUTLINE DESCRIPTIONS OF INTERACTION PROCESSES

The following important processes of radiowave interaction in the lower atmosphere can be identified:

- absorption and dispersion in atmospheric gases (oxygen, water vapour and minor constituents);
- scintillation and scattering from atmospheric turbulence;
- scattering and absorption in populations of hydrometeors (including anisotropic effects, forward scatter, backscatter and scattering at arbitrary angles);
- scattering and absorption in sand and dust particle populations;
- refraction and reflection in stable atmospheric layers;
- thermal emissions from hydrometeors and atmospheric gases.

Also in the upper atmosphere, the following process is of minor but not negligible significance for frequencies above 10 GHz:

- anisotropy and delay in the ionosphere.

In each of the chapters reviewing these processes of interaction, we give a brief description of the characteristics of the physical media, the electromagnetic techniques and modelling assumptions associated with each of these media, followed by a detailed mathematical model of the processes, wherever possible.

1.2 PHYSICAL MODELS AND PREDICTION METHODS

While in some instances a physical process may have an elegant mathematical treatment and description, such a mathematical model may not be directly applicable as a method for prediction of radio system factors. Often semi-empirical techniques are used for such system factor predictions, drawing some general principles from mathematical analysis of the processes.

Tables 1.1–1.6 list the salient features of the media, electromagnetic tools, systems factors and prediction methods for the most important physical processes.

Table 1.1 Absorption and dispersion in atmospheric gases (Chapter 2)

Physical media	Gaseous continua
Media characteristics and modelling assumptions	Atmospheric composition, variability in space and time of pressure, partial pressures and temperature Various reference atmospheres, including ARDC Molecular resonance with coupling between molecules and rotation of nuclei (oxygen) Line broadening from pressure (collisions), Doppler and Zeeman effects in addition to quantum mechanical uncertainty Coupling with earth's magnetic field Dipole resonance effects with polar molecules (e.g. water vapour) but uncertainty in modelling continuum
Electromagnetic modelling and analysis tools	Basic wave equation in homogeneous lossy media
System factors	Attenuation and dispersion on earth–space and upper, middle or lower atmosphere to space paths
Prediction methods	Oxygen: Van Vleck–Weisskopf approximation with coefficients adjusted to fit laboratory measurements; complete computer evaluation for local conditions Water vapour: MPM model Both: CCIR semiempirical model

Table 1.2 Scintillation and scattering by atmospheric turbulence (Chapter 3)

Physical medium	Turbulent eddies in troposphere
Medium characteristics and modelling assumptions	Eddy size distribution Height of layer containing eddies Composition and temperature of atmosphere Velocity structure function, refractive index structure function Locally isotropic and homogeneous turbulence with $r^{2/3}$ structure Turbulent eddies with zero mean drift velocity; weak turbulence
Electromagnetic modelling and analysis tools	Solution of wave equation in locally isotropic medium with λ small compared with turbulence inner scale size Weak scintillations, perturbations assumed smooth compared with λ First Fresnel zone radius lies between inner and outer scale of turbulence
System factors	Scintillation fading, interference from scattering in turbulence, beam spreading
Prediction methods	Prediction of variance of cumulative distribution function (CDF) of scintillation using theoretical frequency scaling laws and aperture smoothing function from Tatarski normalized to an empirical variance for a reference antenna Short-term CDF of amplitude lognormal but long-term γ distributed Prediction of mean variance from monthly averaged wet term refractive index SAR phase error prediction (UCL)

Table 1.3 Scattering and absorption in populations of hydrometeors (Chapter 4)

Physical media	Populations of rain and ice particles (and particulate ice–water mixtures) falling in still or turbulent atmospheres Ice and water clouds, rainstorms
Media characteristics and modelling assumptions	Ice and rain particle composition and temperature Particle size distributions, particle shapes and size relations Particle terminal velocities in still air Collisions and coalescence Rain canting in horizontal wind shear Ice particle oriented effects, aerodynamic and electrostatic Rainstorm macrophysical structure and types Melting region structure Ground rainfall intensity

Table 1.3 (*cont.*)

Electromagnetic modelling and analysis tools	Single scattering: Rayleigh, Mie, Rayleigh–Debye, WKB, spheroidal expansion, point matching, T-matrix, Fredholm integral and unimoment methods for single-particle coherent scattering model, plane wave spectrum treatment Multiple scattering: Rytov solution, transport theory, multiple scattering theory
System factors	Attenuation and dispersion (minimal), cross-polarization, rain scatter interference Sky noise ('thermal emissions' (Table 1.4))
Prediction methods	Attenuation: CCIR model, Crane global models, Leitao–Watson model, Excell model Cross-polarization: XPD–attenuation relationships Clouds: SWD and SIC; Guissard semi-empirical model; Altshuler and Marr model

Table 1.4 Thermal emissions from hydrometeors and atmospheric gases (Chapter 5)

Physical media	Gaseous continua or populations of rain and ice particles
Media characteristics and modelling assumptions	Discussed in sections 5.1 and 5.3 Local thermodynamic equilibrium
Electromagnetic modelling and analysis tools	Transport theory Rayleigh and Mie scattering Radiative transfer theory
System factors	System noise temperature Attenuation Effective medium temperature
Prediction methods	CCIR models Westwater (cloud model) MPM model

Table 1.5 Scattering and absorption in sand and dust particles (Chapter 6)

Physical media	Populations of sand or dust particles in the lower atmosphere
Media characteristics and modelling assumptions	Particle composition, density, temperature and moisture states, size distributions, shapes and size relations Aerodynamic and electrostatic orientation effects Dust storm macrophysical structures
Electromagnetic modelling and analysis tools	Rayleigh scattering (up to 30 GHz) Coherent and incoherent scattering models as for rain
System factors	Attenuation (minimal), depolarization and interference (from bistatic scatter)
Prediction methods	McEwan, Bashir and Connelly (Intelsat) approximations for attenuation, phase shift and depolarization

Table 1.6 Refraction in stable atmospheric layers and reflection from atmospheric layers (Chapter 12)

Physical media	Horizontally stratified atmospheric layers with anomalous vertical refractivity profiles
Media characteristics and modelling assumptions	Vertical gradients of temperature and water vapour concentration Horizontal and temporal variability Various reference atmospheres, including ARDC
Electromagnetic modelling and analysis tools	Refractivity changes smooth compared with wavelength n related to temperature and humidity Modal analysis, ray tracing and parabolic equation method
System factors	Refraction: borseight error–path length variations Reflection: fading and interference at low elevations
Prediction methods	Refraction: use of standard atmospheres; Baars model; Bean and Dutton Reflection:–

2

Absorption and dispersion in atmospheric gases

2.1 INTRODUCTION

When an asymmetrical molecule is placed in an electric field, it has a preferred orientation, and the molecule is said to be polar. Polar molecules tend to rotate to align with the direction of an imposed electric field and in general exhibit more loss than non-polar molecules. Such loss always exhibits resonance and pronounced temperature dependence.

The principal constituents of the lower atmosphere, oxygen and nitrogen, are both electrically non-polar and no absorption occurs as a result of electric dipole resonance. Water is, on the other hand, a polar molecule and so water vapour exhibits absorption due to electric dipole resonance at critical frequencies. If ε'' and ε' (imaginary and real parts of complex permittivity ε) are plotted against frequency for a polar molecule, the critical frequency (or frequencies) shows up as peaks in ε''.

Figure 2.1 illustrates this effect for a polar molecule with one critical frequency. The precise frequency at which the absorption peaks can vary with temperature. The radiowave absorption effects of trace gases (CO, NO, NO_2, N_2O, SO_2, O_3 and other polar molecules in the atmosphere) are negligible

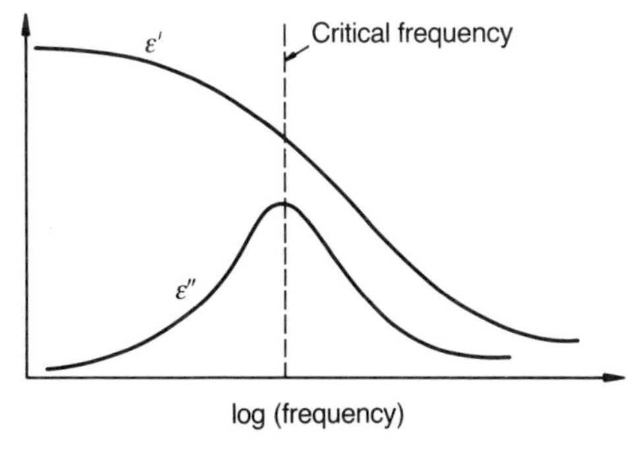

Fig. 2.1 ε' and ε'' plotted against log (frequency) for a substance that has one critical frequency.

compared with the absorption of water and water vapour but, at frequencies above 70 GHz, can contribute significant attenuation in the absence of water vapour. These aspects are dealt with briefly in sections 2.2.5. and 2.5. Below 350 GHz lines are observed at 22.3, 183.3 and 323.8 GHz.

Oxygen is a paramagnetic molecule with a permanent magnetic moment, which can introduce very significant resonance absorption at particular frequencies. An isolated resonance occurs at 118.74 GHz but a large number of lines cluster around 60 GHz. At low pressures (or high altitudes) the individual absorption lines become visible, but at ground atmospheric pressure the lines merge into a broad continuum (pressure broadening) between 52 and 68 GHz.

Oxygen and water vapour are the principal absorbers of radiowave energy in the earth's atmosphere. Nevertheless, in addition to the pronounced resonance absorption lines for these molecules, there is a continuum absorption increasing with frequency in the windows between the resonant frequencies. The origins of this continuum are not fully understood, although they are ascribed in the main to water in some form (section 2.4.3).

2.2 PHYSICAL BACKGROUND OF GASEOUS ATTENUATION (Ramo, Whinnery and Van Duzer, 1984)

2.2.1 General aspects

In this section, a classical molecular physics formulation is used to give a general understanding of the processes involved in radiowave propagation through a gaseous medium. The atmosphere is considered as an inhomogeneous, linear, isotropic and lossy dielectric, which contains gases consisting of atoms or molecules, characterized by a complex permittivity

$$\varepsilon = \varepsilon_0 \varepsilon_r = \varepsilon_0(\varepsilon_r' + j\varepsilon_r'') \tag{2.2.1}$$

in which ε_0 is the permittivity of vacuum, ε_r' is the real part and ε_r'' is the imaginary part of the complex relative permittivity. (Note that a time dependence of $e^{-j\omega t}$ has been assumed in the complex notation.) The atmospheric constituents (atoms and molecules) have dipole moments with a non-zero average alignment. Consequently, a non-zero dipole moment density $\boldsymbol{p}'$ can be defined:

$$\boldsymbol{p}' = \lim_{\Delta V \to 0} \frac{\sum_i \boldsymbol{p}_i}{\Delta V} \tag{2.2.2}$$

where $\boldsymbol{p}_i$ are the elementary dipole moments present in the volume element

ΔV. If higher-order multipoles are neglected, dipole moment density is identical to the electric polarization $\boldsymbol{P}$:

$$\boldsymbol{P} = \varepsilon_0 \chi_e \boldsymbol{E} \tag{2.2.3}$$

where χ_e is the electric susceptibility and $\boldsymbol{E}$ the applied electric field. $\boldsymbol{P}$ is linearly dependent on $\boldsymbol{E}$; this has already been assumed in defining the dielectric as a linear medium. If M like molecules per unit volume are present, the induced polarization may be written as

$$\boldsymbol{P} = M\gamma_t \boldsymbol{E}_{loc} = M\gamma_t g \boldsymbol{E} \tag{2.2.4}$$

where γ_t is the molecular polarizability and g is the ratio between the local electric field $\boldsymbol{E}_{loc}$ acting on the molecule and the applied electric field $\boldsymbol{E}$.

The local electric field of a molecule considered differs from the applied field, because the surrounding molecules influence this field. If the surrounding molecules are assumed to be spherical and sparsely distributed in vacuum, g can be calculated as (Ramo, Whinnery and Van Duzer, 1984)

$$g = \frac{2 + \varepsilon_r}{3}. \tag{2.2.5}$$

The electric flux density $\boldsymbol{D}$ can be defined as

$$\boldsymbol{D} = \varepsilon \boldsymbol{E} = \varepsilon_0 \varepsilon_r \boldsymbol{E} = \varepsilon_0 \boldsymbol{E} + \boldsymbol{P} = \varepsilon_0 (1 + \chi_e) \boldsymbol{E}. \tag{2.2.6}$$

Combining equations (2.2.4), (2.2.5) and (2.2.6) yields

$$\frac{\varepsilon_r - 1}{\varepsilon_r + 2} = \frac{M\gamma_t}{3\varepsilon_0}. \tag{2.2.7}$$

This expression is known as the Clausius–Mossotti relation or, when frequency effects in γ_t are included, the Debye (1929) equation. This equation may be derived in an alternative manner by assuming that each individual small volume element dV acts as an independent scatterer (Rayleigh–Gans theory). For a small (dielectric) sphere the field inside, acting on the molecules, is different from the applied field by a factor of $(2 + \varepsilon_r)/3$ (section 2.5), leading to the same result.

The molecular polarizability γ_t of a gas can be split into three independent parts, each with a different physical background (see for example Ramo, Whinnery and Van Duzer, 1984):

$$\gamma_t = \gamma_e + \gamma_i + \gamma_d \tag{2.2.8}$$

where γ_e is the 'electronic' part, caused by the shift of the electron cloud in each atom relative to its positive nucleus, γ_i is the 'ionic' part, caused by the displacement of positive and negative ions from their neutral positions, and γ_d is the 'dipole' part, caused by permanent dipole moments of individual molecules. The applied electric field tends to align the permanent dipoles against the randomizing forces of molecular collision. Since random motion is a function of temperature, this effect is strongly temperature dependent.

The frequency effects of the electronic and ionic parts can be described as follows. Any displacement of the charge cloud from its central nucleus and any displacement of one ion from another produce a restoring force. The interaction of this restoring force with the inertia of the moving cloud or ion produces resonances as in a mechanical spring–mass system. These resonances in γ_e and γ_i can be represented as

$$\gamma_j = \frac{S_j}{(\omega_j^2 - \omega^2) - \mathrm{j}\omega\Gamma_j} \tag{2.2.9}$$

in which S_j measures the strength of the jth resonance with damping constant Γ_j and frequency ω_j that is related to the restoring force. Because of the greater inertia of the ions in comparison with the electrons of the charge cloud, the ionic resonances occur at lower frequencies.

The frequency effects of the dipole part are different because the force aligning the dipoles in the direction of the applied field is set against the random Brownian movement, which is a thermal effect. It acts as a viscous force and the dynamic response is 'overdamped'. The frequency response of such a system is of the form

$$\gamma_d = \frac{\boldsymbol{P}^2}{3kT(1 - \mathrm{j}\omega\tau)} \tag{2.2.10}$$

where T is the temperature, k Boltzmann's constant, $\boldsymbol{P}$ the permanent dipole moment and τ the relaxation time (i.e. the time for the polarization to fall to $1/\mathrm{e}$ of the original value if the applied field is removed).

The resonances are induced by time-harmonic electromagnetic fields: the electromagnetic energy of the fields is converted to kinetic energy needed for the resonances. Hence a wave propagating through this medium is losing electromagnetic energy, which is equivalent to attenuation. The kinetic energy is, in turn, converted to heat.

2.2.2 Oxygen

The theory presented in section 2.2.1 is quite general, and, as was stated in the beginning of that section, gives only a broad understanding of the processes involved in clear air propagation. As far as the molecule of O_2 is concerned, there are a variety of additional physical mechanisms that must be taken into account for a complete characterization of the attenuation and dispersion with respect to radiowave propagation through the atmosphere.

The presence of a number of resonant absorption lines in the microwave oxygen spectrum gives rise to a complicated atmospheric transfer function (amplitude and phase), the characteristics of which depend on pressure, temperature and water vapour–height profiles and, therefore, on climatic conditions. Spectral lines are only resolvable at high altitudes, namely 15 km or more above sea level. At lower altitudes, pressure broadening of resonances occurs, and the various lines experience a type of mutual interference, which may be explained in terms of a non-linear overlap.

In considering absorption and dispersion caused by the oxygen molecule, we have to take into account its fine structure, related to the coupling between molecular spin and the rotation of the nuclei, and involving energy transitions in the millimetre wave range.

(a) Quantum numbers and energy transitions

The energy level of a diatomic molecular structure is characterized by three quantum numbers.

- The first is the resultant electronic axial quantum number Λ which, for a given molecule and for practical purposes, may be regarded as a constant; in any case transitions involving a change in Λ are responsible for absorption and emission in the optical region only. For the oxygen molecule, $\Lambda = 0$.
- The second is the resulting orbital momentum $\boldsymbol{K}$, which takes into account the rotation of the nuclei. In the oxygen molecule only odd values of the quantum number K are allowed by the Pauli exclusion principle. Quantum jumps between levels with different K values are responsible for absorption and emission in the IR region only.
- The third is the total angular momentum $\boldsymbol{J}$ defined as

$$\boldsymbol{J} = \boldsymbol{K} + \boldsymbol{S}$$

  where $\boldsymbol{S}$ is the molecular spin and the sum must be interpreted in the quantum sense. For the oxygen molecule, the quantum number S is equal to unity, so a given K value corresponds to three different J values, i.e. $J = K - 1,\ K,\ K + 1$. Transitions between these states are possible, but J cannot change by more than one unit. These jumps now are small enough to give millimetre wave resonances. The energy separation between the levels appears to be only slightly dependent on K; in any case, all the corresponding frequencies fall into the 60 GHz band, with the exception of the transition with $K=1$, $J=0 \rightarrow 1$, which corresponds to a frequency separation of about 118 GHz.

(b) Van Vleck–Weisskopf approach

The Van Vleck–Weisskopf theory (Van Vleck and Weisskopf, 1945; Van Vleck, 1947) provides a satisfactory estimate of the absorption and dispersion coefficients due to the oxygen molecule. According to this theory, the oxygen absorption coefficient has the following simple form:

$$\alpha = C_1 p T^{-3} f^2 \sum_K (F_{K^+}\mu_{K^+}^2 + F_{K_-}\mu_{K_-}^2 + F_0\mu_{K_0}^2)\,\mathrm{e}^{-E_K/kT} \qquad (2.2.11)$$

where α (dB km^{-1}) is the absorption coefficient, $C_1 = 1.9779$, a scale factor that allows α to be expressed in decibels per kilometre, p (mbar) is the atmospheric pressure, T (K) is the temperature, f (GHz) is the frequency, K is the orbital angular momentum of the O_2 molecule defined as above, $\sum_K$ indicates summation over the odd values of K (usually $K = 45$ is assumed as the upper limit of the summation; this appears to be quite satisfactory, because all levels with appreciable populations in the range of temperatures encountered in the earth's atmosphere are taken into account),

$$F_{K_-^+} = \frac{\Delta f + \text{non-linear term}}{(f_{K_-^+} - f)^2 + (\Delta f)^2} + \frac{\Delta f + \text{non-linear term}}{(f_{K_-^+} + f)^2 + (\Delta f)^2}$$

is the Lorentzian shape factor with non-linear Rosenkrantz term, in which f_{K^+} and f_{K_-} are resonant frequencies for the transitions of J from K to $K+1$ and from K to $K-1$ respectively (for a fixed K) and Δf is the line width (a factor $1/\pi$ has been incorporated into the scale factor C_1),

$$F_0 = \frac{\Delta f}{f^2 + (\Delta f)^2}$$

is the non-resonant contribution to the absorption coefficient,

$$\mu_{K^+}^2 = \frac{K(2K+3)}{K+1}$$

is the transition matrix element for the transition $J = K \rightarrow K+1$,

$$\mu_{K_-}^2 = \frac{(K+1)(2K-1)}{K}$$

is the transition matrix element for the transition $J = K \rightarrow K-1$,

$$\mu_{K_0}^2 = \frac{2(K^2 + K + 1)(2K + 1)}{K(K + 1)}$$

is the non-resonant transition matrix element ($\Delta J = 0$) and

$$\frac{E_K}{kT} \approx \frac{2.068\,44K(K + 1)}{T}$$

is the Boltzmann factor.

We may use a very similar formulation for the phase dispersion by the oxygen molecule. The only differences lie in the shape factor and, of course, the scale factor C_1. The phase dispersion is then given by

$$\beta = C_2 p T^{-3} f^2 \sum_K (F_{K^+}^{(d)} \mu_{K^+}^2 + F_{K^-}^{(d)} \mu_{K^-}^2)\,\mathrm{e}^{-E_K/kT} \tag{2.2.12}$$

where

$$F_{K_-^+}^{(d)} = \frac{f_{K_-^+} - f + \text{non-linear term}}{(f_{K_-^+} - f)^2 + (\Delta f)^2} - \frac{f_{K_-^+} + f + \text{non-linear term}}{(f_{K_-^+} - f)^2 + (\Delta f)^2}$$

and $C_2 = C_1/8.6858$ is a scale factor which allows β to be expressed in radians per kilometre.

In the shape factors for absorption and dispersion, a suitable parameter Δf has been introduced. In fact $f_{K_-^+}$ may be considered the mean of the frequencies connected with the jump between the two quantum levels and Δf their standard deviation. In terms of a line width for a particular resonance, Δf may be defined as the half-width at half-maximum value of the line centred around the $f_{K_-^+}$ value (in other words, Δf is equivalent to the 'Q factor' of a resonator).

There is a great variety of spectral widths and each must be attributed to a corresponding physical cause. It is possible to classify the line shapes according to their origins and to assign them different values characteristic of their physical causes.

The sources of spectral line broadening that we shall take into account are the following.

- *Natural line width*. This is the frequency broadening due to uncertainty in the energy of an excited quantum level. From Heisenberg's uncertainty principle,

$$\Delta f \geqslant \frac{1}{2\pi}\frac{1}{\Delta t} \tag{2.2.13}$$

where Δf is the frequency uncertainty connected with the jump and Δt is the uncertainty in the lifetimes of the two excited levels involved in the transition process.

In spite of its historical importance, the frequency broadening connected with the uncertainty principle is negligible for our purposes. In fact, sample theoretical computations in the millimetre band always give values smaller than 1 Hz for the natural line width.

- *Pressure broadening*. Interactions between randomly absorbing gas molecules are a major cause of line broadening. These perturbations are generally termed 'pressure broadening'. Although one may differentiate in detail between various mechanisms, all of these effects are related to premature decay of the excited molecular level. A shorter Δt gives, from equation (2.2.13), an increase in Δf. The collision frequency of the air molecules under normal room conditions is, for instance, about 10^9–10^{10} s^{-1}: in this case equation (2.2.13) provides an approximate value of 1 GHz for Δf.

  A more refined theory should evaluate Δf in terms of intermolecular forces. Nevertheless, a complete treatment of pressure broadening and theoretical evaluation of Δf is so complicated that a number of different semi-empirical approaches have been developed in the literature. The most popular is that proposed by Rosenkrantz (1975):

$$\Delta f_{\mathrm{p}} \text{ (GHz)} = 1.16 \times 10^{-3} p \left(\frac{300}{T} \right)^{0.85} \tag{2.2.14}$$

  where p is in millibars.
- *Doppler broadening*. This is a frequency shift in the radiation received from a moving source. When a source emitting monochromatic radiation at frequency f moves towards an observer, the emitted wave trains arrive at a higher frequency. This also affects the radiowave frequency, subject to absorption and dispersion by a gas molecule which is in thermal motion. The Doppler half-width may be expressed in the form

$$\Delta f_{\mathrm{d}} \text{ (GHz)} = 3.581 \times 10^{7} f (T/m)^{1/2} \tag{2.2.15}$$

  where m (u) is the molecular mass of the gas.
- *Zeeman broadening* (Liebe, 1981). When an atom or a molecule is placed in a magnetic field, the energy levels undergo a splitting known as the Zeeman effect. In other words, the third component J_z of the rotational quantum momentum $\boldsymbol{J}$ now becomes essential in the energy quantum level computation. The atmosphere is subject to the earth's magnetic field which, although weak, is capable of generating a splitting, i.e. an O_2 molecular level characterized by a given J value experiences a split into

$2J + 1$ sublevels. The separation energy between such levels is very small (corresponding to 1–2 MHz) and does not affect the millimetre wave region. Nevertheless, this splitting causes broadening of the original unperturbed level known as Zeeman broadening. A good approximation valid for all absorption lines appears to be

$$\Delta f_z \text{ (GHz)} \approx 25.2H \qquad (2.2.16)$$

where H (T) is the earth's magnetic field strength.

For total broadening due to all the above-mentioned phenomena, we have adopted the solution suggested by Liebe (1975) and Liebe, Gimmestad and Hopponen (1977):

$$\Delta f = [(\Delta f_p)^2 + (\Delta f_d + \Delta f_z)^2]^{1/2}. \qquad (2.2.17)$$

In the shape factors for absorption and dispersion another so-called 'non-linear' term has been introduced. If this term is neglected, equations (2.2.11) and (2.2.12) involve a simple sum over all resonant frequencies of the oxygen molecule: we then have a linear addition of the contributions coming from the isolated resonances, each of which is characterized by its Lorentzian shape factor weighted with the Boltzmann distribution law. It should be emphasized, however, that with such a mechanism we suppose that no interactions occur between adjacent resonances. The physical phenomenon is more complicated in that, with increasing pressure, a mutual interaction between the resonances arises, which changes the isolated shape factor. To date there is no completely satisfactory theory able to provide a quantitative description of this phenomenon.

However, Rosenkrantz (1975) has suggested a mechanism that allows good agreement with experimental results to be obtained. It involves introducing a suitable non-linear contribution into the Lorentzian line of shape of absorption and dispersion which may be rewritten as

$$F_{K^{\pm}_{-}} = \frac{\Delta f - y^{\pm}_{K}(f_{K^{\pm}_{-}} - f)p}{(f_{K^{\pm}_{-}} - f)^2 + (\Delta f)^2} + \frac{\Delta f - y^{\pm}_{K}(f_{K^{\pm}_{-}} + f)p}{(f_{K^{\pm}_{-}} + f)^2 + (\Delta f)^2}, \qquad (2.2.18)$$

$$F^{(d)}_{K^{\pm}_{-}} = \frac{f_{K^{\pm}_{-}} - f + y^{\pm}_{K}\Delta f p}{(f_{K^{\pm}_{-}} - f)^2 + (\Delta f)^2} - \frac{f_{K^{\pm}_{-}} + f + y^{\pm}_{K}\Delta f p}{(f_{K^{\pm}_{-}} + f)^2 + (\Delta f)^2}. \qquad (2.2.19)$$

These shape factors (2.2.18) and (2.2.19) are modified by the explicit dependence on pressure p, but the most important quantities are the y_K parameters, introduced by Rosenkrantz.

2.2.3 Water vapour

Water molecules tend to associate through hydrogen bonds having about one-tenth of the strength of a molecular bond. This happens as the electron-rich end of O in the polar molecule H_2O attracts an electron-poor H of a neighbour H_2O. This physical property of water molecules leads to different results according to the phase (solid, liquid, gaseous) of the molecules. In the gaseous phase, singly H-bonded molecules may associate to dimers or aggregate into clusters of preferred sizes (10–50 molecules) under the influence of ion activity.

The number of dry air molecules per unit volume is

$$M_{\mathrm{da}}\ (\text{molecules}\,\mathrm{m}^{-3}) = 2.415 \times 10^{23}\, pt \tag{2.2.20}$$

where p (mbar) is the partial pressure of dry air and $t = 300\ \mathrm{K}/T$ is the relative inverse temperature.

Water vapour is an imperfect gas. From thermodynamic measurements, it is known that there are slightly more H_2O molecules per unit volume than predicted by the ideal gas law. This correction is made by introducing a term $B(t)$. So the number of water vapour molecules per unit volume M_{w} is

$$M_{\mathrm{w}}\ (\text{molecules}\,\mathrm{m}^{-3}) = \left[\frac{2.989 \times 10^{-23}}{\rho} + B(t)\right]^{-1} \tag{2.2.21}$$

with ρ being the water vapour density:

$$\rho\ (\mathrm{g\,m}^{-3}) = 7.219et \tag{2.2.22}$$

with e (mbar) being the partial pressure of water vapour.

Very few values for $B(t)$ have been reported (Bohlander, 1979; Curtiss, Frurip and Blander, 1979). Table 2.1 gives $B(t)$ at various temperatures. The deviations from ideal behaviour

$$\Delta M_{\mathrm{w}} = \frac{M_{\mathrm{w}}}{(M_{\mathrm{w}})_{B=0}} - 1$$

Table 2.1 $B(t)$ at various temperatures

T (K)	t	$B(t)$ ($\times 10^{-27}$ m^3)
250	1.2	−0.9
300	1	−1.9
333	0.9	−2.8

Table 2.2 Deviation of M_w at saturation pressure and various temperatures

T (K)	e (g m^{-3})	Deviation ($\times 10^{-3}$)
250	0.82	0.03
300	25.5	1.6
333	130	12.3

are small, even at saturation. Table 2.2 gives the values of the deviation at various temperatures.

The presence of more H_2O molecules than predicted by the ideal gas law has lead to the postulation of the existence of a dimer molecule $(H_2O)_2$. Its molecular structure, and the millimetre wave spectrum (Viktorova and Zhevakin, 1967, 1971; Bohlander, 1979) are well established; however, its number density under tropospheric conditions is not known.

The dimer number density is expected to depend strongly on temperature since the hydrogen bond strength is rather weak. The dimer density M_d was proposed by Viktorova and Zhevakin (1967) to follow the expression

$$M_d \text{ (molecules m}^{-3}) = M_w^2 \kappa(t). \tag{2.2.23}$$

An approximate equation for the temperature coefficient $\kappa(t)$ was derived from expressions given by Bohlander:

$$\kappa(t) \text{ (m}^3) \approx 1.8 \times 10^{-27} t^{5.6} \tag{2.2.24}$$

which when combined with equations (2.2.21) and (2.2.23) yields

$$M_d \text{ (molecules m}^{-3}) \approx 2.0 \times 10^{18} p^2 t^{5.6} \tag{2.2.25}$$

and for the fractional dimer concentration

$$M_d/M_w \approx 6.1 \times 10^{-5} p t^{5.6}. \tag{2.2.26}$$

There are also other estimates of the fractional dimer concentration suggested by other investigators. These estimations give for M_d/M_w values close to those from equation (2.2.26).

2.2.4 Line shapes at high altitudes

Close to the earth's surface – that is below 30 km – the physical mechanism that dominates spectral line broadening is collisional interaction between molecules. The Van Vleck and Weisskopf profile is then appropriate. However, especially for oxygen, a modified Van Vleck–Weisskopf profile, suggested by Rosenkrantz, has to be used, in order to take into account pressure-induced interference resulting in closely spaced absorption lines.

At higher altitudes, up to 50 km approximately, the line widths become small enough to reduce the Van Vleck–Weisskopf profile to Lorentzian.

At sufficiently low pressures, that is at altitudes above 110 km, collisional broadening is negligible compared with thermal broadening. So, the appropriate line shape function which describes thermal broadening has to be used. This is described by the well-known Doppler line shape function.

Where collisional and thermal broadening become comparable, that is between 50 and 100 km, the resonant frequency in the collision shape expression should be convolved with the Doppler shift probability distribution over all possible Doppler shifts. The convolution is called the Voight profile. A detailed treatment of the applicability of the Voight profile in relation to the calculation of the spectra of various atmospheric constituents in the earth's atmosphere has been given by Papatsoris and Watson (1993). Figure 2.2 illustrates that the Voight profile ought not to be neglected in the calculation of spectra of atmospheric gases at high altitudes.

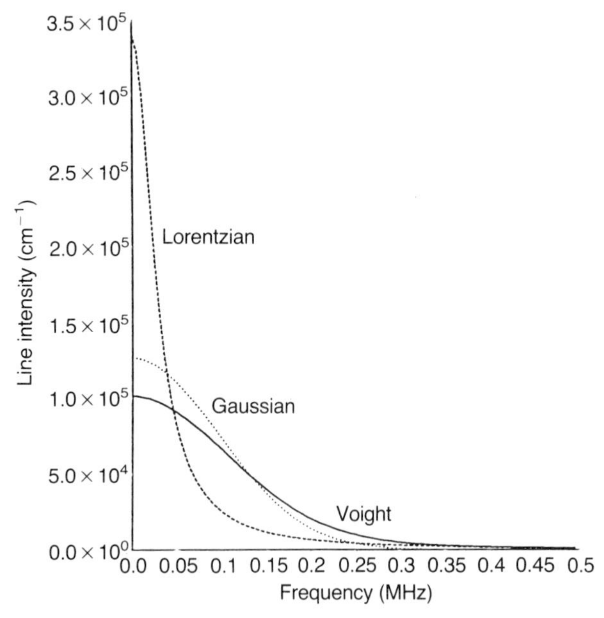

Fig. 2.2 Line shapes for the 184.5 GHz ozone line at 80 km altitude (the wavenumber on the ordinate is proportional to the energy density).

2.2.5 Spectra of trace gases

In the 1–300 GHz frequency range, the clear atmosphere attenuation is due primarily to the absorption lines of water vapour and oxygen. If one were interested only in communication channels in this frequency range, then only oxygen and water vapour would need to be considered.

However, for applications in atmospheric remote sensing and in navigation, we require characterization of other atmospheric constituents, including ozone (O_3), carbon monoxide (CO) and nitrous oxide (N_2O), for completeness and accuracy. The spectral signatures of the most important atmospheric constituents are illustrated in Fig. 2.3 as derived from the US Air Force Geophysical Laboratory (AFGL) HITRAN database (section 2.5).

Calculations of attenuation and dispersion for ozone and nitrous oxide are illustrated in section 2.5.4.

2.3 CURRENT APPROACHES TO MODELLING

2.3.1 Introduction

The calculation of the attenuation and dispersion produced by dry or moist air is very complicated, requiring significant computational effort for each value of temperature, pressure and water vapour concentration. Furthermore, the lack of exact measurements for the preceding atmospheric quantities in most practical cases, the pronounced spatial and time invariability observed for these quantities and the limited accuracy of the available instrumentation mean that a full computer evaluation, encompassing time and spatial variability, may be more time consuming than clearly useful.

In the next sections, approaches to modelling are described, which often use averaged quantities as a practical step to providing answers for system designers. In section 2.5 computer-based methods are also described which take advantage of databases for model atmospheres. The latter may be particularly appropriate in situations where specific atmospheric trajectories must be considered owing to the motions of aircraft or spacecraft.

Many approaches to prediction are based on reference atmospheres. These include

- Air Research and Development Command (ARDC),
- summer mid-latitude (45° N),
- winter mid-latitude (45° N),
- subarctic summer (60° N),
- subarctic winter (60° N), and
- tropical annual average (15° N).

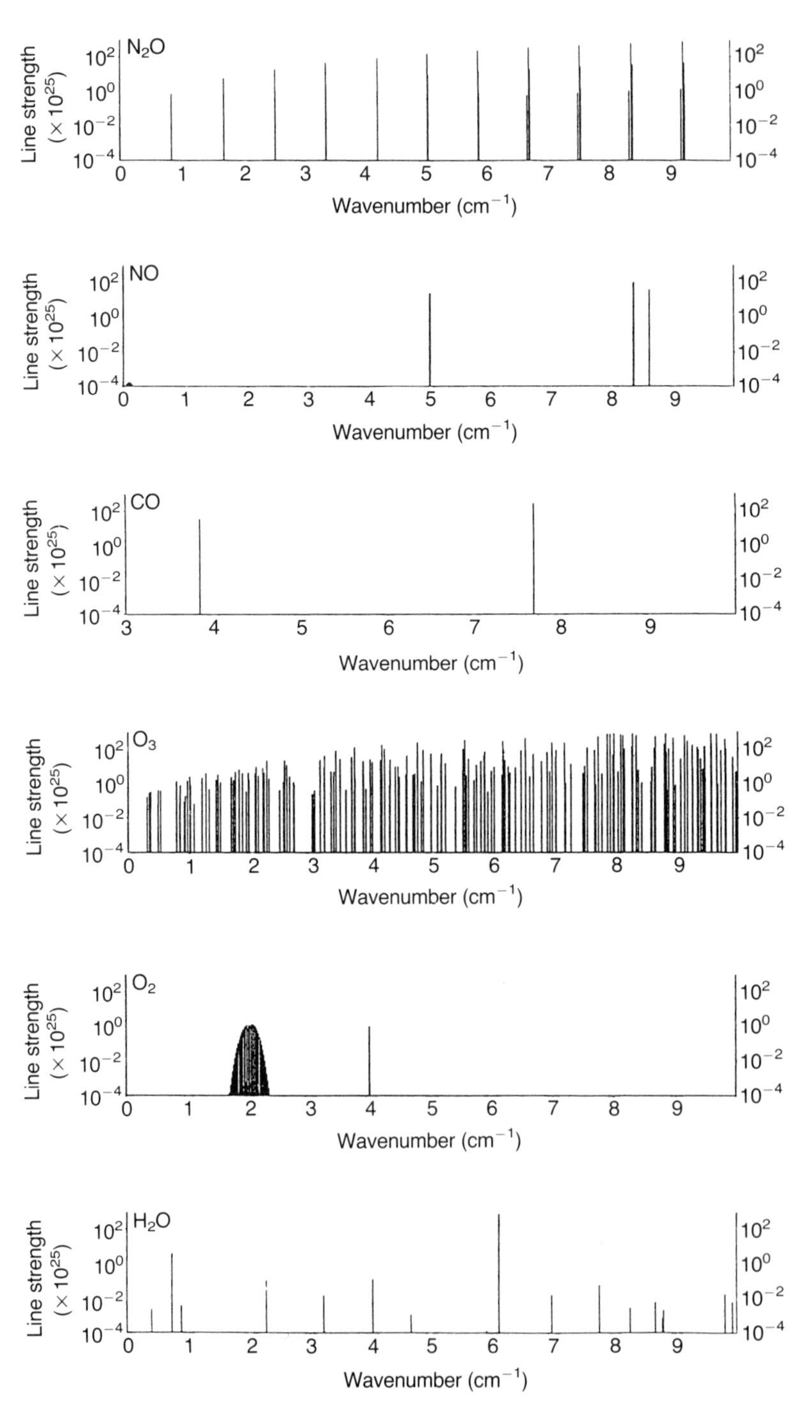

Fig. 2.3 Spectral signatures of the most important atmospheric constituents.

Particularly common is the usage of the ARDC atmosphere, which may be considered as representative of idealized middle-latitude, year-round mean pressure and temperature conditions for the daylight hours, for the range of solar activity between sunspot minimum and sunspot maximum. Figure 2.4 summarizes the characteristics of the ARDC atmosphere.

2.3.2 The CCIR semi-empirical models

(a) General approach

An appropriate technique which may be applied for many practical situations has been adopted by the CCIR (1990) in Report 719. This uses an approximation based on the Van Vleck–Weisskopf line shapes, with coefficients adjusted to fit the results of computer calculations on available measurements (Liebe, 1985). The resulting model takes into account only the attenuation effects of oxygen and water vapour.

The attenuation in the atmosphere over a path length L (km) is then given by

$$A\ (\text{dB}) = \int_0^L \alpha(z)\,\mathrm{d}z \tag{2.3.1}$$

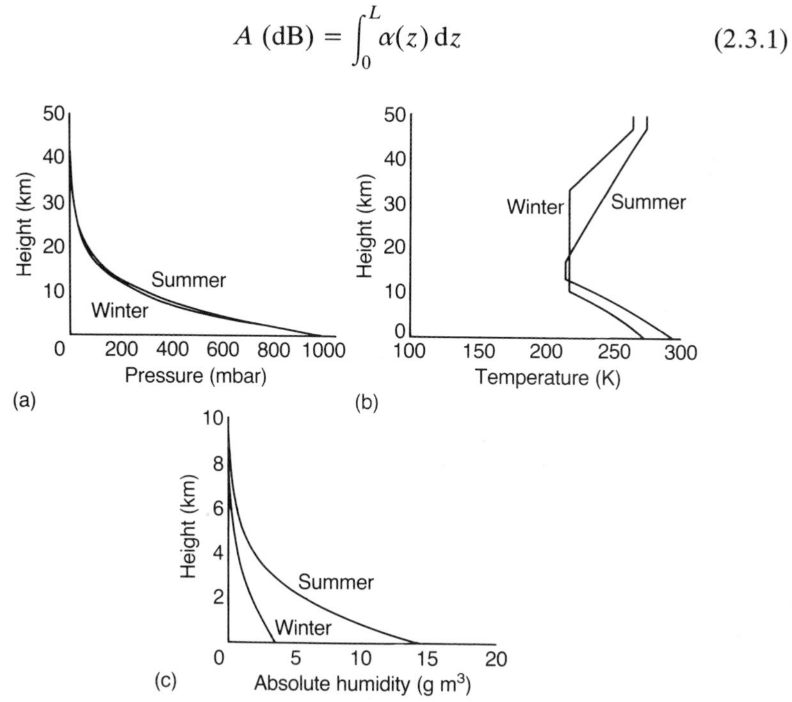

Fig. 2.4 Height profiles of the ARDC atmosphere at 45° N (midlatitude) (Damosso, Stola and Brussaard, 1983): (a) atmospheric pressure–height profile; (b) atmospheric temperature–height profile; (c) atmospheric humidity–height profile.

with

$$\alpha(z)\ (\mathrm{dB\,km^{-1}}) = \alpha_{\mathrm{O}}(z) + \alpha_{\mathrm{w}}(z). \tag{2.3.2}$$

α is the absorption coefficient and is composed of the oxygen absorption α_{O} and the water vapour absorption α_{w}.

The approximation for the oxygen absorption at ground level (1013 mbar) and 15 °C is given by

$$\alpha_{\mathrm{O}}\ (\mathrm{dB\,km^{-1}}) = \left[7.19 \times 10^{-3} + \frac{6.09}{f^2 + 0.227} + \frac{4.81}{(f-57)^2 + 1.50}\right] f^2 \times 10^{-3} \tag{2.3.3}$$

for $f < 57$ GHz and

$$\alpha_{\mathrm{O}}\ (\mathrm{dB\,km^{-1}}) =$$

$$\left[3.79 \times 10^{-7} f + \frac{0.265}{(f-63)^2 + 1.59} + \frac{0.028}{(f-118)^2 + 1.47}\right](f + 198)^2 \times 10^{-3} \tag{2.3.4}$$

for $f > 63$ GHz. It can be seen from these formulae that the resonances of oxygen at 57 GHz, 63 GHz and 118 GHz are modelled as poles in denominators, like the form factors in equation (2.2.11). Between 57 GHz and 63 GHz many overlapping absorption lines are present; these are excluded from the model, but they are given in a figure extracted from the work by Liebe (Fig. 2.5).

The approximation for water vapour used in the CCIR model was given earlier by Gibbins (1986) and predicts the specific attenuation at sea level for a temperature of 15 °C, i.e.

$$\alpha_{\mathrm{w}}\ (\mathrm{dB\,km^{-1}}) = \left[0.050 + 0.0021\rho + \frac{3.6}{(f-22.2)^2 + 8.5} + \frac{10.6}{(f-183.3)^2 + 9} + \frac{8.9}{(f-325.4)^2 + 26.3}\right] f^2 \rho \times 10^{-4} \tag{2.3.5}$$

for $f < 350$ GHz and $\rho < 12\ \mathrm{g\,m^{-3}}$ where ρ is the water vapour density. The absorption lines at 22.3, 183.3 and 323 GHz are again modelled as poles in the denominators.

The far wing effects of absorption peaks at higher frequencies and for various continuous spectra are modelled as constants and factors in formulae

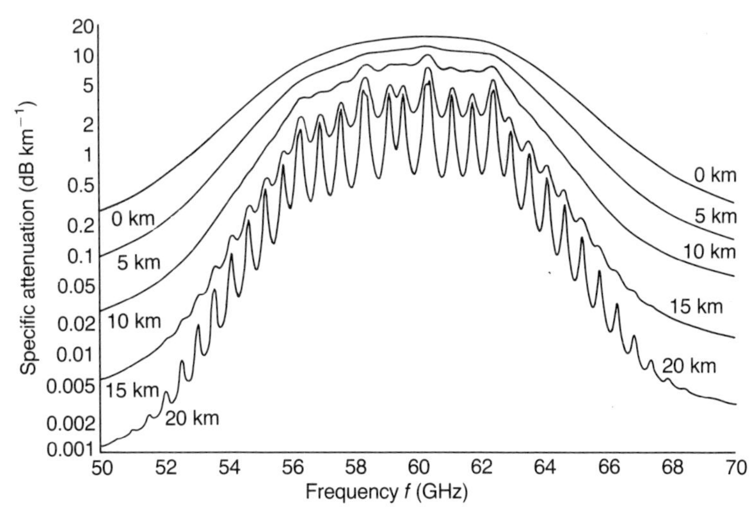

Fig. 2.5 Absorption coefficient in the range 50–70 GHz at various altitudes (CCIR, 1990).

(2.3.3), (2.3.4) and (2.3.5). The CCIR claims that this model evaluates the attenuation due to atmospheric gases for frequencies up to 350 GHz.

The algorithms given above for dry air and water vapour specific attenuation apply at ground level within a pressure range of 1013 ± 50 mbar, at a temperature of 15 °C. Other temperatures may be taken into account by correction factors of $-1.0\%\ °C^{-1}$ from 15 °C for dry air, and $-0.6\%\ °C^{-1}$ from 15 °C for water vapour, valid over the range from -20 °C to $+40$ °C. Note, however, that near resonance lines the temperature correction becomes frequency dependent. A field experiment at 96.1 GHz on a 21.4 km path has confirmed the temperature dependence of attenuation characteristics (Manabe, Debolt and Liebe, 1989).

In applying equation (2.3.5) to water vapour densities higher than about 12 g m^{-3} ($\approx$95% relative humidity at 15 °C) it is important to remember that, except for supersaturation cases (clouds), the water vapour density may not exceed the saturation value valid for the temperature considered. This saturation water vapour density ρ_s may be found in CCIR (1990) Report 563-3. Thus, for high water vapour densities, equation (2.3.5) needs to be corrected for the higher temperatures necessary to sustain such densities.

Overall, taking into account the temperature dependence, the algorithms given by equations (2.3.3), (2.3.4) and (2.3.5) are found to have an accuracy of approximately $\pm 15\%$ over the temperature range from -20 °C to 40 °C and water vapour density range from 0 to 50 g m^{-3}.

The input parameters for the CCIR model are the frequency f and the water vapour density ρ. In most cases the frequency is accurately known. In Table 2.3 typical values of the accuracy of commonly used radiosonde sensors

Table 2.3 Accuracy of radiosonde sensors (Levy and Craig, 1989)

Radiosonde	Accuracy		
	Pressure (mbar)	Temperature (K)	Humidity (%)
AIR AS-3A	3	0.5	5
VAISALA RS 88	0.5	0.2	2

are presented. According to this table the water density at ground level can be measured within a few per cent.

To solve the integral in equation (2.3.1), ρ must be known along the whole path of the radiowave through the atmosphere. Hence the dependence of ρ on height and distance must be known. The vertical dependence can be derived from statistical data of an area in a specific type of climate. The correlation between the point where ρ is measured and a point a few kilometres away is much less. This effect becomes a problem if low elevation angles are used, with relatively long radio paths in the lower atmosphere.

An uncertainty in ρ results in the same uncertainty in the water vapour attenuation α, according to equation (2.3.5). On top of this, for high water vapour densities ($\rho > 12\ \mathrm{g\,m^{-3}}$), equation (2.3.5) is not valid, as stated before. This shows the necessity of knowing ρ along with the whole path with some accuracy.

To account for the change of water vapour and oxygen concentrations along the path of the radiowave, the CCIR gives expressions for the attenuation at different elevation angles. In these expressions separate equivalent heights for oxygen and water vapour components are used. The distance dependence, however, is not accounted for.

In the following the CCIR's suggestions for each case of the elevation angle are described.

(b) Application to terrestrial paths

For a terrestrial path, or for slightly inclined paths close to the ground, the CCIR approach is to write the path attenuation as

$$A\ (\mathrm{dB}) = \alpha L = (\alpha_{\mathrm{O}} + \alpha_{\mathrm{w}})L \tag{2.3.6}$$

where L (km) is the path length.

(c) Application to slant paths (outside the frequency range 57–63 GHz)

For an inclined path or an earth–space path, the expression (2.3.1) must be integrated through the atmosphere to obtain the total path attenuation.

For vertical paths and sea level stations, a first-order approximation is to assume an exponential decay of density with height and to introduce the concept of separate equivalent heights for oxygen and water vapour. These equivalent heights, outside the absorption bands, may be approximated by 6 km for oxygen and by a variable of around 2 km for water vapour, depending on weather conditions, with steep increases at frequencies close to the absorption lines. Using a model atmosphere (Ito, 1987) together with a spectroscopic database (Liebe, 1985), the equivalent height resulting from curve fitting to computer calculations may be expressed as

$$h_O = 6\ \text{km} \qquad \text{for } f < 57\ \text{GHz}, \tag{2.3.7}$$

$$h_O\ (\text{km}) = 6 + \frac{40}{(f-118.7)^2+1} \qquad \text{for } 63\ \text{GHz} < f < 350\ \text{GHz}, \tag{2.3.8}$$

$$h_w\ (\text{km}) = h_{w0}\left[1 + \frac{3.0}{(f-22.2)^2+5} + \frac{5.0}{(f-183.3)^2+6} + \frac{2.5}{(f-325.4)^2+4}\right] \qquad \text{for } f < 350\ \text{GHz} \tag{2.3.9}$$

with

$$A\ (\text{dB}) = \alpha_O h_O + \alpha_w h_w \tag{2.3.10}$$

where h_O is the oxygen equivalent height, h_w is the water vapour equivalent height and h_{w0} is the water vapour equivalent height in the window regions $h_{w0} = 1.6$ km in clear weather and $h_{w0} = 2.1$ km in rain). The equivalent height for water vapour was determined at a ground level temperature of 15 °C. For other temperatures, h_w may be corrected by 0.1% °C^{-1} or 1% °C^{-1} in clear weather or during rain respectively in the window regions and by 0.2% °C^{-1} or 2% °C^{-1} in the absorption bands, between −20 °C and +40 °C.

The concept of equivalent height is based on the assumption of an exponential atmosphere specified by a scale height to describe the decay in density with altitude. Note that scale heights for both dry air and water vapour may vary with latitude, season and/or climate, and that water vapour distributions in the real atmosphere may deviate considerably from the exponential, with corresponding changes in equivalent heights.

As an example, the total one-way zenith attenuation through the atmosphere is shown in Figs 2.6 and 2.7 for clear weather, calculated for the US standard atmosphere, for a dry atmosphere and for an atmosphere with an exponential water vapour layer. These results may be scaled by considering the difference between curves A and B in Fig. 2.6 to different ground level water vapour densities within the range 0–15 g m^{-3} with an accuracy of about ±15% and using the same temperature dependences as given here and in Gibbins (1986). Greater accuracy can be obtained from Liebe's model at any

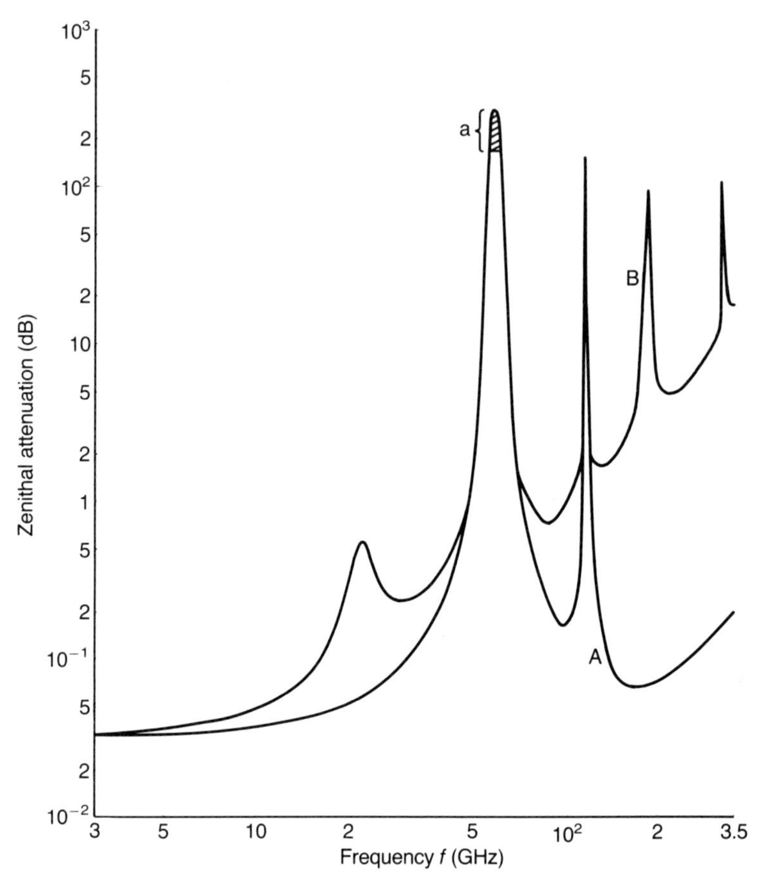

Fig. 2.6 Zenithal attenuation through the atmosphere for clear weather, calculated for the ARDC atmosphere ($p = 1013$ mbar; $T = 15$ °C at ground level): curve A, for a dry atmosphere; curve B, with an exponential water vapour layer of 7.5 $\mathrm{g\,m^{-3}}$ at ground level, and a scale height of 2 km, corresponding to an equivalent height of 1.6 km in the window regions. The region around 60 GHz is shown in more detail in Fig. 2.7.

frequency up to 350 GHz, when the meteorological parameters, especially the water vapour density profile, are known to high accuracy.

Approximation for elevation angles $> 10°$ In this case, it is sufficient to apply a cosecant law. Hence the total attenuation is given by the relation

$$A\ (\mathrm{dB}) = \frac{h_\mathrm{O}\alpha_\mathrm{O} + h_\mathrm{w}\alpha_\mathrm{w}}{\sin \epsilon} \tag{2.3.11}$$

where ϵ is the elevation angle.

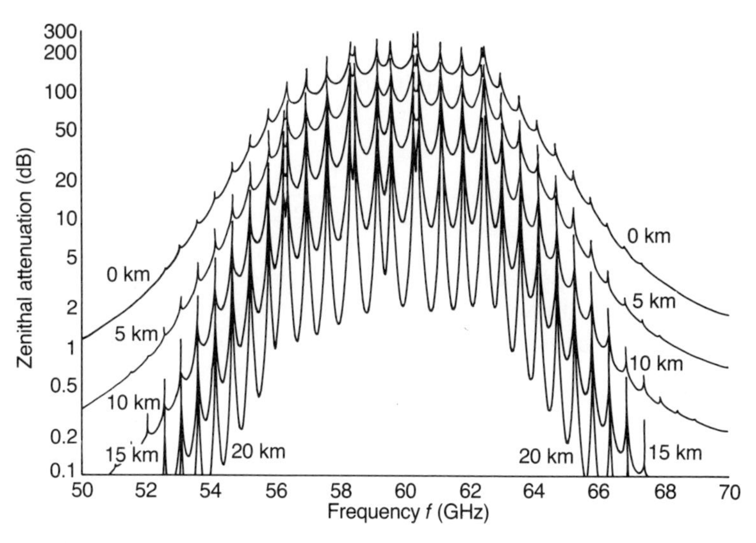

Fig. 2.7 Zenithal oxygen attenuation in the region around 60 GHz for different initial heights and ARDC atmosphere (CCIR, 1990).

These formulae are applicable to cases of inclined paths between a satellite and an earth station situated at sea level. To determine the attenuation on an inclined path between a station situated at altitude h_1 and another at a higher altitude h_2, the values h_O and h_w in equation (2.3.11) must be replaced by the following h'_O and h'_w values:

$$h'_O = h_O(e^{-h_1/h_O} - e^{-h_2/h_O}), \tag{2.3.12}$$

$$h'_w = h_w[1 - e^{(h_1-h_2)/h_w}], \tag{2.3.13}$$

it being understood that the value ρ of the water vapour concentration used in equation (2.3.5) is the value corresponding to altitude h_1 of the lower station in question.

Elevation angles between 0° and 10° In this case, the relations (2.3.11), (2.3.12) and (2.3.13) must be replaced by more accurate formulae allowing for the real length of the atmospheric path. This leads to the following relation:

$$A\ (\mathrm{dB}) = \frac{a_e^{1/2}}{\cos\epsilon}\{\alpha_O h_O^{1/2} F[(a_e/h_O)^{1/2}\tan\epsilon] + \alpha_w h_w^{1/2} F[(a_e/h_w)^{1/2}\tan\epsilon]\} \tag{2.3.14}$$

where a_e is the effective earth radius including refraction, given in Report 718 (CCIR, 1990), expressed in kilometres (a value of 8500 km is generally acceptable for the immediate vicinity of the earth's surface) and $F(x)$ is a function defined by

$$F(x) = \frac{1}{0.661x + 0.339(x^2 + 5.51)^{1/2}}. \tag{2.3.15}$$

Equation (2.3.14) is applicable to cases of inclined paths between a satellite and an earth station situated at sea level. To determine the attenuation values on an inclined path between a station situated at altitude h_1 and another at a higher altitude h_2, the relation (2.3.14) must be replaced by the following:

$$A = \frac{{a_e}^{1/2}}{\cos\epsilon}\{\alpha_O {h_O}^{1/2}[F(x_1)\,\mathrm{e}^{-h_1/h_O} - F(x_2)\,\mathrm{e}^{-h_2/h_O}] + \alpha_w {h_w}^{1/2}[F(x_1') - F(x_2')\,\mathrm{e}^{-(h_2-h_1)/h_w}]\} \tag{2.3.16}$$

where

$$x_i = \cos\epsilon\left[\sin\epsilon\tan^2\epsilon\left(\frac{a_e}{h_O}\right)^{1/2} + \left(\frac{a_e}{h_O}\sin^2\epsilon + 2\frac{h_i}{h_O} + \frac{h_i^2}{2a_e h_O}\right)^{1/2}\right] \text{ for } i = 1, 2,$$

$$x_i' = \cos\epsilon\left[\sin\epsilon\tan^2\epsilon\left(\frac{a_e}{h_w}\right)^{1/2} + \left(\frac{a_e}{h_w}\sin^2\epsilon + 2\frac{h_i}{h_w} + \frac{h_i^2}{2a_e h_w}\right)^{1/2}\right] \text{ for } i = 1, 2,$$

it being understood that the value ρ of the water vapour concentration used in equation (2.3.5) is the value corresponding to altitude h_1 of the lower station in question.

If h_2 is greater than about 1 km, a_e should be recomputed using equation (4) of Report 718 (CCIR, 1990) as it decreases rapidly with altitude towards the actual radius of the earth (6370 km).

Elevation angles < 0° This could be the case of a radio path between two satellites, for example. Most of the attenuation occurs in the portion of the path which is closest to the earth. In order to compute the attenuation, find the point on the path closest to the surface of the earth corresponding to a height, h_3, above the earth, again computed on the basis of an effective earth radius. Find the attenuation for the two path segments, namely from h_3 to h_1 and from h_3 to h_2. Note that the elevation angle of these two paths is zero at point h_3. The total attenuation is the sum of the attenuations for each partial path.

2.3.3 Liebe's semi-empirical model for line-by-line calculation

For atmospheres other than the simple one considered in CCIR's model and for any frequency below about 350 GHz (and especially between 50 and 70 GHz), it is more appropriate to determine the gaseous attenuation by means of a line-by-line calculation, in which the particular atmospheric conditions, pressure, temperature and humidity, can be specified along any path through the atmosphere.

It should be noted, however, that the accuracy of this procedure is strongly dependent on the accuracy with which the meteorological input parameters are known. The distribution of water vapour in the atmosphere, in particular, can be subject to much uncertainty, as noted earlier.

Below are given details of the microwave propagation model (MPM) of Liebe (1985) using a 'reduced line base'. This model, which takes into account only the most important resonance lines of the oxygen and water vapour absorption spectrum, enables the calculation of gaseous attenuation at frequencies up to 350 GHz at any temperature, pressure and humidity. For frequencies higher than 350 GHz, the full line base should be employed (Liebe, 1985). (Note, also, that some of the constants given in this section are those revised by Liebe in private communication, rather than those in Liebe, 1985.)

The complex refractivity N, which is related to the refractive index n, can be written in real and imaginary parts as

$$N(f)\ (\text{ppm}) = (n-1)10^6 = N_0 + N'(f) + \mathrm{j}N''(f) \qquad (2.3.17)$$

The real frequency-dependent part $N'(f)$ is the cause of dispersive effects; the imaginary part $N''(f)$ causes absorption. The gaseous absorption coefficient is related to the latter as

$$\alpha\ (\text{dB km}^{-1}) = 10^6(4\pi/c)10\log \mathrm{e}\, fN''(f) = 0.1820 \times 10^6 fN''(f) \qquad (2.3.18)$$

where f is in gigahertz.

The frequency dependence of $N''(f)$ can be expressed as

$$N''(f) = \sum_i S_i F_i(f) + N''_{\mathrm{d}}(f) + N''_{\mathrm{w}}(f). \qquad (2.3.19)$$

S_i is the strength of the ith line, F_i is the line shape factor and the sum extends over all the lines; $N''_{\mathrm{d}}(f)$ and $N''_{\mathrm{w}}(f)$ are dry and wet continuum spectra.

The line strength is given by

$$S_i = a_{1i} p t^3\, \mathrm{e}^{\,a_{2i}(1-t)} \text{ for oxygen lines} \qquad (2.3.20)$$

and

$$S_i = b_{1i} e t^{3.5} \mathrm{e}^{b_{2i}(1-t)} \text{ for water vapour lines} \tag{2.3.21}$$

where p (mbar) is the dry air pressure, e (mbar) is the water vapour partial pressure (total barometric pressure $P = p + e$) and $t = 300\ \mathrm{K}/T$. The coefficients a_{1i} and a_{2i} for oxygen are given in Table 2.4 and those for water vapour, b_{1i} and b_{2i}, are given in Table 2.5.

The line shape factor is given by

$$F_i(f) = \frac{f}{f_i}\left[\frac{\Delta f - s(f_i - f)}{(f_i - f)^2 + (\Delta f)^2} + \frac{\Delta f - s(f_i + f)}{(f_i + f)^2 + (\Delta f)^2}\right] \tag{2.3.22}$$

where f_i is the line frequency and Δf is the width of the line:

$$\Delta f = a_{3i}(pt^{0.8} + 1.1et) \text{ for oxygen,} \tag{2.3.23}$$

$$\Delta f = b_{3i}(pt^{0.6} + 4.80et^{1.1}) \text{ for water vapour} \tag{2.3.24}$$

and s is a correction factor which arises from interference effects in the oxygen lines (Rosenkrantz, 1975) given by

$$s = a_{4i} p t^{a_{5i}} \quad \text{for oxygen} \tag{2.3.25}$$

and

$$s = 0 \qquad \text{for water vapour} \tag{2.3.26}$$

The spectroscopic coefficients are given in Tables 2.4 and 2.5. The line shape functions (2.3.22) are modified versions of those by Van Vleck and Weisskopf, which are used in equation (2.2.11).

The dry air continuum arises from the non-resonant Debye spectrum of oxygen below 10 GHz and a pressure-induced nitrogen attenuation above 100 GHz.

$$N''_{\mathrm{d}}(f) = fpt^2\left\{\frac{6.14 \times 10^{-5}}{d[1 + (f/d)^2][1 + (f/60)^2]} + 1.4 \times 10^{-11}(1 - 1.2 \times 10^{-5} f^{1.5})pt^{1.5}\right\} \tag{2.3.27}$$

where f is in gigahertz and d is the width parameter for the Debye spectrum:

$$d\ (\mathrm{GHz}) = 5.6 \times 10^{-4}(p + 1.1e)t^{0.8}. \tag{2.3.28}$$

Table 2.4 Spectroscopic parameters for oxygen lines

f_i (GHz)	a_{1i} ($\times 10^{-7}$ kHz mbar^{-1})	a_{2i}	a_{3i} ($\times 10^{-4}$ GHz mbar^{-1})	a_{4i} ($\times 10^{-4}$ mbar^{-1})	a_{5i}
51.5034	6.08	7.74	8.90	5.60	1.8
52.0214	14.14	6.84	9.20	5.50	1.8
52.5424	31.02	6.00	9.40	5.70	1.8
53.0669	64.10	5.22	9.70	5.30	1.9
53.5957	124.70	4.48	10.00	5.40	1.8
54.1300	228.00	3.81	10.20	4.80	2.0
54.6712	391.80	3.19	10.50	4.80	1.9
55.2214	631.60	2.62	10.79	4.17	2.1
55.7838	953.50	2.12	11.10	3.75	2.1
56.2648	548.90	0.01	16.46	7.74	0.9
56.3634	1344.00	1.66	11.44	2.97	2.3
56.9682	1763.00	1.26	11.81	2.12	2.5
57.6125	2141.00	0.91	12.21	0.94	3.7
58.3269	2386.00	0.62	12.66	−0.55	−3.1
58.4466	1457.00	0.08	14.49	5.97	0.8
59.1642	2404.00	0.39	13.19	−2.44	0.1
59.5910	2112.00	0.21	13.60	3.44	0.5
60.3061	2124.00	0.21	13.82	−4.13	0.7
60.4348	2461.00	0.39	12.97	1.32	−1.0
61.1506	2504.00	0.62	12.48	−0.36	5.8
61.8002	2298.00	0.91	12.07	−1.59	2.9
62.4112	1933.00	1.26	11.71	−2.66	2.3
62.4863	1517.00	0.08	14.68	−4.77	0.9
62.9980	1503.00	1.66	11.39	−3.34	2.2
63.5685	1087.00	2.11	11.08	−4.17	2.0
64.1278	733.50	2.62	10.78	−4.48	2.0
64.6789	463.50	3.19	10.50	−5.10	1.8
65.2241	274.80	3.81	10.20	−5.10	1.9
65.7648	153.00	4.48	10.00	−5.70	1.8
66.3021	80.09	5.22	9.70	−5.50	1.8
66.8368	39.46	6.00	9.40	−5.90	1.7
67.3696	18.32	6.84	9.20	−5.60	1.8
67.9009	8.01	7.74	8.90	−5.80	1.7
118.7503	945.00	0.00	15.92	−0.13	−0.8

Table 2.5 Spectroscopic parameters for water vapour lines

f_i (GHz)	b_{1i} ($\times$kHz mbar^{-1})	b_{2i}	b_{3i} ($\times 10^{-4}$ GHz mbar^{-1})
22.235	0.0109	2.143	27.84
183.310	0.2300	0.653	31.64
325.153	0.1540	1.515	29.70

The wet continuum, $N''_w(f)$, is included to account for the fact that measurements of water vapour attenuation are generally in excess of those predicted using the theory described by equations (2.3.14)–(2.3.28), plus a term to include the effects of high-frequency water vapour lines not included in the reduced line base:

$$N''_w(f) = 1.18 \times 10^{-8}(p + 30.3et^{6.2})fet^{3.0} + 2.3 \times 10^{-10}pe^{1.1}t^2f^{1.5}. \quad (2.3.29)$$

Note that if the full line base, given by Liebe, is employed, the second term in equation (2.3.29) is not required. Note also that the spectroscopic parameters given in Tables 2.4 and 2.5 together with numerical constants in the above equations have been altered from those given by Liebe to allow pressures to be given in millibars.

The dispersion coefficient β is related to the real part of $N(f)$ as follows:

$$\beta\,(\text{rad km}^{-1}) = (2\pi/c)fN'(f) = 0.0210fN'(f). \quad (2.3.30)$$

The most recent expressions for $N'(f)$ are given by Liebe and Hufford (1989), as well as an expression for the frequency-independent part of $N(f)$, N_0. Generally, $N'(f)$ is less than 1% of N_0 as given in CCIR's Report 563, at frequencies above 350 GHz.

To calculate gaseous attenuation along a vertical path through the atmosphere, it generally suffices to divide the atmosphere into horizontal layers 1 km thick, assigning to each layer a pressure, temperature and water vapour partial pressure, for example from the US Standard Atmosphere, and to sum the attenuations for each layer. For elevation angles other than the zenith, the total one-way attenuation through the atmosphere is considered by means of a path extension factor, s, by which the attenuation is multiplied.

For a stable, spherically symmetric and horizontally stratified atmosphere, and at elevation angles between 10° and 90°, the increase in path length through the atmosphere follows a cosecant law:

$$s = 1/\sin\epsilon. \quad (2.3.31)$$

At elevation angles below about 10°, however, the refractive index of the atmosphere and the curvature of the earth's surface must be considered, and the path length extension for each layer of the atmosphere should be determined from

$$s = \left\{1 - \left[\frac{N_i(a_e + h_i)}{N_j(a_e + h_j)}\cos\epsilon\right]^2\right\}^{1/2} \quad (2.3.32)$$

where $N_{i,j}$ is the complex refractivity N at altitude $h_{i,j}$ and $a_e = 6370$ km is the radius of the earth. It should be noted that equation (2.3.32), which is the spherical form of Snell's law, is only valid when N is purely real. In the cases at hand, N'' is small compared with $N_0 + N'$.

To take account properly of refractive bending at very low elevation angles, ϵ should be recalculated along the path according to

$$\cos \epsilon_j = \frac{N_i(a_e + h_i)}{N_j(a_e + h_j)} \cos \epsilon_i \tag{2.3.33}$$

2.4 WEAKNESSES IN CURRENT APPROACHES TO MODELLING

2.4.1 Applicability of current spectroscopic models

Liebe's model consists of a synthesis of theoretically and experimentally derived data in order to calculate the attenuation and dispersion produced by the propagation of a radiowave through the atmosphere. Despite the fact that CCIR recommends the use of this model (if the physical input parameters are accurately known), the model seems to suffer from certain weaknesses as a result of insufficient understanding of various physical processes (section 2.4.3). However, it gives good insight into the dependence of the attenuation on the physical input parameters, such as pressure, temperature and water vapour concentration. Furthermore, the results from the model tend to be very susceptible to variations in the input data. It is therefore necessary to have accurate input data to obtain reliable results.

The classical spectroscopic predictions thus seem to be in good agreement with experimental data near to the absorption peaks for both main gaseous radiowave absorbers, O_2 and H_2O. Furthermore, the classical spectroscopic theory models successfully the oxygen molecule behaviour not only near the absorption peaks but in the windows as well. However, for water vapour, the theory proves to be insufficient to explain the observed measurements. In fact, this is not necessarily a weakness of the spectroscopic theory itself, but it has rather to do with a variety of complicated physical processes which take place in the formation of certain water vapour patterns.

A more detailed critique on the latest water vapour studies is found in the next two sections. The aim is to illustrate the difficulties encountered and to explain why the classical spectroscopic theory fails to predict the measured values.

2.4.2 Water vapour studies

The experimental data are presented in terms of the absorption coefficient α, as a function of f, T, absorbing gas partial pressure p_a and foreign gas partial pressures p_i, $i = 1, 2, \ldots$. A general empirical form for the continuum absorption coefficient used to represent the data is

$$\alpha(f, T, p_1, p_2, \ldots, p_a) = \frac{p_a}{RT}\left[\sum_i C_i(f, T)p_i + C_s(f, T)p_a\right] \quad (2.4.1)$$

where C_s is the self-broadening coefficient of the absorbing gas, C_i is the foreign broadening coefficient due to the ith type of foreign gas and R is the gas constant. The equation can conveniently be written for water vapour as the absorbing gas, and nitrogen and oxygen as the foreign gases, to obtain

$$\alpha(f, T)\,(\text{km}^{-1}) = \frac{C_1(f, T)}{RT} p_a[p_1 + F(f, T)p_2 + B(f, T)p_a] \quad (2.4.2)$$

where $F = C_2/C_1$ and $B = C_s/C_1$ are the dimensionless broadening coefficients. Near line centre, B has the value of 5 for water vapour relative to nitrogen. In the real atmosphere, the effects of oxygen broadening must also be included. The dimensionless broadening coefficient F accounts for oxygen. However, many laboratory experiments ignore the effects of oxygen and use only nitrogen as the broadening gas along with the absorbing gas.

Figure 2.8 shows continuum absorption for 10–1000 GHz (total absorption minus local lines). The full line represents an empirical formula given by Gaut and Reifenstein (1971):

$$\alpha_{\text{cont}}\,(\text{km}^{-1}) = 1.08 \times 10^{-6} \rho t^{2.1}\left(\frac{P}{101}\right) f^2 \quad (2.4.3)$$

where ρ (g m^{-3}) is the water vapour density, $t = 300\text{ K}/T$, P (kPa) is the total pressure and f is in gigahertz. The plotted points indicate experimental data. The formula correctly demonstrates the frequency dependence of the continuum but not the temperature and pressure dependence. More recent work by Liebe (1987) uses a continuum formula, fitted to experimental data at 138 GHz, of the form

$$\alpha_{\text{cont}}\,(\text{dB km}^{-1}) = 1.61 \times 10^{-5} f^2 t^3 e(p + 31.6t^{7.5} e) \quad (2.4.4)$$

where f is in gigahertz and e and $p = P - e$ are in kilopascals.

Comparing equation (2.4.4) with (2.4.2) it is seen that $B = 31.6t^{7.5}$ and $F = 0$. This means a strong dependence on the water vapour partial pressure ($B \gg 5$), while B is inversely dependent on T. On the basis of the work by Liebe and Layton (1983), the parameter B decreases with f increasing from 110 to 833 GHz (Table 2.6). The continuum calculated using equation (2.4.4) is smaller than that calculated using equation (2.4.3), because of the improved local line modelling.

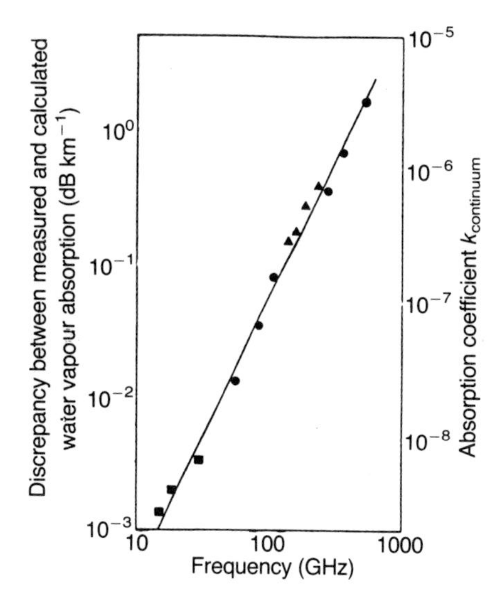

Fig. 2.8 Discrepancy between the measured and calculated water vapour absorption (▲, ●, ■) and empirical correction term given by equation (2.4.3) (——). The plotted points are from Becker and Autler (1946) (▲), Frenkel and Woods (1966) (●) and Burch, Grynvak and Pembrook (1971) (■) ($T = 300$ K; $P = 101$ kPa; $\rho = 1.0\,\mathrm{g\,m^{-3}}$; $k_{\text{continuum}} \propto \nu^2$).

Table 2.6 Experimental frequency dependence of B; $T = 300$ K

f (GHz)	B
110	32
138	31.6
213	20
833	7.4

As equation (2.4.4) indicates, B is not only a function of frequency but a strong function of temperature as well. Although Liebe chooses to represent his data in a power law form, a comprehensive study at 213 GHz by Llewellyn-Jones (1980) shows that the data fit an Arrhenius plot with the functional form

$$C_s(f, T) \propto e^{5\times10^4 K/T}. \tag{2.4.5}$$

Also at this frequency, a strong negative dependence on T is observed. Figure 2.9 illustrates the experimental results of Llewellyn-Jones.

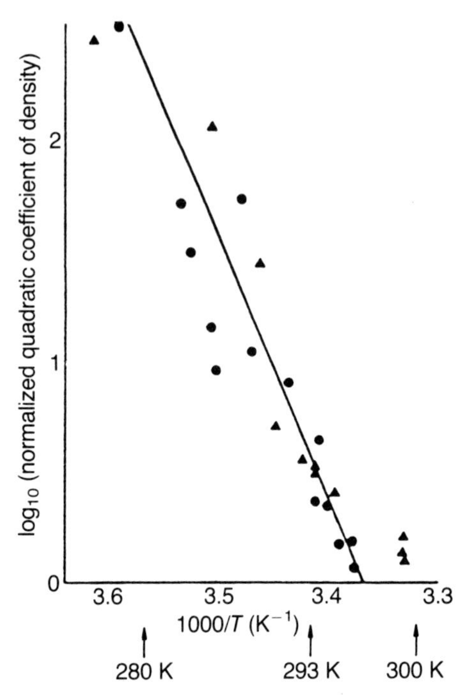

Fig. 2.9 The temperature dependence of the observed continuum absorption due to water vapour (absorption coefficients normalized to values at 293 K): ▲, observations at frequencies in the range 115–126 GHz; ●, observations at 213 GHz.

Other minor sources of continuum absorption in this window region arise from the blue wing of the non-resonant O_2 spectrum and collision-induced absorption of the pure rotational band of N_2.

In summary, the millimetre wave water vapour continuum absorption falls off as f^2, has an enhanced self-broadening contribution that grows with decreasing f and has a strong negative temperature dependence.

2.4.3 Excess water vapour absorption (EWA)

(a) Introduction to EWA

The differences which have been observed between the classical theoretical spectroscopic predictions and experimental data obtained from both laboratory and field measurements, regarding the absorption produced by water vapour in the absorption windows (24–48, 72–112, 128–160, 200–260 GHz), have established the existence of excess water vapour absorption to account for those discrepancies. In Liebe's model the abbreviation EWA is used to describe the term N''_w in equation (2.3.19).

Two schools of thought have evolved to explain EWA:

- a molecular approach searching for water polymers $(H_2)_n$ and their spectra in the atmosphere (sizes of $n = 2$ and $n = 3$ are possibilities);
- liquid water uptake by submicron aerosol particles under conditions of high relative humidity ($> 85\%$).

Each conjecture is supported by some experimental evidence, but appears to be contradicted by other results. There is still no universally accepted physical explanation. Recent observations appear to show absorptions that exceed any of the well-known semi-empirical corrections for the upper millimetre wave and IR regions (Emonns and de Zafra, 1990), although a recent theoretical interpretation (Ma and Tipping, 1990) attempts an explanation in terms of the accumulated absorption from the far wings of allowed transitions.

(b) Definition of EWA

Window attenuation, both model and experimental (previous section), is fitted to expressions in decibels per kilometre of the form

$$\alpha\ (\text{dB km}^{-1}) = 0.1820 f N'' = C p^x t^y (f/30)^z \tag{2.4.6}$$

where x, y and z are the proper exponents of a particular absorption model and f is in gigahertz. To give a clear understanding about the factors that contribute in EWA, one could say that the millimetre wave window attenuation α might very well be a combination of up to five different contributions:

$$\alpha = \alpha_l\ (\text{local lines}) + \alpha_f\ (\text{far wings}) + \alpha_a + \alpha_d + \alpha_x.$$

The last three contributions address the existence of EWA. α_a has to deal with the aerosol liquid water attenuation, α_d with the dimer spectrum, and α_x is a correction term in order to bridge the differences observed between various theoretical assumptions and experimental evidence.

(c) Evidence for EWA

A significant number of experiments have been conducted in order to characterize the continuum absorption of water vapour. These use a wide variety of techniques, falling into two broad categories: (a) measurements within the Earth's atmosphere (or field measurements) and (b) laboratory measurements using either single transit or multi-pass cells (resonators).

A comprehensive review of earlier work is given in the paper by Liebe

(1980) and a more up-to-date summary of measurements in the IR and millimetre wave regions is reported by Hinderling, Sigrist and Kneubühl (1987). Table 2.7 summarizes some of the laboratory studies as given by Liebe (1980). Where the resonator technique has been used the Q factor is indicated in the table. The reader is referred to the paper by Liebe (1980) for further information on individual studies.

Field measurements may be carried out to give three types of data:

- attenuation (giving specific attenuation, in dB km^{-1}) for horizontal line-of-sight paths,
- total zenith attenuation, and
- sky noise, often used to deduce the attenuation value.

Most of the measurements that have been reported were performed at a single frequency; a few were carried out over a broad band. There are several difficulties and weaknesses with such measurements, specifically:

- problems of absolute calibration;
- large scatter (from ±10% to ±30%) in data due to unspecified weather along the path;

Table 2.7 Summary of laboratory studies of continuum H_2O absorption

		Experimental conditions			
Frequency	Foreign gas	ρ (gm^{-3})	T (K)	L (m)	Resonator $Q \times 10^3$
18–31 GHz	air	0–40	318		800
22 GHz	N_2	0–50	312		45
22, 24 GHz		0–20	297	(30)	16
31, 62 GHz		0–35	280–325	(> 100)	> 200
117–120 GHz		0–25	295	150	
170–300 GHz		0–20	295		100
213 GHz	N_2	0–60	270–320	(40)	
210–300 GHz		3–6	273–333	28	
450–960 GHz		1	295	2	
890–965 GHz		0–35	293, 323	10–60	
300–1500 GHz			290–355	5–103	
9–18 THz			283, 329	133	
12–36 THz					
14–27 THz		2–20	293–313	500	
21–38 THz		14	296–388	1185	
28–33 THz	N_2, O_2	0–20	289–301		
75–86 THz		14	298	21	

- scarcity of data at high humidities (> 90%);
- lack of simultaneous recordings of the integrated water vapour density;
- absence of data on integrated liquid water content and visibility in cloud-free air;
- difficulty of fitting data empirically to surface-based meteorological variables.

The horizontal and zenith path data reviewed by Liebe are summarized in Table 2.8, which includes comparisons with model calculations using the US Standard Atmosphere (Liebe and Rosich, 1978). Once again for details of the studies the reader is referred to Liebe (1980).

As expected, in the windows (W1–W5) absorption increases with the water vapour density or path-integrated liquid water content *w*. Also, as observed earlier, measurements in the 100 to 117 GHz range revealed considerably higher values of water vapour absorption than were predicted by H_2O line shape theory.

2.5 COMPUTATIONS BASED ON LINE-BY-LINE SUMMATIONS

2.5.1 Introduction

Line-by-line summations are necessarily demanding on computer time, especially if an integral along a radio path is required; nevertheless, in some applications they represent the only practical approach. In this section, comments are given on some of the available computer codes and their associated databases. Some of the results of calculations are also shown. The computations of Liebe (for which a computer code is also available) have already been described (section 2.3.3).

2.5.2 FASCOD

A model and computer code, FASCOD (fast atmospheric signature code) has been developed at the AFGL for the calculation of radiance and transmittance with particular applicability to the earth's atmosphere (Clough *et al.*, 1981).

The program is applicable to spectral regions from the microwave to the visible, and performs its calculations based on a line-by-line summation technique. All the spectroscopic information concerning the seven principal atmospheric absorbers (H_2, CO_2, O_3, N_2O, CO, CH_4 and O_2) and 21 additional molecular species resides in an external database, which is selectively accessed according to the user's requirements.

Table 2.8 Summary of reported EHF radio path attenuation measurements and comparison with model calculations (abstracted from Liebe, 1980)[a]

	Field data			Model (Liebe and Rosich, 1978)	
Horizontal paths at sea level					
Frequency (GHz)	α (dB/km) RH = 0	RH = 100%	Water vapour data fit $\rho(\mathrm{gm}^{-3})$	α (dB km^{-1}) RH = 0	RH = 100%
80	–	0.510	$0.041(2)\rho$	0.010	0.56
130.000	–	1.540	$0.12(3)\rho$	0.020	1.28
171.000	0.200	4.800	$0.375\rho + b\rho^2$	0.010	4.93
250.000	–	4.900	0.38ρ	0.000	4.57
304.000	0.000	41.000	3.2ρ	0.000	8.16
337.000	0.600	18.400	$0.8\rho + 0.05\rho^2$	0.000	16.89
345.000	0.000	8.400	$0.45\rho + 0.016\rho^2$	0.000	13.50
350	–	17.900	1.4ρ	0.000	14.00

Zenith paths from sea level[b]

	Attenuation (dB)		Water vapour data fit		Attenuation (dB)	
	Dry term	Total	ρ_0 (gm^{-3})	w (cm)	Dry term	Total
15.000	0.055	0.106	0.004ρ		0.050	0.11
15.000	0.046*	0.085*	0.003ρ		0.050	0.11
20.600	0.11(2)	0.52(2)			0.090	0.56
22.000 W0	0.33(3)	1.150		$0.27(2)w$	0.065	0.88
22.200	0.200	0.900	0.70		0.066	0.90
22.235	0.110	0.720	$0.043(8)\rho$		0.066	0.91
31.700	0.13(2)	0.35(2)		$0.032w^2 - 0.025w$	0.120	0.35
35.00	0.168*	0.300*	0.010ρ		0.150	0.39

35.000	W1	0.170	0.340	0.013(2)ρ		0.150	0.39
35.00		0.100	0.250			0.150	0.39
36.00		0.15(2)	0.640	0.038(6)$\rho + b\rho^2$		0.160	0.41
80.000		0.5(2)	1.400			0.450	1.37
90.000		0.170	0.940	0.6(2)0		0.250	1.38
91.000		0.320	1.510		$0.2w + 0.06w^2$	0.240	1.40
95.000	W2	0.41(4)	1.500		0.35(2)w	0.220	1.48
95.000		0.250	1.350			0.220	1.48
110.000		0.6(2)	2.180	0.17(3)ρ		0.450	2.10
111.000		0.97(17)	2.400		0.45(14)w	0.470	2.20
118.000		10.000	12.000			14.110	16.45
123.000		1.67(6)	3.400		0.56(4)w	1.110	3.29
150.000	W3	0.28(25)	4.000		1.19(14)w	0.100	3.62
210.000			9.200	0.72ρ		0.070	7.63
225.000		0.050	6.900	0.54(30)ρ			
230.000	W4	0.000	4.800	0.37(5)ρ		0.050	7.82
240.000			7.900	0.62		0.040	8.32
300.000			13.400	1.05ρ		0.020	17.40
345.000	W5		23.000	1.18(8)ρ		0.010	31.00
411.000			50.000	4ρ			104.00
667.000			150.000	12ρ			

[a]Figures in parentheses give the standard deviation from the mean in terms of the least significant digit.
[b]An asterisk (*) indicates data extrapolated for a tangential path.

In Figure 2.10, a comparison of a FASCOD calculation with experimental data obtained by Rice and Ade (1979) in the millimetre spectral regions is shown. The data were taken with an interferometer measuring downwelling radiance at the ground from which atmospheric transmittance was inferred. The calculation was performed with a two-layer atmosphere and the water amount is that measured in the Rice and Ade experiment.

2.5.3 Attenuation and dispersion calculations in the oxygen absorption band (50–70 GHz)

In order to obtain an outline of the attenuation and dispersion behaviours for the whole oxygen spectrum, computations have been performed (Damosso, Stola and Brussaard, 1983) of zenithal attenuation and phase shift for the ARDC atmosphere and from various initial heights up to the upper atmosphere. Figures 2.11 and 2.12 show such results. It can be seen from Fig. 2.11 that attenuation from 10 km to the upper atmosphere exceeds 10 dB even in windows between peaks for frequencies between about 56 GHz and about 64 GHz. The corresponding behaviour of the phase–frequency characteristic appears to be sufficiently smooth (with an approximate linear–parabolic trend) outside the 56–64 GHz region. Near the resonance peaks, however, its behaviour is as expected: very oscillatory, with abrupt changes of slope.

These kinds of calculations have been performed for each reference atmosphere previously defined. For brevity, comparisons are presented for a segment of the oxygen absorption band only, which is one of the most interesting portions of the spectrum, for an initial height of 10 km and for two

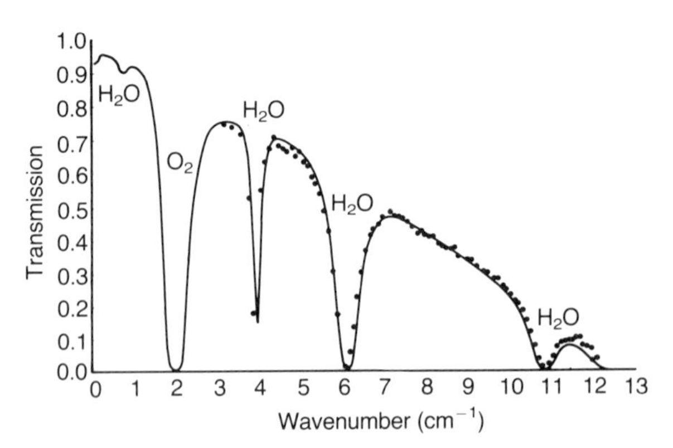

Fig. 2.10 Spectral transmittance inferred from a radiance measurement looking up from the ground: ●, data of Rice and Ade (1979); ——, two-layer FASCOD1B calculation.

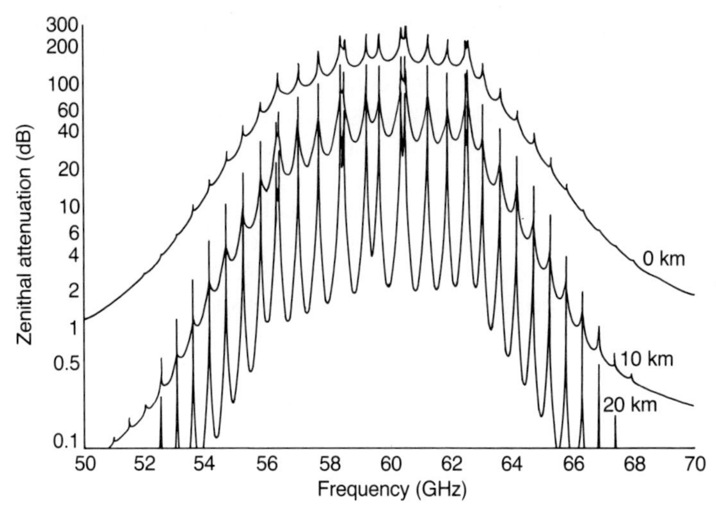

Fig. 2.11 Attenuation in the 50–70 GHz band, up from various initial heights.

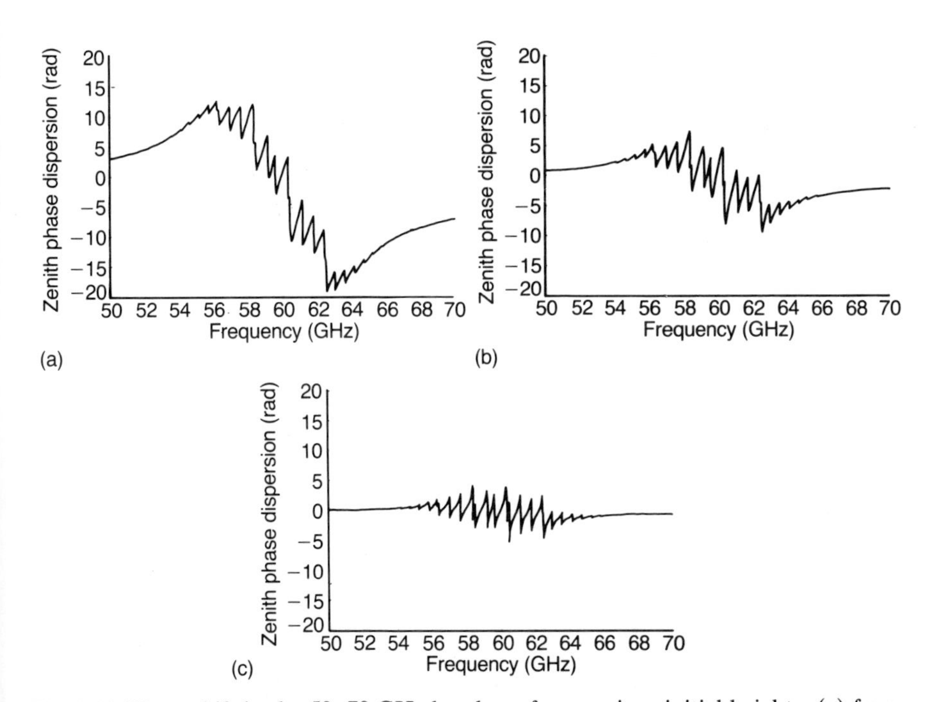

Fig. 2.12 Phase shift in the 50–70 GHz band, up from various initial heights: (a) from 0 km; (b) from 10 km; (c) from 20 km.

extreme conditions (Figs. 2.13 and 2.14). Comparing these figures it can be seen that no dramatic differences show up in absorption or phase shift between the two atmospheres. The conclusion is that the influence of climatic conditions, for heights above 10 km at least, is only limited. This will be due to the low water vapour content and the reduced height dependence of pressure.

At high altitudes (above 50 km) these calculations may not completely represent the line shape profiles, owing to the omission of the Voight profile and Zeeman splitting effects (next section). A more complete treatment, including the Voight profile, is reported in Papatsoris and Watson (1993).

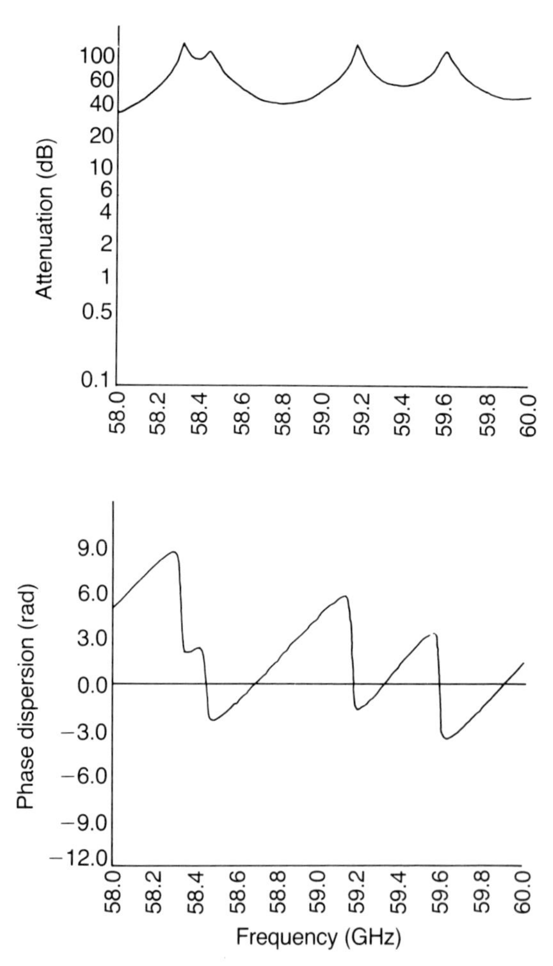

Fig. 2.13 Attenuation and phase shift in the frequency range 58–60 GHz for the 15° N average tropical atmosphere (initial height, 10 km).

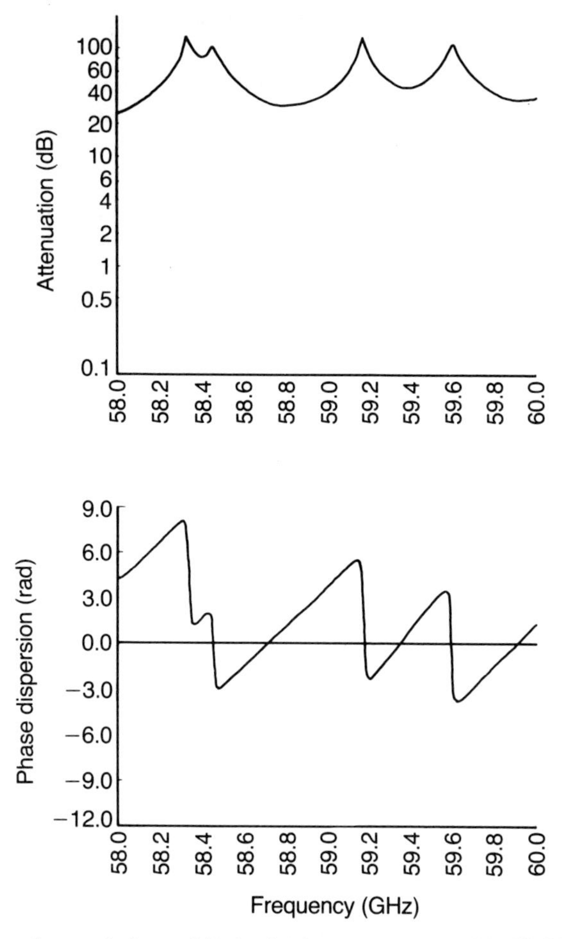

Fig. 2.14 Attenuation and phase shift in the frequency range 58–60 GHz for the 50° N winter subarctic atmosphere (initial height, 10 km).

2.5.4 Further computations

As a result of perceived shortcomings in the FASCOD model, a more comprehensive computer based prediction method is under development (called APM and described in Papatsoris and Watson, 1993).

Again a line-by-line summation technique is employed to provide a complete picture of the attenuation and dispersion of atmospheric gases. The program uses a source of spectroscopic information the AFGL's HITRAN 1986 (high-resolution transmission molecular absorption) database, and is applicable to spectral regions from the microwave to the submillimetre.

Several reference atmospheres may be selected, including user-specified models.

Compared with FASCOD, which performs radiance and transmittance calculations, the APM model calculates attenuation (decibels) and dispersion (radians), being oriented towards the radio system engineer. Also, Voight line shapes are evaluated, using the fitted algorithm suggested by Kielkopf (1973), and the effects of Zeeman splitting are to be included.

Figures 2.15 and 2.16 show the specific attenuation and dispersion calculated for ozone from the APM model, while Figs. 2.17 and 2.18 illustrate some calculations for nitrous oxide. Finally, Fig. 2.19 shows the absorption and dispersion for the lower tail of the 60 GHz absorption band from a 10 km initial height (from Papatsoris and Watson, 1993).

2.6 CONCLUDING REMARKS

At millimetre wavelengths, absorption and dispersion by atmospheric gases can have a significant effect on the propagation of electromagnetic radiation. Close to the earth's surface, absorption is dominated by the water vapour lines at 22 and 183 GHz and oxygen lines at 60 and 118 GHz. Nevertheless, most of the minor atmospheric constituents also have spectral signatures in

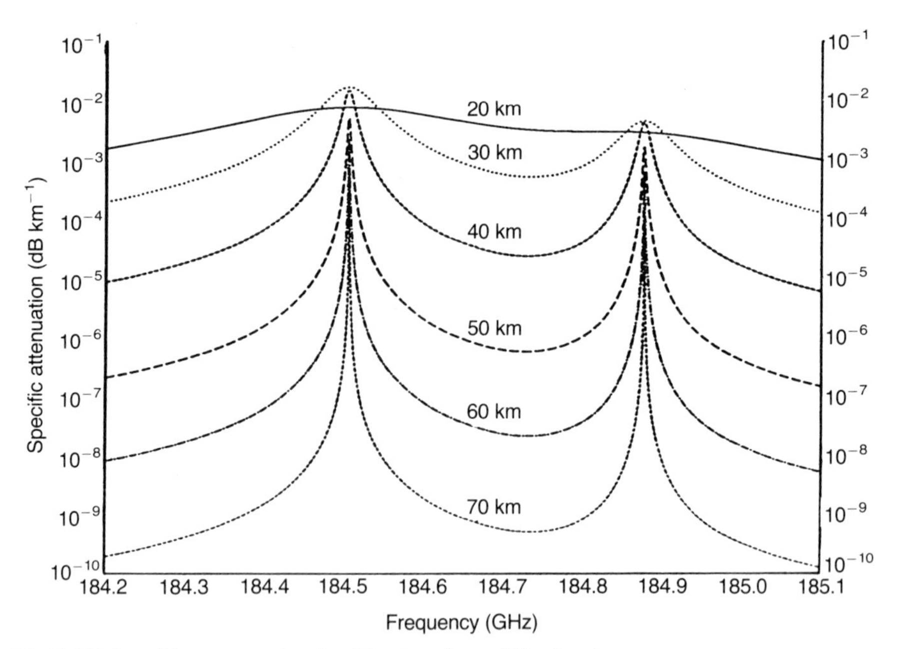

Fig. 2.15 Specific attenuation for O_3 at various altitudes.

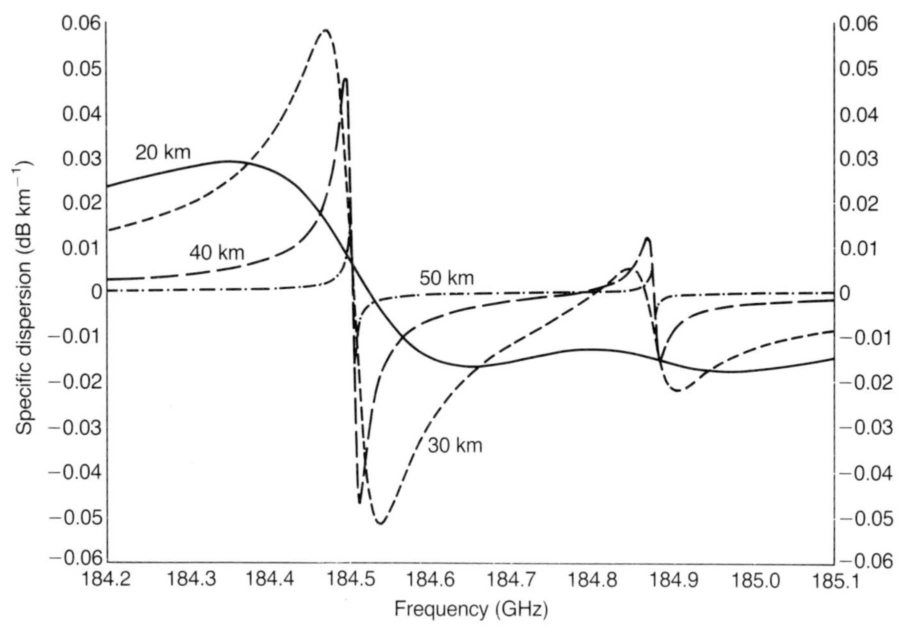

Fig. 2.16 Specific dispersion for O_3 at various altitudes.

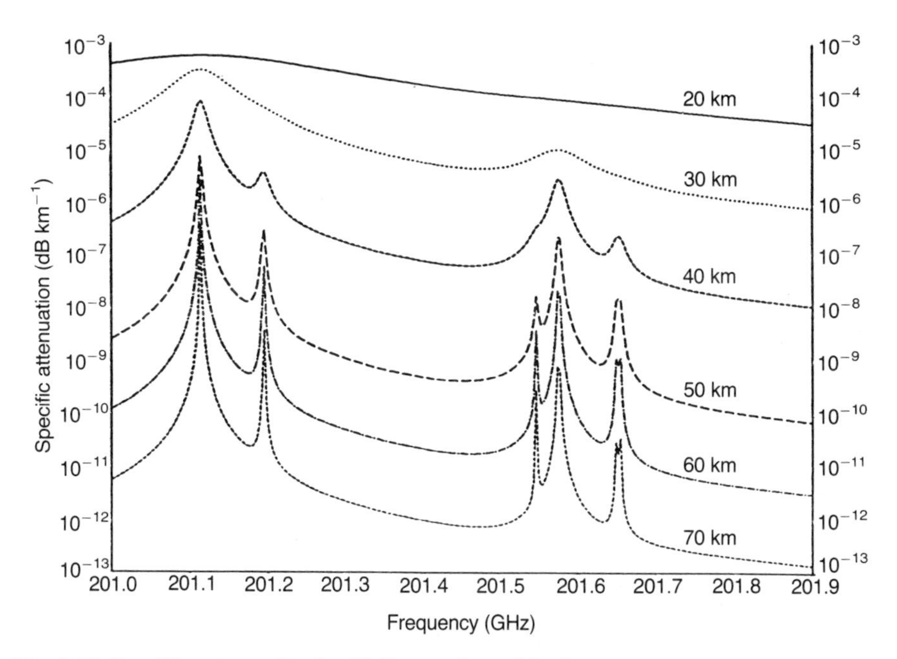

Fig. 2.17 Specific attenuation for N_2O at various altitudes.

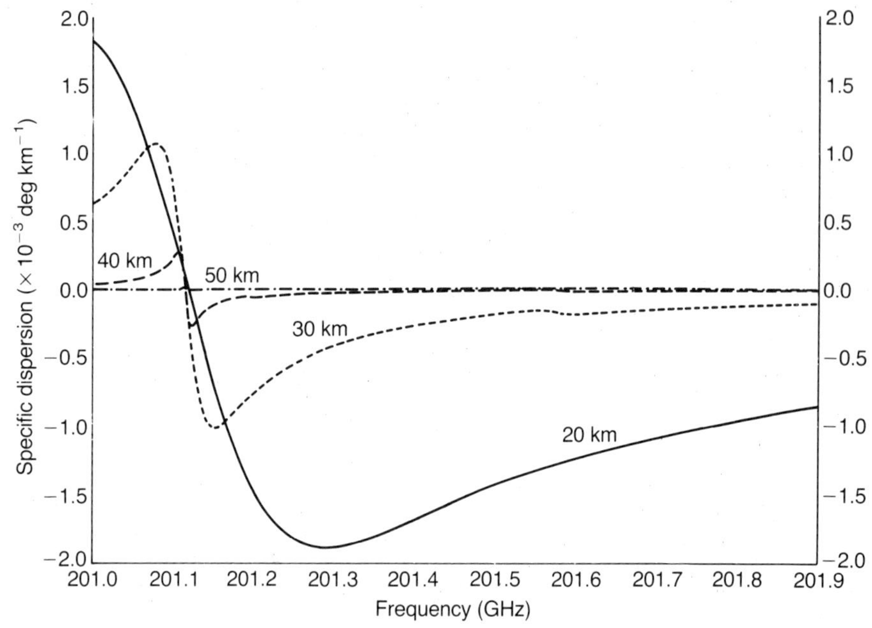

Fig. 2.18 Specific dispersion for N_2O at various altitudes.

the millimetre wave range, which, although negligible on a communication link close to the earth's surface, become important for applications in remote sensing and navigation.

While spectroscopic models appear to give adequate (although computationally difficult) predictions of line shapes near to resonance peaks, experimental evidence reveals the inability of such models to predict the resultant attenuation in the continuum between lines, especially in relation to the water vapour continuum.

Currently, more than two-thirds of the water vapour contribution to the calculation of clear air attenuation is described only by empirical formulae, lacking both physical insight and general applicability. Research to uncover the true nature of the millimetre wave continuum thus remains a serious challenge.

REFERENCES

Altshuler, E.E. and Marr, R.A. (1988) A comparison of experimental and theoretical values of atmospheric absorption at the longer millimeter wavelengths. *IEEE AP-S*, **36**(10), 1471.

Becker, G.E. and Autler, S.H. (1946) Water vapour absorption of electromagnetic radiation in the cm wave-length range. *Phys. Rev.*, **70**, 300–7.

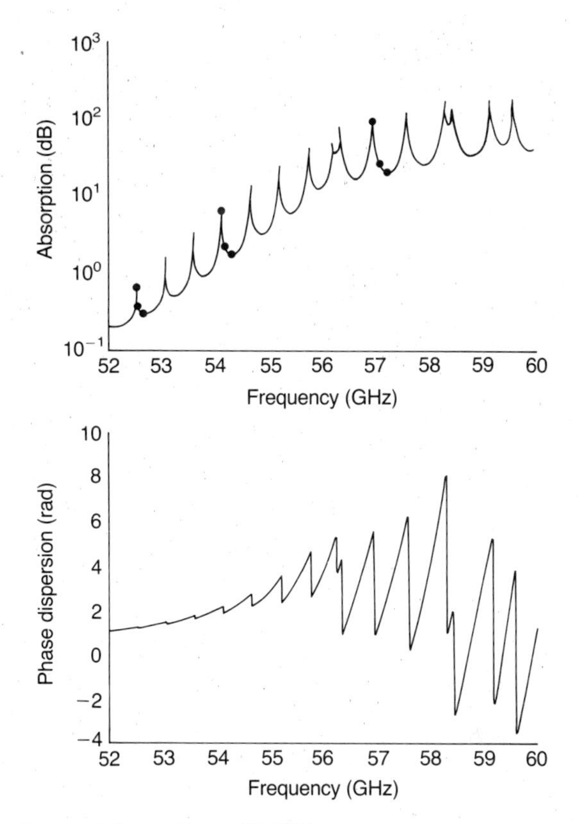

Fig. 2.19 Absorption and dispersion at 60 GHz.

Bohlander, R.A. (1979) Spectroscopy of water vapour, PhD Thesis, Department of Physics, Imperial College, London.

Burch, D., Grynvak, D. and Pembrook, J. (1971) Continuous absorption in the 8–14 μm range by atmospheric gases. *Philco–Ford Rep. 19*, 628–69-c-0263.

CCIR (1990) Propagation in non-ionized media, in *Recommendations and Reports of the CCIR*, Vol. V.

Clough, S.A., Kneizys, F.K., Rothman, L.S. and Gallery, W.O. (1981) Atmospheric spectral transmittance and radiance: FASCODIB. *TR-81-0269*, Air Force Geophysics Laboratory, Hauscom–Air Force Base, MA 01731.

Curtiss, L.A., Frurip, D.J. and Blander, M. (1979) Studies of molecular association in H_2O and D_2O vapours by measurement of thermal conductivity. *J. Chem. Phys.*, **71**(6), 2703–11.

Damosso, E., Stola, L. and Brussaard, G. (1983) Characterisation of the 50–70 GHz band for space communications, *ESA J.*, **7**, 23–43.

Debye, P. (1929) *Polar Molecules*, Dover Publications, New York, pp. 89–95.

Emonns, L.K. and de Zafra, R.L. (1990) Observations of strong inverse temperature dependence for opacity of atmospheric water vapour in the mm continuum near 280 GHz. *Int. J IR MMW*, **11**(4), 469–88.

Frenkel, L. and Woods, D. (1966) The microwave absorption of H_2O vapour and its mixtures with other gases between 100 and 300 GHz. *Proc. IEEE*, **54**(4), 498–505.

Gaut, N.E. and Reifenstein, E.C. (1971) *Environ. Res. Tech. Rep. 13*, Lexington, MA.

Gibbins, C.J. (1986) Improved algorithms for the determination of specific attenuation at sea level by dry air and water vapour in the frequency range 1–350 GHz. *Radio Sci.*, **21**(6), 945–54.

Hinderling, J., Sigrist, M.W. and Kneubühl, F.K. (1987) Laser-photoacoustic spectroscopy of water-vapour continuum and line absorption in the 8 to 14 μm atmospheric window. *Infrared Phys.*, **27**(2), 63–120.

Ito, S. (1987) A method for estimating atmospheric attenuation on earth–space paths in fair and rainy weather. *IEICE Trans.*, **J70-B**(11), 1407–14.

Kielkopf, J.F. (1973) New approximation to the Voight function with applications to spectral line profile analysis. *J. Opt. Soc. Am.*, **63**, 987.

Levy, M.F. and Craig, K.H. (1989) Assessment of anomalous propagation predictions using minisonde refractivity data and the parabolic equation method, in AGARD Conference Proc. 453, San Diego, CA.

Liebe, H.J. (1975) Molecular transfer characteristics of air between 40 and 140 GHz. *IEEE Trans. Microwave Theory Techn.*, **23**(4), 380–6.

Liebe, H.J. (1980) In *Atmospheric Water Vapour* (eds A. Deepack, T.D. Wilkerson and L.H. Ruhnke), Academic Press, New York, pp. 143–201.

Liebe, H.J. (1981) Modelling attenuation and phase of radio waves in air at frequencies below 1000 GHz. *Radio Sci.*, **16**(6), 1183–99.

Liebe, H.J. (1985) An updated model for millimetre wave propagation in moist air. *Radio Sci.*, **20**(5), 1069–89.

Liebe, H.J. (1987) A contribution to modeling atmospheric millimeter-wave properties. *Frequenz*, **41**(1/2), 31–6.

Liebe, H.J. and Hufford, G.A. (1989) Modelling millimeter-wave propagation effects in the atmosphere, AGARD Conf. Proc. CP 454 on Atmospheric propagation, Copenhagen, Oct. 1989.

Liebe, H.J. and Layton, D.H. (1983) Proc. URSI Commission F, 1983 Symp., Louvain.

Liebe, H.J. and Rosich, R.K. (1978) *Modeling of EHF propagation in clear air. IEEE Conf. Space Instrumentation for Atmospheric Observation*, El Paso, Texas, 5/1–15.

Liebe, H.J., Gimmestad, G.G. and Hopponen, J.D. (1977) Atmospheric oxygen microwave spectrum – experiment versus theory. *IEEE Trans. Antennas Propag.*, **25**(3), 327–35.

Llewellyn-Jones, D.T. (1980) In *Atmospheric Water Vapour* (eds A. Deepack, T.D. Wilkerson and L.H. Ruhnke), Academic Press, New York, p. 255.

Ma, Q. and Tipping, R.H. (1990) Water vapour continuum in the millimetre spectral region. *J. Chem. Phys.*, **93**(9), 6127–39.

Manabe, T., Debolt, R.O. and Liebe, H.J. (1989) Moist air attenuation at 96 GHz over a 21 km line-of-sight path. *IEEE Trans. Antennas Propag.*, **37**(2), 262.

Papatsoris, A.D. and Watson, P.A. (1993) Calculation of absorption and dispersion spectra of atmospheric gases at millimeter-wavelengths. *IEE Proc. H*, **140**(6), 461–8.

Ramo, S., Whinnery, J.R. and Van Duzer, T. (1984). *Fields and Waves in Communication Electronics*, 2nd edn, Wiley, New York, pp. 666–72.

Rice, D.P. and Ade, P.A.R. (1979) Absolute measurements of the atmospheric transparency at short/millimeter wavelengths. *Infrared Phys.*, **19**, 575.

Rosenkrantz, P.W. (1975) Shape of the 5 mm oxygen band in the atmosphere. *IEEE Trans. Antennas Propag.*, **23**(4), 498–506.

Van Vleck, J.H. (1947) The absorption of microwaves by oxygen. *Phys. Rev.*, **71**(7), 413–24.

Van Vleck, J.H. and Weisskopf, V.F. (1945) On the shape of collision broadened lines. *Rev. Mod. Phys.*, **17**(2–3), 227–36.

Viktorova, A.A. and Zhevakin, S.A. (1967) The water-vapour dimer and its spectrum. *Sov. Phys. Dokl.*, **11**(12), 1059–68.

Viktorova, A.A. and Zhevakin, S.A. (1971), Band spectrum of a dimer of water vapour. *Sov. Phys. Dokl.*, **15**(9), 836–55.

3
Scintillation

3.1 INTRODUCTION

Scintillation is the phenomenon of signal variation due to small-scale refractive index variations in the atmosphere. These refractive index variations are the result of temperature, humidity and pressure irregularities called atmospheric turbulence. In the optical range, the influence of the temperature fluctuations is dominant, resulting in, for instance, the twinkling of a star. In the microwave region, where the humidity fluctuations are more important, the result is random variation in signal strength and phase received on a satellite–earth link as well as degradation in performance of large antennas, notably in synthetic-aperture radars.

3.2 PHYSICAL MODELLING

3.2.1 **Basic assumptions** (Tatarski, 1961)

The basis for the theory of scintillation is the fluid dynamics model for velocity fluctuations in a turbulent medium, proposed first by Kolmogorov. Two scales of turbulence can be identified, the outer and the inner scales.

- *The outer scale.* If kinetic energy is introduced into a fluid at some large scale l_1 (the input range), corresponding to a Reynolds number much larger than the critical value, it results in a turbulent flow. (The Reynolds number is the ratio of the production rate and the dissipation rate of kinetic energy.) At that large scale, the dissipation is negligible and the energy is transferred successively in eddies of decreasing size (the inertial range). The dependence of the magnitude of velocity fluctuations on eddy size follows from energy conservation considerations. In the Kolmogorov model it is assumed that the time needed to introduce kinetic energy into a volume V is some characteristic time of the order $t = l/\Delta v$, where l is the linear dimension and Δv is the velocity fluctuation (variation with respect to the mean).

- *The inner scale.* Finally, a scale size l_2 is reached where the production rate and dissipation rate are of the same order, and the enegy is dissipated into heat (dissipation range).

l_1 and l_2 are the outer and inner scale of turbulence respectively. l_2 is in the order of a millimetre, while l_1 ranges from a few tens of metres to a kilometre. It is assumed that at any size l between the inner and outer scales (the inertial subrange), the eddies are isotropic. Furthermore, in the lower atmosphere l_1 is generally assumed to be of the order $h/3$ where h is the height (Fante, 1980).

The spatial extent of the velocity variations is modelled by the **structure function**. The structure function of the velocity variations $\mathcal{F}_v(r)$ is defined as

$$\mathcal{F}_v(r) = \overline{[v(r) - v(0)]^2} \tag{3.2.1}$$

where $v(0)$ and $v(r)$ are the velocity at a point 0 and r respectively and $\bar{x}$ represents the mean of x. It follows from the Kolmogorov theory that

$$\mathcal{F}_v(r) \propto \begin{cases} r^{2/3} & \text{for } l_2 \ll r \ll l_1 \text{ (inertial range)} \\ r^2 & \text{for } r \ll l_2 \text{ (dissipation range).} \end{cases} \tag{3.2.2}$$

3.2.2 Refractive index structure function

Temperature, pressure and humidity variations result in refractive index variations to which the theory of velocity variations was applied by Tatarski in order to derive an estimate of signal fluctuations. In this theory, the structure function of refractive index fluctuations has the form

$$\mathcal{F}_n(r) = \begin{cases} C_n^2 \, r^{2/3} & \text{for } l_2 \ll r \ll l_1 \\ C_n^2 \, l_2^{2/3} \, (r/l_2)^2 & \text{for } r \ll l_2 \end{cases} \tag{3.2.3}$$

where C_n^2 is called the **structure constant**.

In practice, measurements of the microwave refractive index are carried out by a single sensor, either a refractometer or a combined temperature–pressure–humidity sensor. In order to derive the structure function $\mathcal{F}_n(r)$ or the structure constant C_n^2 from these measurements, Taylor's hypothesis of 'frozen turbulence' is invoked: it is assumed that the temporal measurements in one point are equivalent to spatial samples, since the turbulent structure is a floating, frozen structure.

The relationship between C_n^2 and the measured samples is then given by

$$C_n^2 = \overline{[n(t) - n(t + \mathrm{d}t)]^2}/(v \, dt)^{2/3} \tag{3.2.4}$$

where $n(t) - n(t + \mathrm{d}t)$ is the difference between the refractive index at some time t and $t + \mathrm{d}t$, v is the wind velocity and $v\,\mathrm{d}t$ values should be taken between the inner scale and outer scale of turbulence. Alternatively, C_n^2 can be calculated from the inertial range of the refractive index spectrum (Herben and Kohsiek, 1984).

For elevated propagation paths, the height profile of C_n^2 is important. Tatarski suggested the following two profiles:

$$C_n^2(h) = C_{n0}^2\,\mathrm{e}^{-h/h_0}$$
$$C_n^2(h) = C_{n0}^2/[1 + (h/h_0)^2] \tag{3.2.5}$$

where h and h_0 are measured from the surface of the earth and C_{n0}^2 is the value of C_n^2 at ground level. For heights below a few hundred metres, use is often made of the relation (Herben and Kohsiek, 1984)

$$C_n^2(h) = C_n^2(h_0)\,(h/h_0)^{-b} \tag{3.2.6}$$

where $b = 4/3$ for moderately up to highly unstable atmospheres (that is, atmospheres with relatively large upward heat fluxes, and moderate to low wind speeds), while $b = 2/3$ for a neutral atmosphere (no vertical heat flux, which mostly occurs at night). Note that the choice of h_0 is free in equation (3.2.6).

Cole, Ho and Mavrokoukoulakis (1978) performed measurements from which they calculated values of C_n^2, on a dual link at 36 and 110 GHz, over a 4.1 km path at $h \approx 50$ m above central London. The mean values they obtained were in the range from 0.5×10^{-14} to 4×10^{-14} m$^{-2/3}$.

Using the spectral representation of the refractive index structure function, Tatarski derived for the three-dimensional spectrum of the refractive index fluctuations F_n the formula

$$F_n(\kappa) = 0.033\,C_n^2\,\kappa^{-11/3} \tag{3.2.7}$$

where κ is the 'spatial wavenumber', which is related to the size l of the fluctuations by $\kappa = 2\pi/l$.

However, in practice a slower decrease has been found for low spatial wavenumbers, and a more rapid decrease was observed for high spatial wavenumbers, leading to the von Kármán representation (Ishimaru, 1978) given by

$$F_n(\kappa) = 0.033\,C_n^2\frac{\mathrm{e}^{-(\kappa/\kappa_2)^2}}{(\kappa^2 + l_1^{-2})^{11/6}} \tag{3.2.8}$$

where $\kappa_2 = 5.91/l_2$ is the spatial wavenumber corresponding to a length near

l_2. It should be noted, however, that in some papers this length is taken equal to l_2 (e.g. Fante, 1975), i.e. $\kappa_2 = 2\pi/l_2$.

More recent experiments have also shown a fairly large variation in the slope of the spectrum. This has led to the generalized von Kármán spectrum (Haddon and Vilar, 1986)

$$F_n(\kappa) = \alpha(n)\frac{\mathrm{e}^{-(\kappa/\kappa_2)^2}}{(\kappa^2 + l_1^{-2})^{n/2}} \tag{3.2.9}$$

with

$$\alpha(n) = \frac{\Gamma(n-1)}{4\pi^2}\sin\left[\frac{\pi(n-3)}{2}\right]C_n^2 \text{ for } 3 \leqslant n \leqslant 5.$$

For $n = 11/3$, $\alpha = 0.033\ C_n^2$.

Using this generalization, measurements of spectra of refractive index fluctuations may be used to derive values of the exponent n, the function $\alpha(n)$ and the structure constant C_n^2.

For modelling purposes it is generally assumed that C_n^2 has a particular distribution with height, while $\alpha(n)$ is constant or slowly varying with height and n is constant.

3.3 ELECTROMAGNETIC MODELLING

3.3.1 Phase structure function

Using a small perturbation approximation (Rytov method and Born approximation (Tatarski, 1961)), the phase variations can be obtained for plane wave propagation through a homogeneous medium. It follows that, assuming a Kolmogorov spectrum, the phase structure function $\mathcal{F}_\mathrm{p}(r)$ of the wave is given by (Tatarski, 1961):

$$\mathcal{F}_\mathrm{p}(r) = \begin{cases} 1.72\ C_n^2 l_2^{-1/3}\ k^2\ L\ r^2 & \text{for } r \ll l_2 \\ 1.46\ C_n^2\ k^2\ L\ r^{5/3} & \text{for } l_2 \ll r \ll (\lambda L)^{1/2} \\ 2.91\ C_n^2\ k^2\ L\ r^{5/3} & \text{for } r \geqslant (\lambda L)^{1/2} \end{cases} \tag{3.3.1}$$

where L is the length of the turbulent part of the path, r is the distance between the observation points in the transverse plane and $k = 2\pi/\lambda$ with λ the wavelength. $(\lambda L)^{1/2}$ is the size of the first Fresnel zone of the turbulent part of the path. For an inhomogeneous path, the product $C_n^2 L$ has to be replaced by the integral

$$\int_0^L C_n^2(z)\,\mathrm{d}z.$$

Equations (3.3.1) are valid only for the case when $l_2 < (\lambda L)^{1/2} < l_1$. For the other cases, $\mathcal{F}_\mathrm{p}(r)$ cannot as easily be expressed (Tatarski, 1961).

3.3.2 Log-amplitude variations

The fluctuations in amplitude of the received signal due to refractive index variations are generally expressed in terms of the variable $g = \ln(1 + \Delta E/\bar{E})$ where ΔE and $\bar{E}$ are the (zero-mean) amplitude fluctuations and the mean amplitude respectively; g is expressed in nepers.

For a homogeneous path of turbulent length L, assuming a point receiver, the variance σ_g^2 of the 'log-amplitude variation' g may be calculated from

$$\sigma_g^2 = 2\pi k^2 L \int_0^\infty \kappa F_n(\kappa)\left[1 - \frac{\sin(\kappa^2 L/k)}{\kappa^2 L/k}\right]\mathrm{d}\kappa. \tag{3.3.2}$$

To interpret this equation, we note that the diameter of the first Fresnel zone at a distance L from the receiver is given by $(\lambda L)^{1/2} = (2\pi L/k)^{1/2}$. When the size of the eddies becomes larger than the size of the first Fresnel zone, their contribution to the signal fluctuations decreases, as expressed by a high-pass filter function

$$f_L(\kappa) = 1 - \frac{\sin[\kappa(L/k)^{1/2}]^2}{[\kappa(L/k)^{1/2}]^2}$$

in the integration of the spectrum.

Applying the generalized von Kármán spectrum of equation (3.2.9) we obtain

$$\sigma_g^2 = \alpha(n)\pi^2 k^{(6-n)/2} L^{n/2}\Gamma\left(-\frac{n}{2}\right)\sin\left(\frac{n\pi}{4}\right) \tag{3.3.3}$$

for $l_2 < (\lambda L)^{1/2} \ll l_1$. For $n = 11/3$ this results in

$$\sigma_g^2 = 0.307 C_n^2 k^{7/6} L^{11/6} \tag{3.3.4a}$$

for $l_2 < (\lambda L)^{1/2} \ll l_1$. For sizes outside the inertial subrange, the following expressions were derived (Tatarski, 1961; Cole, Ho and Mavrokoukoulakis, 1978):

$$\sigma_g^2 = \begin{cases} 2.46 l_2^{-7/3} C_n^2 L^3 & \text{for } (\lambda L)^{1/2} < l_2 \quad (3.3.4b) \\ \overline{(\Delta n)^2} L l_t k^2 & \text{for } l_1 \ll (\lambda L)^{1/2} \quad (3.3.4c) \end{cases}$$

where $\overline{(\Delta n)^2}$ is the mean square of the refractive index fluctuations and l_t is called the **integral scale** of turbulence and is of the same order of magnitude as the outer scale of turbulence l_1. To convert σ_g^2 from nepers squared to decibels squared, all the equations in this section should be multiplied by $(20 \log e)^2 \approx 75.44$.

Simultaneous experiments made at 36 and 110 GHz (Cole, Ho and Mavrokoukoulakis, 1978) (Fig. 3.1) confirmed the validity of the $k^{7/6}$ wavenumber dependence when the condition $l_2 < (\lambda L)^{1/2} \ll l_1$ is valid ($(110/36)^{7/6} = 3.68$) and the k^2 wavenumber dependence when $l_1 \ll (\lambda L)^{1/2}$ ($(110/36)^2 = 9.34$), under homogenous path conditions.

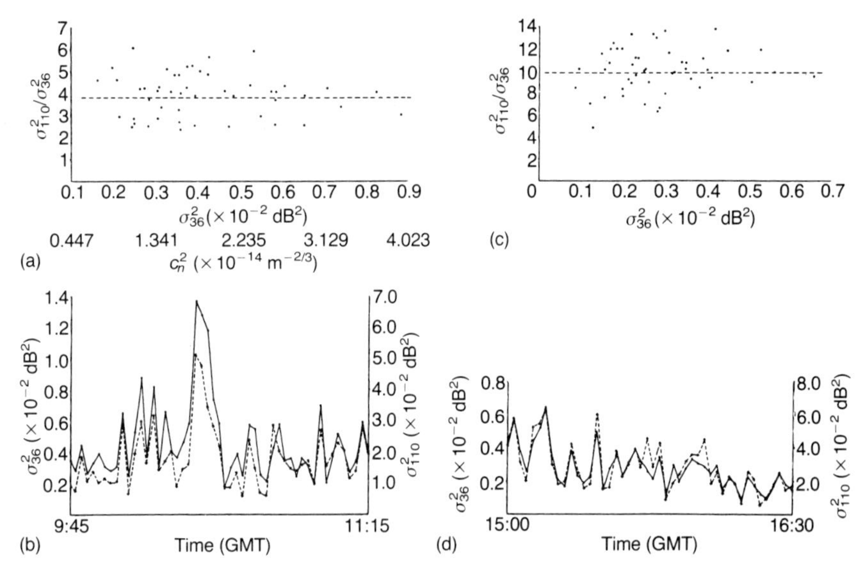

Fig. 3.1 Results of experiments by Cole, Ho and Mavrokoukoulakis (1978). (a) Ratio of variances of log-amplitude fluctuations at 110 GHz and 36 GHz ($\sigma_{110}^2/\sigma_{36}^2$; mean value, 3.83) against variance of log-amplitude fluctuations at 36 GHz (σ_{36}^2) and value of the refractive index structure parameter C_n^2 under the condition $L_0 > (\lambda L)^{1/2}$. (b) Variances of log-amplitude fluctuations at 110 GHz (σ_{110}^2, ---) and 36 GHz (σ_{36}^2, ——) against time under the condition $L_0 > (\lambda L)^{1/2}$. (c) Ratio of variances of log-amplitude fluctuations at 110 GHz and 36 GHz ($\sigma_{110}^2/\sigma_{36}^2$; mean value, 9.81) against variance of log-amplitude fluctuations at 36 GHz (σ_{36}^2) under the condition $L_0 < (\lambda L)^{1/2}$. (d) Variances of log-amplitude fluctuations at 110 GHz (σ_{110}^2, ---) and 36 GHz (σ_{36}^2, ——) against time under the condition $L_0 < (\lambda L)^{1/2}$.

For an inhomogeneous propagation path, the above formulae become

$$\sigma_g^2 = \begin{cases} 7.37 l_2^{-7/3} \int_0^L C_n^2(z) z^2 \, dz & \text{for } (\lambda L)^{1/2} \ll l_2 \\ 0.56 k^{7/6} \int_0^L C_n^2(z) z^{5/6} \, dz & \text{for } l_2 < (\lambda L)^{1/2} \ll l_1 \\ l_t k^2 \int_0^L \overline{(\Delta n)^2} \, dz & \text{for } l_1 \ll (\lambda L)^{1/2}. \end{cases} \tag{3.3.5}$$

Frequency (f) and elevation angle (ϵ) scaling of σ_g^2 can easily be obtained from the expressions presented above. For instance, if $l_2 < (\lambda L)^{1/2} \ll l_1$ and $n = 11/3$ it appears from equation (3.3.4a) that

$$\frac{(\sigma_g^2)_1}{(\sigma_g^2)_2} = \left(\frac{f_1}{f_2}\right)^{7/6} \left(\frac{\sin \epsilon_2}{\sin \epsilon_1}\right)^{11/6}. \tag{3.3.6}$$

All the results given above are valid for point receivers only. For an antenna of diameter D, some correction is necessary to account for the fact that as the antenna diameter increases the incident wave front fluctuations become less correlated across the aperture. As the antenna output is a spatial average of the various incident rays, the antenna acts as a low-pass filter for the scattered components corresponding to wavenumbers greater than about $1/D$. Hence the observed signal fluctuations are smaller than those obtained from a point antenna. This subject has been covered by various authors, Using a simplified model for the antenna pattern, and assuming $n = 11/3$, Haddon and Vilar (1986) found that the normalization to the point receiver is given by

$$\frac{\sigma_g^2(D)}{\sigma_g^2(0)} = 3.8637(x^2 + 1)^{11/12} \sin\left[\frac{11}{6} \arctan\left(\frac{1}{x}\right)\right] - 7.0835 x^{5/6}$$

$$\approx 1 - 7.0835 x^{5/6} \text{ for } x \ll 1 \tag{3.3.7}$$

where $x = 0.0584 kD^2/L$ is a measure of the ratio between the effective antenna diameter D and the Fresnel zone size $((2\pi L)^{1/2}/k)$. The effective diameter is quoted to be approximately 75% of the physical diameter for typical reflector antennas.

In general the effect of antenna smoothing may be described by a smoothing function $s(x)$:

$$\sigma_g^2(D) = \sigma_g^2(0)\, S_n^2(x) \tag{3.3.8}$$

where $S_n^2(x)$ is a function of the exponent n in the von Kármán spectrum.

The frequency scaling of σ_g^2 then has the following form:

$$\frac{\sigma_g^2(k_2)}{\sigma_g^2(k_1)} = \left(\frac{k_2}{k_1}\right)^{(6-n)/2} \frac{S_n(x_2)}{S_n(x_1)}. \tag{3.3.9}$$

3.3.3 Probability density function (PDF)

Because g can be expressed as the sum of a large number of uncorrelated terms, it follows from the central limit theorem that g is normally distributed with zero mean and variance σ_g^2 as given in the previous section. This short-term PDF was verified experimentally by various authors (e.g. Herben, 1982, 1983; Figs 3.2 and 3.3). For long time periods σ_g^2 itself is a random variable, because C_n^2 varies slowly over time. Moulsley and Vilar (1982)

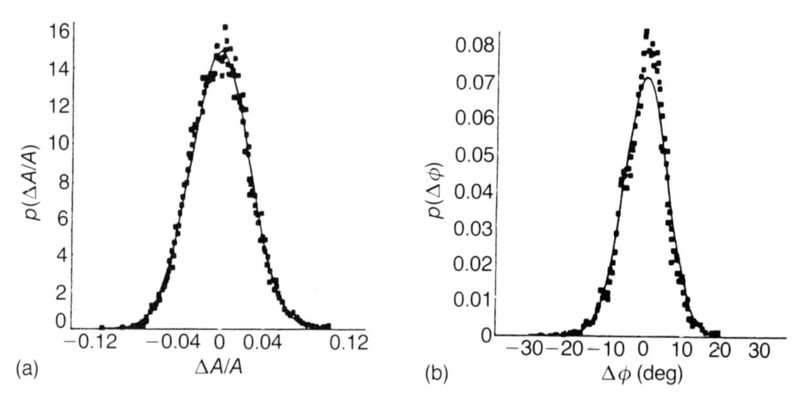

Fig. 3.2 Probability density functions for scintillation event on September 22, 1981 (line of sight; $f = 30$ GHz) (Herben, 1982): (a) amplitude variation; (b) phase scintillation; ■, measured values; ——, Gaussian distribution.

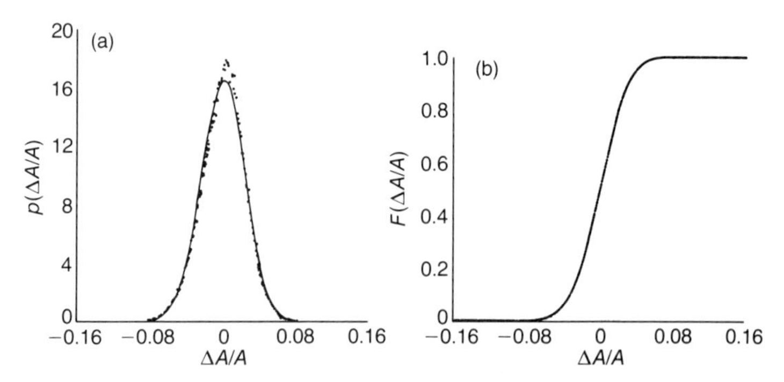

Fig. 3.3 (a) Probability density function and (b) cumulative distribution for scintillation event on June 15, 1981 (Orbital Test Satellite; $f = 11.5$ GHz) (Herben, 1983): ●, measured values; ——, Gaussian distribution.

found from experimental observations that $\ln(\sigma_g^2)$ is Gaussian distributed with mean $\ln(\sigma_m^2)$ and standard deviation σ_{gg}. This leads to the following long-term PDF $p(g)$ for g:

$$p(g) = \frac{1}{2\sigma_{gg}\sigma_m\pi}\int_{-\infty}^{+\infty}\exp\left(-\frac{s^2}{2\sigma_{gg}^2}-\frac{g^2}{2\sigma_m^2}e^{-s}-\frac{s}{2}\right)ds \qquad (3.3.10)$$

with $s = \sigma_g^2/\sigma_m^2$ (Fig. 3.4).

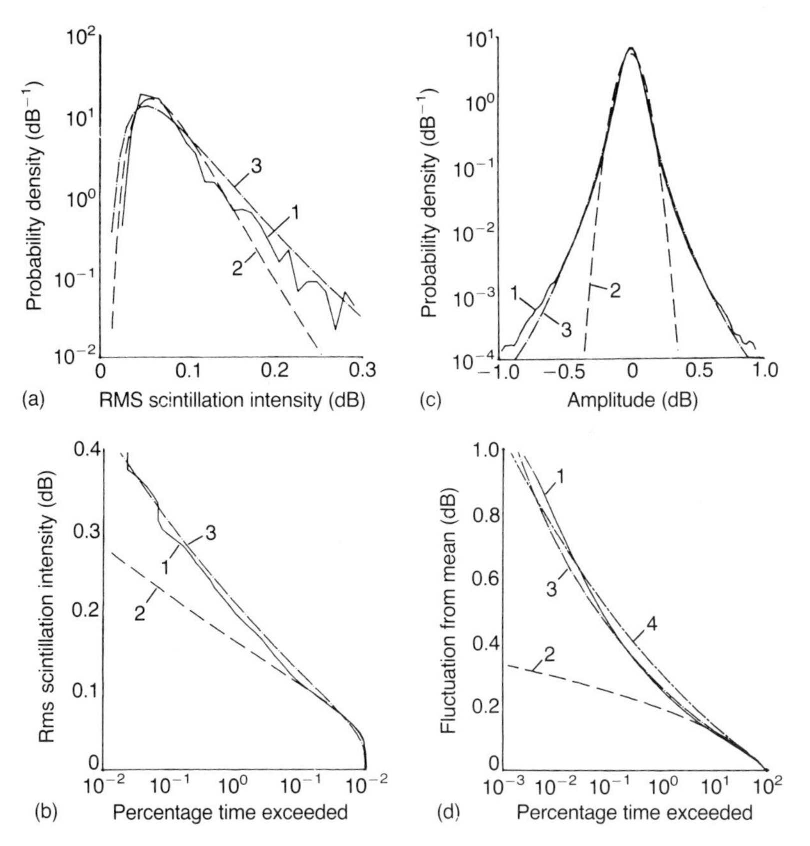

Fig. 3.4 Results of experiments by Moulsley and Vilar (1982). (a) Probability density of rms scintillation intensity measured over 10 min intervals (curve 1) and probability densities corresponding to models with $\sigma_{gg} = 0.77$ dB, $\sigma_m = 0.067$ dB (curve 2) and $\sigma_{gg} = 1.0$ dB, $\sigma_m = 0.067$ dB (curve 3). (b) Cumulative distributions corresponding to curves 1–3 in (a). (c) Comparison of probabilities of amplitude derived from scintillations in bandwidth 0.01–28 Hz (curve 1), Gaussian distribution with standard deviation 0.076 dB (curve 2) and measured distribution of rms values shown in (a) (curve 3). (d) Cumulative distributions of fluctuations from the mean values derived from the distributions shown in (c) (curves 1–3) and distribution derived from a model with $\sigma_{gg} = 1.0$ dB, $\sigma_g = 0.09$ dB (curve 4).

3.3.4 Spectral density function

The spectrum of scintillation noise has the appearance of band-limited white noise with a flat part $W_g^0(\omega)$ and a roll-off $W_g^\infty(\omega)$, $\omega = 2\pi f$ representing the Fourier frequency in the scintillation spectrum. Using the Kolmogorov refractive index spectrum $F_n(k) = 0.033C_n^2 k^{-11/3}$, Ishimaru (1978) derived the following expressions for the two asymptotes:

$$\begin{aligned} W_g^0(\omega) &= 2.765\sigma_g^2/\omega_t && \text{for } \omega \to 0 \\ W_g^\infty(\omega) &= 7.13\sigma_g^2\left(\frac{\omega}{\omega_t}\right)^{-8/3}/\omega_t && \text{for } \omega \to \infty \end{aligned} \tag{3.3.11}$$

with σ_g^2 as given by equation (3.3.4) for $l_2 \ll (\lambda L)^{1/2} \ll l_1$ and $\omega_t = v_t(k/L)^{1/2}$ with v_t the wind velocity component transverse to the propagation path. The two asymptotes are valid as long as ω/ω_t is between $(\lambda L)^{1/2}/l_1$ and $(\lambda L)^{1/2}/l_2$. Therefore, the departure of the actual spectrum from the asymptotes indicates the effects of the outer and inner scales of turbulence. Note that the two asymptotes meet at the frequency $\omega = 1.43v_t(k/L)^{1/2}$, which means that if v_t increases, the scintillation noise power is distributed over a wider frequency band.

These results of Ishimaru, which are valid for a point receiver, were modified by Haddon and Vilar (1986) in order to take the low-pass filtering effect of a circular aperture with diameter D into account. They found that the normalization to the point receiver is given by

$$\frac{W_g^0(\omega, D)}{W_g^0(\omega, 0)} = \frac{14}{3}x^{4/3} - 2(1 + x^2)^{7/6}\sin\left(\frac{7}{3}\arctan\frac{1}{x}\right) \quad \text{for } l_2 < D < l_1$$

$$\frac{W_g^\infty(\omega, D)}{W_g^\infty(\omega, 0)} = 1.053\left(\frac{\omega_s}{\omega}\right)e^{-(\omega/\omega_s)^2} \tag{3.3.12}$$

where $x = 0.0584kD^2/L$ as in equation (3.3.7) and $\omega_s = 4.1391v_t/D$ can be interpreted as the scintillation frequency above which the spectrum becomes dominated by aperture smoothing effects and rolls off faster. This reduction of the high-frequency spectral components was observed during propagation experiments by Cox, Arnold and Hoffman (1981) and Herben (1983) (Figs 3.5 and 3.6). The difference between the curves of the antennas with different aperture diameters, 0.6 m and 7 m in Fig. 3.5 and 0.5 m and 8 m in Fig. 3.6, clearly shows the aperture smoothing effect.

A particularly interesting case of an inhomogeneous path is a slant path through a thin turbulent layer. This case was studied by Haddon and Vilar (1986), leading to a generalization of equation (3.3.12), with the distance of

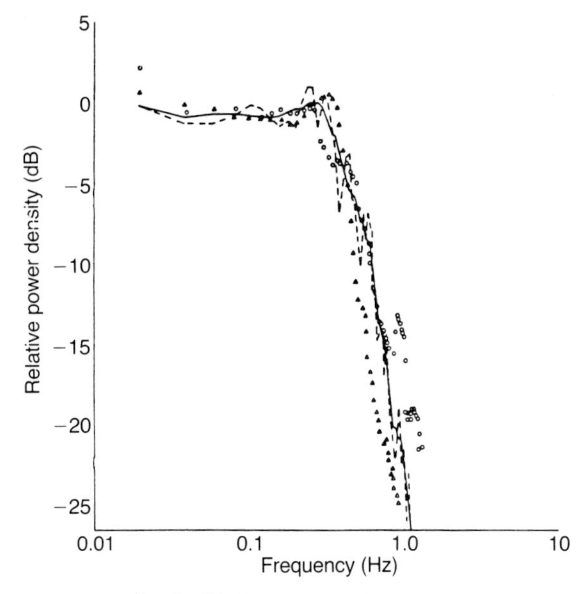

Fig. 3.5 Power spectrum of scintillation events from two days (Cox, Arnold and Hoffman, 1981) (Crawford Hill; 28 GHz; elevation, 41.5°).

Date	Antenna	Data treatment	Symbol
August 1, 1979	7 m	Seven-point average	▲
August 3, 1979	7 m	Seven-point average	——
August 3, 1979	7 m	Unsmoothed	- - -
August 3, 1979	0.6 m	Seven-point average	○

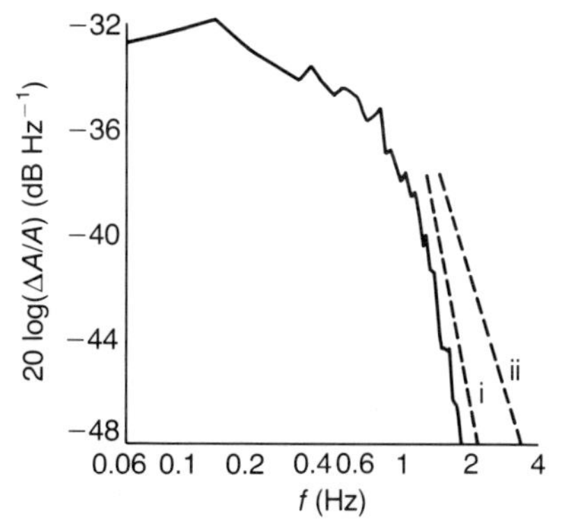

Fig. 3.6 Amplitude scintillation power density spectrum for events in Fig. 3.3 (Herben, 1983); ——, measured spectrum; curve i (– – –), slope of theoretical spectrum with an antenna aperture diameter of 8 m and $(\Delta E/\bar{E})^3 \propto f^{-5.02}$; curve ii (– – –), slope of theoretical spectrum with an antenna aperture diameter of 0.5 m and $(\Delta E/\bar{E})^3 \propto f^{-2.83}$.

the turbulent layer as an additional parameter. Using multifrequency observations of scintillation and fitting the model spectra to the measurements in this way, a tool may be developed for the derivation of the physical model parameters.

3.4 PREDICTION METHOD

3.4.1 CCIR model

CCIR Report 718-2 (1990) contains a prediction method for the calculation of the standard deviation of signal fluctuations due to scintillation, which starts with equation (3.3.5) with $l_2 \ll (\lambda L)^{1/2} \ll l_1$. The '7/6 law' for the frequency dependence is applied in the CCIR procedure, which is largely based on the model by Karasawa, Yamada and Allnutt (1988). An empirical estimation of the intensity of the fluctuations, equivalent to giving a C_n^2 profile, is represented by a factor σ_{ref}, which is a function of ground refractivity. Aperture averaging by the antenna is taken into account using an averaging function (Haddon and Vilar, 1986) which depends on the diameter of the antenna in terms of wavelength as well as the length of the turbulent part of the path. The resulting prediction has the form

$$\sigma_{pre} = \sigma_{ref} f^{7/12} \zeta(x)/(\sin \epsilon)^{1.2} \tag{3.4.1}$$

where

$$\sigma_{ref} = 3.6 \times 10^{-3} + 1.03 \times 10^{-4} \, N_{wet},$$

$$\zeta(x) = \{3.86(x^2 + 1)^{11/12} \sin[(11/6) \arctan(1/x)] - 7.08x^{5/6}\}^{1/2},$$

$$x = 0.0584 \, D_{eff}^2 k/L,$$

$$D_{eff} = D\eta^{1/2},$$

$$L = 2h/[(\sin^2 \epsilon + 2h/a_e)^{1/2} + \sin \epsilon],$$

and σ_{pre} is the (predicted) monthly average value of the standard deviation of the amplitude variation, f is in gigahertz, ϵ is the apparent elevation angle, N_{wet} is the wet term of ground refractivity, averaged over one month, k is the wavenumber, L is the length of the effective turbulent path, D is the antenna diameter, η is the antenna efficiency, h is the height of turbulence and $a_e = 8.5 \times 10^6$ m is the effective earth radius including refraction.

Under the assumption that the distribution of the standard deviation itself is approximated by a Γ distribution, the long-term cumulative distribution function of fading due to scintillation is predicted by

$$A(P) = [-0.061(\log P)^3 + 0.072(\log P)^2 - 1.71(\log P) + 3.0]\ \sigma_{\text{pre}} \quad (3.4.2)$$

for $0.01 < P < 50$, with P the percentage of time. The model is stated to be based on measured data at $\epsilon = 4°–32°$ and $D = 3–36$ m; Karasawa, Yamada and Allnutt (1988) showed the method to become problematic for $\epsilon < 5°$. The frequency range is 7–14 GHz. One of the stated advantages of the model is that it is no longer necessary to consider the various regions of validity of the Tatarski theory (geometrical optics, e.g. diffraction region); the model is 'more practical'.

3.4.2 Extension to higher frequencies

In order to extend the CCIR prediction model to higher frequencies, it is necessary to examine how this model was derived from the theory described in section 3.3 and which assumptions were made. (This takes more information than was present in the CCIR report.) Then it must be checked whether these assumptions are still valid for the higher frequencies or not. In the latter case, the prediction method should be modified accordingly. This question is a subject for further investigation.

3.5 SCINTILLATION AT FREQUENCIES IN ATMOSPHERIC ABSORPTION REGIONS

For millimetre waves in an absorption region, the turbulence-induced amplitude fluctuations will be superimposed on slow signal fluctuations due to the gaseous absorption mechanism. The results are a departure from the short-term Gaussian PDF and a change in the scintillation spectrum at low scintillation frequencies. This was investigated theoretically by Ott and Thompson (1978) and observed experimentally by Medeiros Filho, Jayasuriya and Cole (1983). Results are shown in Fig. 3.7; in particular, note the first part of the curve, which should be horizontal without the absorption mechanisms. Very recently, similar observations were made on satellite-to-earth links at frequencies within an atmospheric window (Sarma and Herben, 1989; Basili *et al.*, 1990) (Figs 3.8 and 3.9). In those cases the additional slow signal fluctuations were presumably caused by slowly varying cloud attenuation.

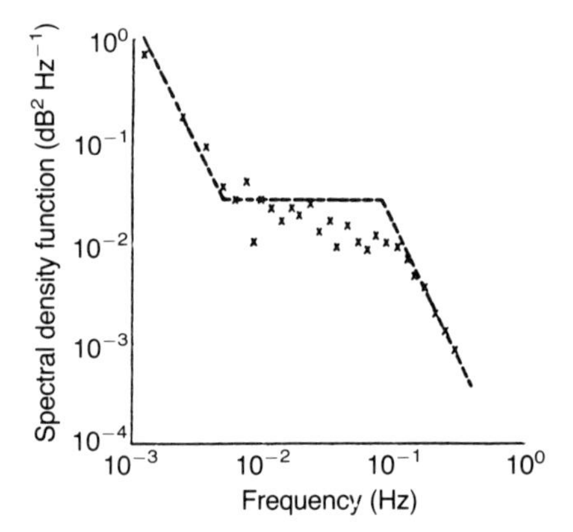

Fig. 3.7 Spectral density function of log-amplitude scintillations at 55.5 GHz for a wind speed of 1.2 m s^{-1} (Medeiros Filho, Jayasuriya and Cole, 1983): – – –, spectrum predicted by theory, for the meteorological conditions under which the experimental spectrum was obtained.

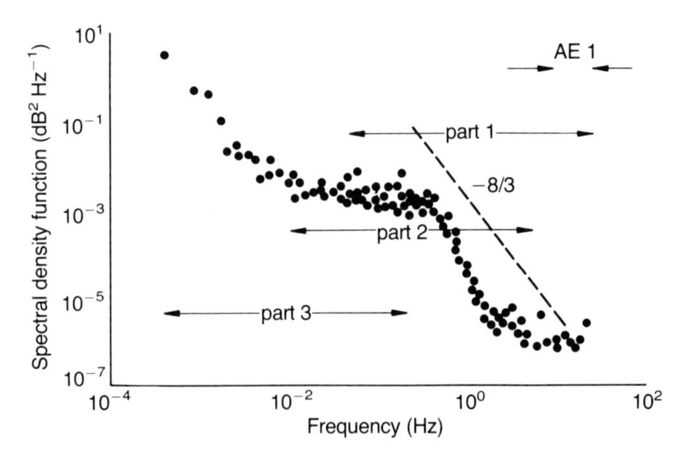

Fig. 3.8 Spectral density function of amplitude scintillations at 11.5 GHz (February 11, 1988; starting time, 12:50 GMT) (Sarma and Herben, 1989): part 1, due to first data set; part 2, due to second data set; part 3, due to third data set; AE1, aliased region where error is greater than the ripple of the filter. Line with slope of −8/3 is also shown.

3.6 RAIN SCINTILLATION

During rain, the turbulence-induced amplitude fluctuations are superimposed on slow signal fluctuations due to the absorption mechanism and fast signal fluctuations caused by incoherent scattering by the falling raindrops. The results are a departure from the short-term Gaussian PDF and a change in the

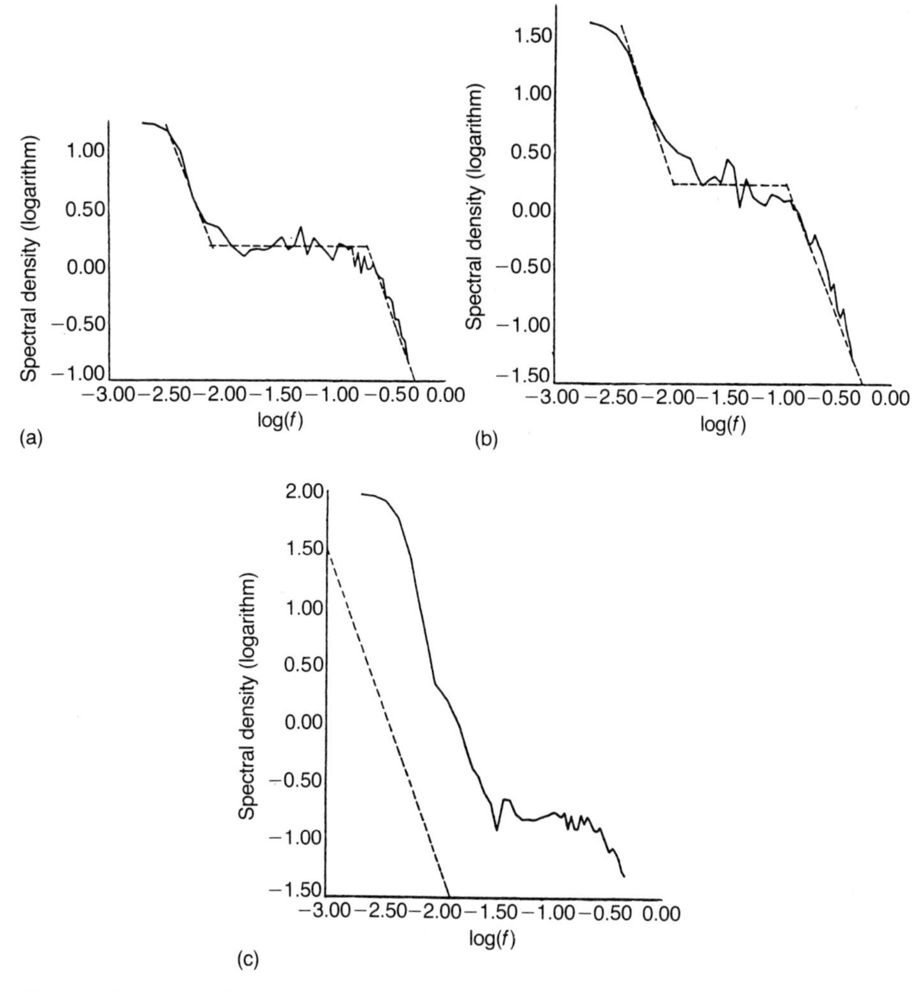

Fig. 3.9 Spectra of fluctuations for event on August 8, 1979, 12:00–14:00 GMT (Basili *et al.*, 1990): (a) field amplitude received by 3 m antenna; (b) field amplitude received by 17 m antenna; (c) atmospheric brightness temperature. Lines with slope of −8/3 are also shown.

scintillation spectrum at both the low and high scintillation frequencies, as observed experimentally by Herben (1984) (Fig. 3.10). Rain scintillation was investigated theoretically by Capsoni, Mauri and Paraboni (1977), Herben (1984) and Haddon and Vilar (1986) (Fig. 3.11). Capsoni, Mauri and Paraboni found some interesting properties of rain scintillation, such as that the power fluctuations are proportional to path length and inversely proportional to antenna gain.

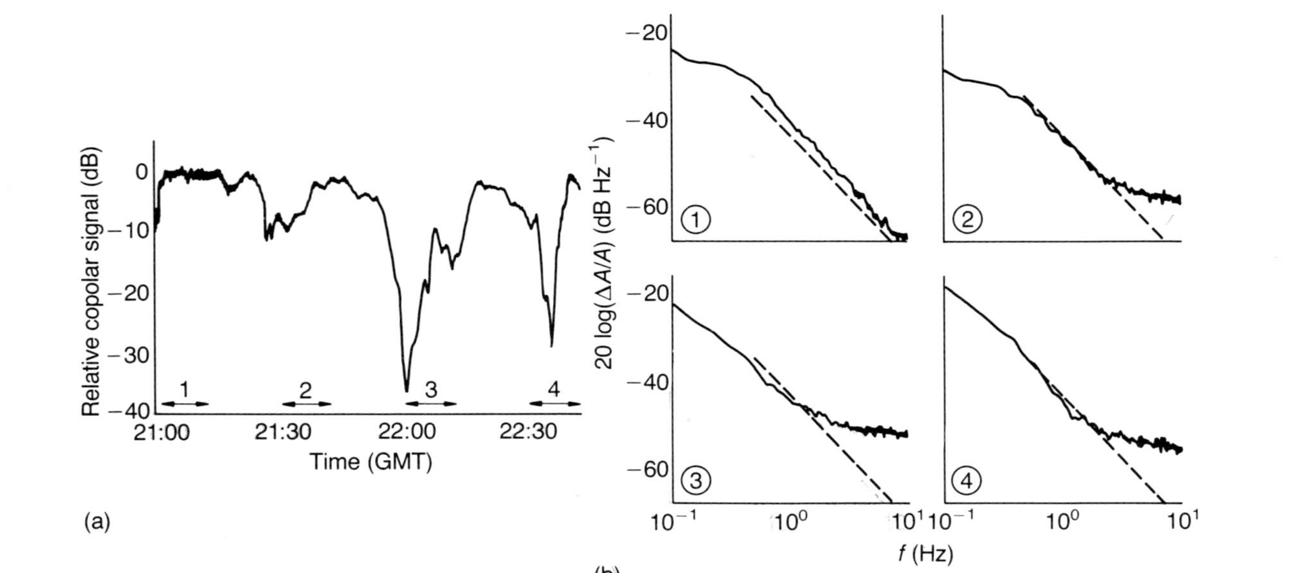

Fig. 3.10 Results from Herben (1984). (a) Copolar signal for the rain event on September 25, 1982. (b) Amplitude scintillation power density spectra for time periods 1–4 of the event in (a): ——, measured spectra; – – –, slope of theoretical and measured spectra obtained for scintillation caused by tropospheric turbulence ($(\Delta A/A)^2 \propto f^{-2.83}$). (c) Wet-snow event on December 8, 1981: (A) relative copolar signal as a function of time; (B) amplitude spectrum for record i; (C) phase spectrum for record i; curves ii, slope of the (theoretical as well as measured) spectra for turbulence-induced scintillation; curves iii, shape of the theoretical spectrum for rain-induced scintillation.

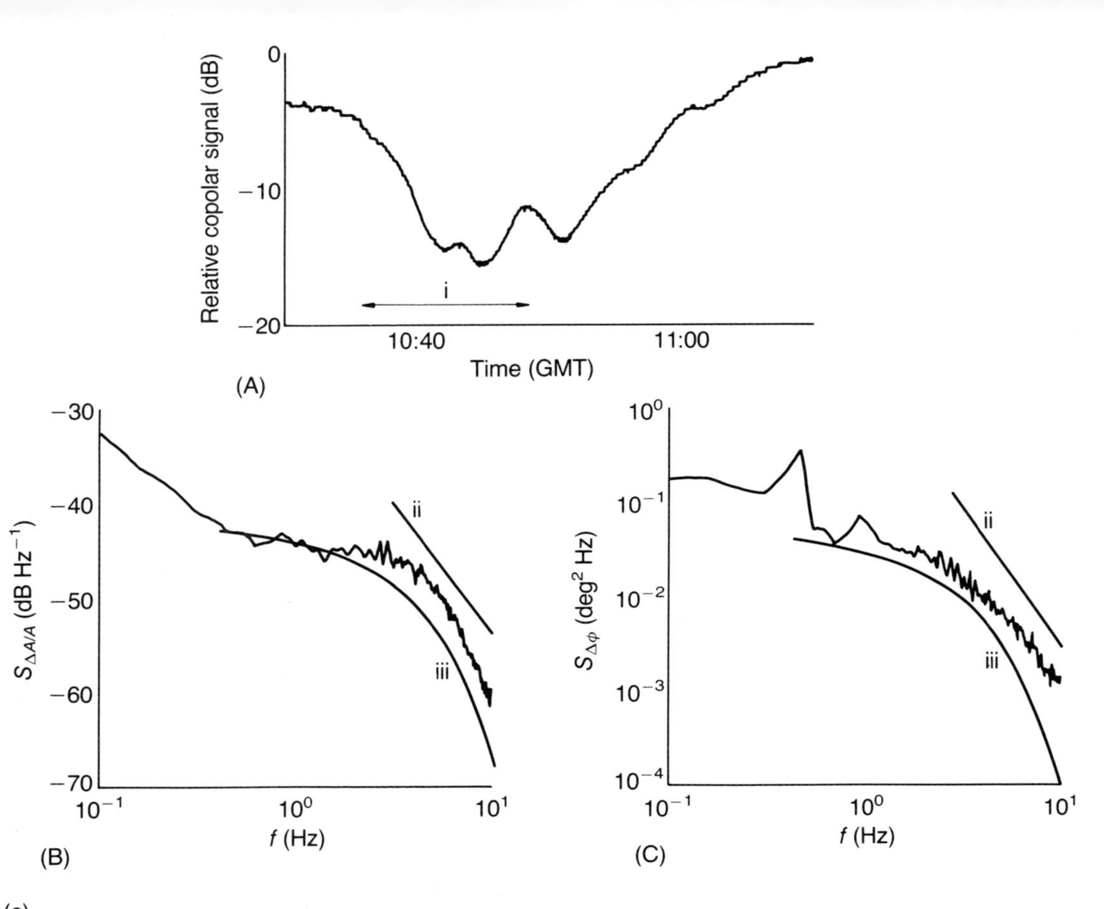

Relative copolar signal (dB)
0
−10
−20
i
10:40
11:00
Time (GMT)
(A)
$S_{\Delta A/A}$ (dB Hz^{-1})
−30
−40
−50
−60
−70
ii
iii
10^{-1}
10^{0}
10^{1}
f (Hz)
(B)
$S_{\Delta\phi}$ (deg^2 Hz)
10^{0}
10^{-1}
10^{-2}
10^{-3}
10^{-4}
ii
iii
10^{-1}
10^{0}
10^{1}
f (Hz)
(C)

(c)

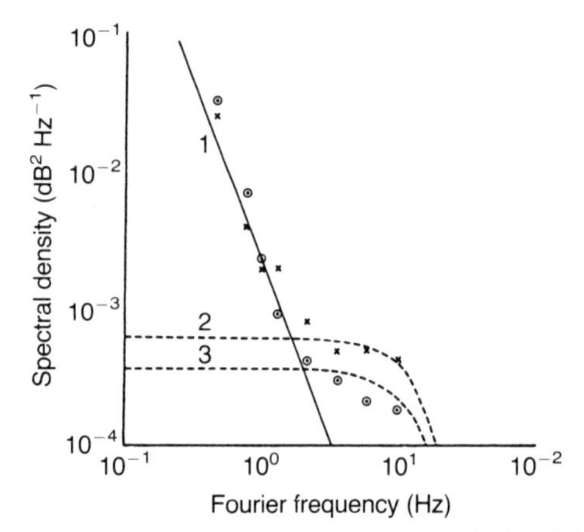

Fig. 3.11 $W_g(\omega)$ from rain events at 30 GHz ($L = 8.2$ km) (Haddon and Vilar, 1986). Measured values: ×, mean attenuation of 26.9 dB and $R = 20.6\ \mathrm{mm\,h^{-1}}$; ⊙, mean attenuation of 18.9 dB and $R = 14.4\ \mathrm{mm\,h^{-1}}$. Theoretical values: curve 1, $C_n^2 = 2.5 \times 10^{-14}\ \mathrm{m^{-2/3}}$, $v = 10\ \mathrm{m\,s^{-1}}$; curve 2, $R = 20.6\ \mathrm{mm\,h^{-1}}$, $v = 10\ \mathrm{m\,s^{-1}}$; curve 3, $R = 14.4\ \mathrm{mm\,h^{-1}}$, $v = 10\ \mathrm{m\,s^{-1}}$.

REFERENCES

Basili, P., Ciotti, P., d'Auria, G., Ferrazzoli, P. and Solimini, D. (1990) Case study of intense scintillation events on the OTS path. *IEEE Trans. Antennas Propag.*, **38**(1), 107–13.

Capsoni, C., Mauri, M. and Paraboni, A. (1977) Incoherent effects in electromagnetic propagation through rain. *Ann Telecommun.*, **32**, 407–14.

CCIR (1990) Effects of tropospheric refraction on radiowave propagation. Report *718-3*.

Cole, R.S., Ho, K.L. and Mavrokoukoulakis, N.D. (1978) The effect of the outer scale of turbulence and wavelength on scintillation fading at millimeter wavelengths. *IEEE Trans. Antennas Propag.*, **26**(5), 712–15.

Cox, D.C., Arnold, H.W. and Hoffman, H.H. (1981) Observation of cloud produced amplitude scintillation on 19 and 29 GHz earth–space paths. *Radio Sci.*, **16**(5), 885–907.

Fante, R.L. (1975) Electromagnetic beam propagation in turbulent media. *Proc. IEEE*, **63**(12), 1669–92.

Fante, R.L. (1980) Electromagnetic beam propagation in turbulent media: an update. *Proc IEEE*, **68**(11), 1424–43.

Haddon, J. and Vilar, E. (1986) Scattering induced microwave scintillations from clear air and rain on earth space paths and the influence of antenna aperture. *IEEE Trans. Antennas Propag.*, **34**(5), 646–57.

Herben, M.H.A.J. (1982) Amplitude and phase scintillation measurements on 8.2 km line-of-sight path at 30 GHz. *Electron. Lett.*, **18**(7), 287–9.

Herben, M.H.A.J. (1983) Amplitude scintillations on the OTS-TM/$\overline{\text{TM}}$ beacon. *Arch. Elekt. Übertrag.*, **37**(3/4), 130–2.

Herben, M.H.A.J. (1984) The influence of tropospheric irregularities on the dynamic behaviour of microwave radio systems. PhD Thesis, Eindhoven University of Technology.

Herben, M.H.A.J. and Kohsiek, W. (1984) A comparison of radio wave and *in situ* observations of tropospheric turbulence and wind velocity. *Radio Sci.*, **19**(4), 1057–68.

Ishimaru, A. (1978) *Wave Propagation and Scattering in Random Media*, Vol. 2, Academic Press, New York.

Karasawa, Y., Yamada, M. and Allnutt, J.E. (1988) A new prediction method for tropospheric scintillation on earth–space paths. *IEEE Trans. Antennas Propag.*, **36**(11), 1608–14.

Medeiros Filho, F.C., Jayasuriya, D.A.R. and Cole, R.S. (1983) Tropospheric effects on line-of-sight links at 36 GHz and 55 GHz. *Proc. IEE F*, **130**(7), 679–87.

Moulsley, T.J. and Vilar, E. (1982) Experimental and theoretical statistics of microwave amplitude scintillations on satellite down-links. *IEEE Trans. Antennas Propag.*, **30**(6), 1099–106.

Ott, R.H. and Thompson, M.C. (1978) Atmospheric amplitude spectra in an absorption region. *IEEE Trans. Antennas Propag.*, **26**(2), 329–32.

Sarma, A.D. and Herben, M.H.A.J. (1989) Variable cut-off frequency filter for wide scintillation bandwidth measurements. *Arch. Elekt. Übertrag.* **43**(6), 378–81.

Tatarski, V.I. (1961) *Wave Propagation in a Turbulent Medium*, Dover Publications, New York.

4

Theory of scattering and absorption

4.1 INTRODUCTION

Attenuation, depolarization and volume scattering of radiowaves due to atmospheric particles (notably liquid water drops, ice particles and dust) are important phenomena which can limit severely the performance of telecommunication systems. To date, most communication systems use frequencies below 20 GHz. The demand for bandwidth for new services, however, forces the telecommunication engineer to consider the development of new systems using higher frequencies.

The phenomena mentioned are all aggregate effects of scattering by the population of particles which is encountered along the wave propagation path. Therefore, the basic theory underlying various models for attenuation, depolarization and volume scattering is the theory for single-particle scattering.

The propagation effects may be modelled by volumetric integration of scattering by individual particles, taking into account mutual influences where and to the extent that this is necessary.

Particle scattering effects become more severe with higher frequencies. This is aggravated by the increasing effect of small particles, such as liquid droplets, which are present in great numbers in the atmosphere. Clouds and fog consist mainly of small drops, but in rain small drops are also present in great numbers. In the discussion of the theory of single and multiple scattering, the emphasis will be on liquid water droplets in the atmosphere. The adaptation of the theory to other types of particles is straightforward.

In view of the increasing importance of small drops, the basic scattering theory must be reviewed in order to find the limitations of its applicability to millimetre wave problems. Therefore, the basic single-particle scattering theory is presented first. Secondly, models for the effects of random distributions of particles are presented; in particular the single scattering and first-order multiple scattering assumptions are discussed, leading to models for the coherent and incoherent fields that result from propagation of a plane wave through a polydispersion of atmospheric particles. The emphasis will be on the application to continuous wave (CW) waves for transmission of information (line-of-sight propagation). Scattering of electromagnetic pulses and problems of radar systems are not treated extensively.

The theory of single and multiple scattering has been discussed by many authors. Some discussions have led to different interpretations of formulae and different ways to derive them. A notorious example is the interpretation of the formulae in multiple scattering theory. At this moment, no reference can be found that treats the basic scattering theory as applied to communication systems completely in a consistent way. Also, most textbooks refer to others when it comes to some of the basics of the scattering theory (Ishimaru, 1978). This makes it difficult to understand the scattering theory. Therefore, in this chapter the focus will be on the understanding and the interpretation of the theory, together with all the assumptions that led to the formulae.

The most important assumption in the models will be that the small drops have spherical shapes. This assumption is valid for drops with a radius up to 1 mm (Pruppacher and Pitter, 1971). In general, the assumption that rain droplets are spherical in shape is reasonable for moderate rain. Only in heavy rain do the droplets have the shape of an oblate ellipsoid and fall with a certain canting angle. The assumption for all models is that the medium surrounding the drops (the air) is non-conducting; this means that there is no power loss in this medium.

Where possible, the theory as presented in this chapter follows Ishimaru (1978). Also several of the illustrations are after this book. An attempt has been made to standardize notation in the equations where possible, in order to bring out equivalences.

4.2 THEORY OF SINGLE-PARTICLE SCATTERING

4.2.1 Introduction

Concerning the single-particle scattering models, at first four basic analytical models will be discussed: Mie theory, Rayleigh scattering, the Born approximation and the WKB interior wavenumber approximation. Secondly, two numerical models are presented: the point-matching method and the Fredholm integral equation method. The assumption in these models will be that the drops are axially symmetric. This assumption is valid for all liquid water drops in the atmosphere.

4.2.2 Definition of quantities

In this section some of the most important quantities of single-particle scattering are introduced.

In scattering theory, the total (complex) electromagnetic field around the scattering (i.e. diffracting) object is split into an incident and a scattered field:

$$\boldsymbol{E}_{\mathrm{t}} = \boldsymbol{E}_{\mathrm{i}} + \boldsymbol{E}_{\mathrm{s}}$$
$$\boldsymbol{H}_{\mathrm{t}} = \boldsymbol{H}_{\mathrm{i}} + \boldsymbol{H}_{\mathrm{s}}. \tag{4.2.1}$$

From here on, in general only the electric field will be discussed. The incident field is taken to be a linearly polarized plane wave propagating in the $\boldsymbol{u}_{\mathrm{i}}$ direction with an amplitude of E_{i}:

$$\boldsymbol{E}_{\mathrm{i}}(\boldsymbol{r}) = E_{\mathrm{i}}\boldsymbol{u}_{\mathrm{e}}\,\mathrm{e}^{\mathrm{j}k_0\boldsymbol{u}_{\mathrm{i}}\cdot\boldsymbol{r}} \tag{4.2.2}$$

where $\boldsymbol{r}$ is the vector from the origin to the observation point, k_0 is the free-space wavenumber, and $\boldsymbol{u}_{\mathrm{e}}$ is a unit vector defining the polarization of the wave. Note that for the time dependence the convention $V(r, t) = \mathrm{Re}\,[U(r)\,\mathrm{e}^{-\mathrm{j}\omega t}]$ has been used, where V is the (real) field considered and U is its complex representation. The factor $\mathrm{e}^{-\mathrm{j}\omega t}$ will be omitted from here on. The sources of the plane wave are not considered. The scattered field is mostly taken to be a spherical wave, which is valid at a large distance from the scatterer:

$$\boldsymbol{E}_{\mathrm{s}}(\mathrm{r}) = E_{\mathrm{i}}\boldsymbol{f}(\boldsymbol{u}_{\mathrm{s}}, \boldsymbol{u}_{\mathrm{i}})\frac{\mathrm{e}^{\mathrm{j}k_0 r}}{r} \tag{4.2.3}$$

where r is the distance from the object to the observation point. The function $\boldsymbol{f}(\boldsymbol{u}_{\mathrm{s}}, \boldsymbol{u}_{\mathrm{i}})$ is called the scattering amplitude; it is a function of the direction of propagation $\boldsymbol{u}_{\mathrm{i}}$ of the incident field and of the direction $\boldsymbol{u}_{\mathrm{s}}$ from the object to the observation point (Fig. 4.1). This function represents the amplitude, phase and polarization of the scattered wave in the far-field zone. In order to describe the most general case, including polarization, properly, $\boldsymbol{f}(\boldsymbol{u}_{\mathrm{s}}, \boldsymbol{u}_{\mathrm{i}})$ should be defined as a tensor. For the description and understanding of the general theory, however, this is not required (section 4.6 should be referred to).

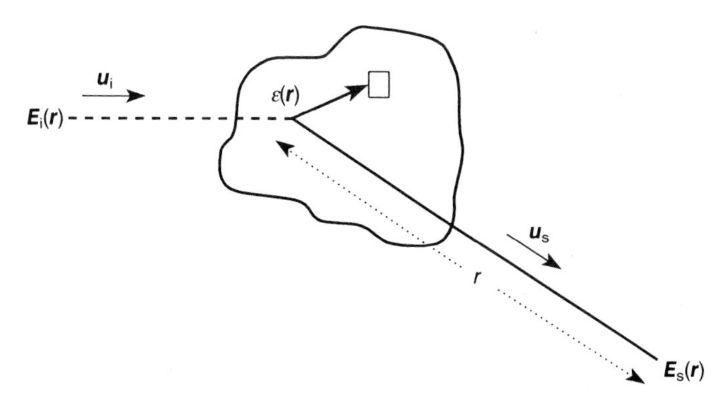

Fig. 4.1 Scattering configuration.

The scattering amplitude may be obtained directly from exact calculations of the scattered field. Alternatively, a general integral expression can be derived for the scattering amplitude in terms of the total field $\boldsymbol{E}$ inside the particle:

$$f(\boldsymbol{u}_s, \boldsymbol{u}_i) = \frac{k_0^2}{4\pi}\int_V \{-\boldsymbol{u}_s \times [\boldsymbol{u}_s \times \boldsymbol{E}(\boldsymbol{r}')]\}[\varepsilon_r(\boldsymbol{r}') - 1]\,e^{-jk_0\boldsymbol{r}'\cdot\boldsymbol{u}_s}\,dV' \quad (4.2.4)$$

(section 4.2.5(c)). Approximations for the field inside the particle then produce the approximated solution for the scattering function.

The far-field zone is mostly taken at the following distance:

$$r > 2l^2/\lambda \quad (4.2.5)$$

where l is the diameter of the object and λ is the wavelength. This condition can be derived geometrically; its interpretation is that at the distance r the spherical wave must be 'locally' plane (i.e. the wave front must not differ much from the wave front of a real plane wave). This far-field condition will also be encountered in multiple scattering theory.

It should be mentioned that the incident wave is defined everywhere in space. However, the total field in the direct neighbourhood of the particle can differ much from a plane wave. This is due to the summation of the incident and the scattered fields.

Finding a solution for a scattering problem means finding a scattered field that, together with the incident field matches on the surface of the scattering object with its internal field. Therefore, the scattered field reflects all of the properties of the scattering particle. This idea can help us to understand the forward scattering theorem.

4.2.3 The forward scattering or extinction theorem

The power density of the incident wave is uniform in space. Therefore, the power loss of the incident wave can be calculated by multiplying the uniform power density with a cross-section, having the dimension of area. Since the scattered field reflects all the properties of the scatterer, it should not be surprising that absorption as well as scattering by an object can be related to the scattered field only. The total power removed from the incident field by absorption and scattering can be represented by the total or extinction cross-section. The total cross-section can be related to the behaviour of the scattered wave in only the forward direction. Because of this, this relation is called the forward scattering theorem.

The time-averaged energy flow density (power density) is represented by the averaged Poynting vector $\bar{\boldsymbol{S}}$. This Poynting vector can be split into separate terms using the separation of the total field into the incident and scattered fields (equations (4.2.1) and (4.2.2)), resulting in

$$\bar{S} = \overline{S_i} + \overline{S_s} + \overline{S_d} \tag{4.2.6}$$

where

$$\overline{S_i} = \tfrac{1}{2}\,\mathrm{Re}\,(\boldsymbol{E}_i \times \boldsymbol{H}_i^*) \tag{4.2.7}$$

$$\overline{S_s} = \tfrac{1}{2}\,\mathrm{Re}\,(\boldsymbol{E}_s \times \boldsymbol{H}_s^*) \tag{4.2.8}$$

$$\overline{S_d} = \tfrac{1}{2}\,\mathrm{Re}\,(\boldsymbol{E}_i \times \boldsymbol{H}_s^* + \boldsymbol{E}_s \times \boldsymbol{H}_i^*). \tag{4.2.9}$$

Consider a large sphere of radius a embodying the scatterer. The net power flow through the surface of this sphere can be calculated by integrating the radial component of $\bar{S}$ over the sphere. If the scatterer is a lossless dielectric, the power flow will be zero. If it is lossy (permittivity is complex), the net power flow will represent the absorption rate P_a of energy by the scatterer (Fig. 4.2):

$$-P_a = P_i + P_s + P_d \tag{4.2.10}$$

where P_a represents the integral of the radial component of $\bar{S}$. The medium surrounding the particle is assumed to be lossless, so the net flow of the energy of the incident field is zero for every closed surface, i.e. $P_i = 0$, and thus

$$P_a + P_s = -P_d = -\tfrac{1}{2}\,\mathrm{Re} \iint_S (\boldsymbol{E}_i \times \boldsymbol{H}_s^* + \boldsymbol{E}_s \times \boldsymbol{H}_i^*) \cdot \boldsymbol{u}_n \,\mathrm{d}S \tag{4.2.11}$$

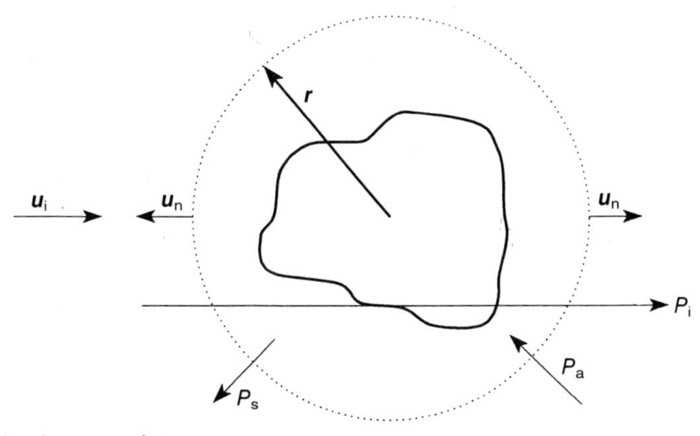

Fig. 4.2 Stationary points.

where $\boldsymbol{u}_n$ is the unit vector normal to the surface S and directed outwards and the dot represents the inner product. Using equation (4.2.4) and the Maxwell equations yields

$$(\boldsymbol{E}_i \times \boldsymbol{H}_s^*) \cdot \boldsymbol{u}_n = \left(\frac{\varepsilon}{\mu}\right)^{1/2} E_i E_s \boldsymbol{u}_e \cdot \boldsymbol{f}^*(\boldsymbol{u}_s, \boldsymbol{u}_i) \frac{e^{-jk_0 a}}{a} e^{jk_0 \boldsymbol{u}_i \cdot \boldsymbol{u}_n} \quad (4.2.12)$$

$$(\boldsymbol{E}_s \times \boldsymbol{H}_i^*) \cdot \boldsymbol{u}_n =$$

$$\left(\frac{\varepsilon}{\mu}\right) E_i E_s \{\boldsymbol{u}_n \cdot \boldsymbol{u}_i [\boldsymbol{f}(\boldsymbol{u}_s, \boldsymbol{u}_i) \cdot \boldsymbol{u}_e] - \boldsymbol{u}_n \cdot \boldsymbol{u}_e [\boldsymbol{u}_i \cdot \boldsymbol{f}(\boldsymbol{u}_s, \boldsymbol{u}_i)]\} \frac{e^{jk_0 a}}{a} e^{-jk_0 a \boldsymbol{u}_i \cdot \boldsymbol{u}_n}. \quad (4.2.13)$$

These two equations must be integrated over the surface of the large sphere. If the radius of the sphere is chosen large enough, the integrals can be evaluated by the method of the stationary phase. The basic idea behind this method is to find those points on the surface of the scattering object where the phase of the fields does not change significantly. Only the contribution of these points will be taken into account. The total contribution of all the other points is zero, because the phase difference between them will cancel their individual contributions. The restriction for this cancellation to hold is that the amplitude function does not change much as the phase function changes sign. This is the case for equations (4.2.12) and (4.2.13), as long as a is taken large enough.

The points where the phase is constant are called the stationary points; they can be found intuitively. In the phase terms (the exponents) of equations (4.2.12) and (4.2.13), the vector dot product determines the change of phase at the surface of the large sphere. The phase is nearly constant if the vector $\boldsymbol{u}_n$ normal to the sphere is parallel to the vector $\boldsymbol{u}_i$ of the incident wave (Fig. 4.2). This results in two solutions:

$$\boldsymbol{u}_n = \boldsymbol{u}_i, \quad \boldsymbol{u}_n = -\boldsymbol{u}_i. \quad (4.2.14)$$

If the stationary phase method is applied properly, the absorbed and scattered energy is given by

$$P_s + P_a = \frac{2\pi}{k_0}\left(\frac{\varepsilon}{\mu}\right)^{1/2} E_i^2 \, \mathrm{Im}[\boldsymbol{u}_e \cdot \boldsymbol{f}(\boldsymbol{u}_i, \boldsymbol{u}_i)]. \quad (4.2.15)$$

The stationary point associated with backscattering ($\boldsymbol{u}_n = -\boldsymbol{u}_i$) gives a purely imaginary contribution to the integral. For the power balance, only the real parts are needed (equation (4.2.9)), and therefore this contribution is absent in equation (4.2.15). In the solution for the stationary point associated with forward scattering, the real part of the scatter function is eliminated owing to the complex conjugate. Therefore, only the imaginary part is of importance (equation (4.2.15)). We may thus conclude that the imaginary part of the contribution of the stationary point associated with forward scattering ($\boldsymbol{u}_n = \boldsymbol{u}_i$) determines the total absorbed and scattered power. Explicitly, the rate at which energy is removed from the incident field is proportional to the imaginary part of the forward scattering amplitude ($\boldsymbol{u}_n = \boldsymbol{u}_i$) in the direction of the electric vector of the incident field.

The ratio of the rate of removal of energy ($P_a + P_s$) to the rate of the incident energy (P_i) on a unit cross-sectional area of the scatterer is defined as the extinction or total cross-section:

$$\sigma_t = \frac{P_s + P_a}{P_i} = \frac{4\pi}{k_0} \operatorname{Im}[\boldsymbol{u}_e \cdot \boldsymbol{f}(\boldsymbol{u}_i, \boldsymbol{u}_i)]. \qquad (4.2.16)$$

In a similar way the scattering and absorption cross-sections may be defined as $\sigma_s = P_s/P_i$ and $\sigma_a = P_a/P_i$. Evidently

$$\sigma_t = \sigma_s + \sigma_a. \qquad (4.2.17)$$

The forward scattering or extinction theorem is an exact relation, but care must be taken when applying it to approximations of the scattered field. In the case of an approximation (e.g. Rayleigh scattering) it is better to calculate for each solution the energy flow through a surface that embodies the scatterer. For Mie scattering solutions, the theorem can be applied without problems, because the solution is exact. The forward scattering or extinction theorem will be encountered in the form shown in equation (4.2.16) in the theory of multiple scattering.

4.2.4 Analytical models for scattering by a spherical particle

(a) Introduction

In this section, four models will be presented. They give an analytical solution (closed expression) for the scattering of a plane wave by a spherical particle. Therefore they are called analytical models. The four analytical models presented here are Mie theory, Rayleigh scattering, the Born approximation and the WKB interior wavenumber approximation. In these models the direction of propagation of the incident field will be chosen to be the positive z direction. This is not a restriction because of the spherical symmetry of the scattering particle.

(b) Mie theory

Mie (1908) was the first to obtain the exact solution for the scattering of a plane wave incident on a homogeneous non-magnetic sphere of any diameter and of any composition placed in a homogeneous, non-conducting, non-magnetic and isotropic medium.

Mie solved the problem using a representation of the fields in potentials and then solving the boundary value problem using spherical expansion functions. Nowadays, this way of solving a boundary value problem is a common procedure in applied mathematics. An excellent description of the procedure and the solution can be found in Born and Wolf (1986). Here only the solution will be presented.

The incident field is a plane wave propagating in the z direction with its polarization in the x direction (Fig. 4.3):

$$\boldsymbol{E}_{\mathrm{i}} = E_{\mathrm{i}}\boldsymbol{u}_x\,\mathrm{e}^{\mathrm{j}k_0 z}. \tag{4.2.18}$$

The sphere with constant permittivity $\varepsilon = m^2\varepsilon_0$ and radius a is placed in the origin of the reference frame. The surrounding medium has constants ε_0 and μ_0; this is appropriate for a raindrop in the atmosphere.

The scattered fields E_ψ and E_θ far from the sphere are given by

$$E_\psi = -E_{\mathrm{i}}\frac{\mathrm{j}\,\mathrm{e}^{\mathrm{j}k_0 r}}{k_0 r}S_1(\theta)\sin\psi, \quad E_\theta = E_{\mathrm{i}}\frac{\mathrm{j}\,\mathrm{e}^{\mathrm{j}k_0 r}}{k_0 r}S_2(\theta)\cos\psi \tag{4.2.19}$$

$$S_1(\theta) = \sum_{n=1}^{\infty}\frac{2n+1}{n(n+1)}(a_n\kappa_n + b_n\tau_n) \tag{4.2.20}$$

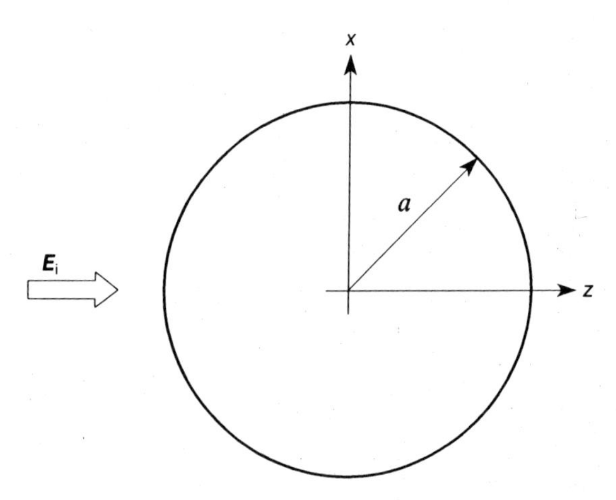

Fig. 4.3 Scattering geometry for Mie scattering.

$$S_2(\theta) = \sum_{n=1}^{\infty} \frac{2n+1}{n(n+1)}(a_n\tau_n + b_n\kappa_n) \tag{4.2.20}$$

$$\kappa_n = \frac{P_n^1(\cos\theta)}{\sin\theta}, \qquad \tau_n = \frac{\mathrm{d}}{\mathrm{d}\theta}P_n^1(\cos\theta) \tag{4.2.21}$$

$$a_n = \frac{\psi_n(\alpha)\psi_n'(\beta) - m\psi_n(\beta)\psi_n'(\alpha)}{\zeta_n(\alpha)\psi_n'(\beta) - m\psi_n(\beta)\zeta_n'(\alpha)} \tag{4.2.22}$$

$$b_n = \frac{m\psi_n(\alpha)\psi_n'(\beta) - \psi_n(\beta)\psi_n'(\alpha)}{m\zeta_n(\alpha)\psi_n'(\beta) - \psi_n(\beta)\zeta_n'(\alpha)} \tag{4.2.23}$$

$$\beta = k_0 ma, \qquad \alpha = k_0 a \tag{4.2.24}$$

$$\psi_n(x) = xj_n(x) = (\pi x/2)^{1/2} J_{n+1/2}(x) \tag{4.2.25}$$

$$\zeta_n(x) = xh_n^{(1)}(x) = (\pi x/2)^{1/2} H_{n+1/2}^{(1)}(x) \tag{4.2.26}$$

where $P_n^1(\cos\theta)$ is the associated Legendre function of the first kind, $J_n(x)$ is the Bessel function of the first kind and $H_n^{(1)}(x)$ is the Hankel function of the first kind. For this exact solution, a number of approximations can be derived. The most important approximation is for $k_0 a < 1$, meaning that the radius a of the particle is small compared with the wavelength. In this case the Mie solution may be approximated by the so-called Rayleigh scattering function. Rayleigh scattering will be treated separately, because this approximation can also be derived on more physical arguments instead of the pure mathematical way of approximating the cylindrical functions of Mie theory. Another approximation results in the Rayleigh–Debye solution, which will also be treated separately.

Mie theory has been the subject of many mathematical studies for some time, resulting in a number of rapidly converging expansion functions and approximations. In particular, it is worth mentioning that, based on the same Mie theory, solutions were also obtained for spheroidal and arbitrarily shaped objects (Yeh, 1964, 1969).

(c) Rayleigh scattering

The assumption for Rayleigh scattering is that $k_0 a < 1$, meaning that the radius a of the particle is small compared with the wavelength. How small this radius should be can be investigated by comparing the exact Mie solution with the Rayleigh solution. Kerker (1969) has done this analysis; he concludes that the upper limit of the radius can be taken to be $a = 0.05\lambda$, resulting in an error of less than 4% for a single scatterer. For a medium containing more scatterers another limit could be possible, as averaging effects and/or multiplying effects become important.

For a dielectric sphere that is small compared with the wavelength, the field inside the particle is chosen to be the solution of the equivalent dielectric electrostatic problem. This field is given by (Fig. 4.4)

$$\boldsymbol{E} = \frac{3}{\varepsilon_r + 2}\boldsymbol{E}_i. \tag{4.2.27}$$

A solution for the scattering problem can now be obtained using the integral expression (4.2.4) for the scattering amplitude in terms of the field inside the particle:

$$f(\boldsymbol{u}_s, \boldsymbol{u}_i) = \frac{k_0^2}{4\pi}[-\boldsymbol{u}_s \times (\boldsymbol{u}_s \times \boldsymbol{u}_e)]V\frac{3(\varepsilon_r - 1)}{\varepsilon_r + 2} \tag{4.2.28}$$

where $V = 4\pi a^3/3$ is the volume of the scatterer. It can be proven that the angular dependence of the scattering function is identical to that of an electric dipole. Another property of Rayleigh scattering is that the intensity $|f(\boldsymbol{u}_s, \boldsymbol{u}_i)|^2$ of the scattered wave is inversely proportional to the fourth power of the wavelength (k_0^4) and directly proportional to the square of the volume of the scatterer (V^2).

In the Rayleigh approximation, the field inside the scattering particle is not modelled properly. It has already been said that, for an exact solution, the scattered field reflects all the properties of the object. In an approximate solution this is no longer true. Problems arise if general electromagnetic

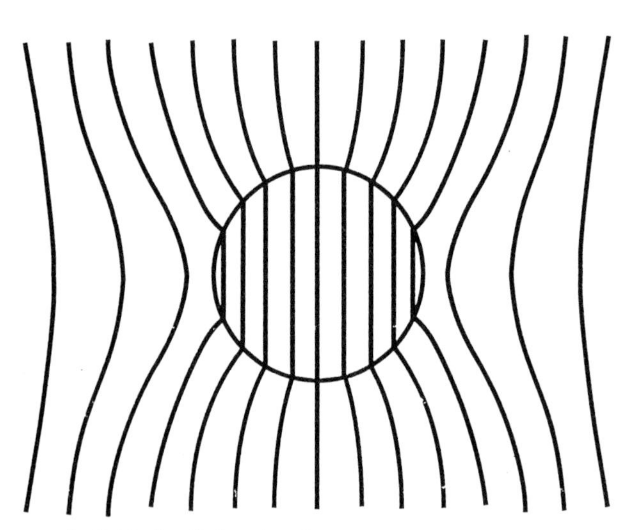

Fig. 4.4 Electrostatic field of a dielectric sphere.

theorems are applied to this approximated scattered field. For example, the forward scattering theorem fails for the Rayleigh approximation. This remark also holds for the following approximation.

(d) Rayleigh–Debye scattering (Born approximation)

This approximation is valid if

$$(\varepsilon_r - 1)k_0 a \ll 1. \tag{4.2.29}$$

This means that either the relative dielectric constant ε_r is close to unity or the size a of the scatterer is small. In both cases, the influence of the scatterer on the total field is small and the field inside the scatterer can be approximated by the incident field:

$$\boldsymbol{E}(\boldsymbol{r}) \approx \boldsymbol{E}_i(\boldsymbol{r}). \tag{4.2.30}$$

For a homogeneous sphere of radius a, the scattering amplitude then becomes (equation (4.2.4))

$$\boldsymbol{f}(\boldsymbol{u}_s, \boldsymbol{u}_i) = \frac{k_0^2}{4\pi}[-\boldsymbol{u}_s \times (\boldsymbol{u}_s \times \boldsymbol{u}_e)](\varepsilon_r - 1)VF(k_1) \tag{4.2.31}$$

where

$$F(k_1) = \frac{3}{k_1^3 a^3}[\sin(k_1 a) - k_1 a \cos(k_1 a)] \tag{4.2.32}$$

$$k_1 = k_0|\boldsymbol{u}_i - \boldsymbol{u}_s|. \tag{4.2.33}$$

(e) WKB interior wavenumber approximation

The letters WKB denote Wentzel, Kramers and Brillouin, who contributed to this approximation (Saxon, 1955). The restrictions for this approximation are

$$(\varepsilon_r - 1)k_0 a \gg 1, \quad \varepsilon_r - 1 < 1. \tag{4.2.34}$$

Therefore, it provides results for which the Rayleigh and Born approximations are not applicable.

The field inside the scatterer is approximated by a field propagating in the direction of the incident field (because $\varepsilon_r - 1 < 1$) but with the wavenumber (propagation constant) of the medium of the scatterer. The transmission coefficient at the surface is approximated by that of a plane wave for normal incidence.

With these approximations, the scattering amplitude for a sphere with radius a becomes

$$f(\boldsymbol{u}_s, \boldsymbol{u}_i) = \frac{k_0^2}{4\pi}[-\boldsymbol{u}_s \times (\boldsymbol{u}_s \times \boldsymbol{u}_e)]VS(\boldsymbol{u}_s, \boldsymbol{u}_z) \tag{4.2.35}$$

where

$$S(\boldsymbol{u}_s, \boldsymbol{u}_z) = V^{-1}\int_V 2[n(\boldsymbol{r}') - 1]\,\mathrm{e}^{\,\mathrm{j}k_0 z_1 + \mathrm{j}k_0(z - z_1) - \mathrm{j}k_0 \boldsymbol{r}' \cdot \boldsymbol{u}_s}\,\mathrm{d}V' \tag{4.2.36}$$

where $n(\boldsymbol{r})$ is the refractive index in the particle, V is the volume of the particle, $\boldsymbol{u}_z$ is the unit vector in the positive z direction and z_1 is the coordinate z of the surface of the particle.

(f) Conclusion

All models described above give also solutions for other shapes than spheres; those solutions are omitted here. The assumptions and ideas behind the methods have been emphasized, and they are the same for other shapes. Table 4.1 gives an overview of the methods described, including the ranges of validity. Section 4.2.6 presents results of further analysis of the limits of validity for spherical particles.

4.2.5 Numerical models

(a) Introduction

In this section, two numerical models are presented: the point-matching method and the Fredholm integral equation method. The point-matching method is very well described by Morrison and Cross (1974); for the Fredholm integral equation method Holt (1980) provides an introduction.

Table 4.1 Overview of solutions to scatter problems

Solution	Restrictions
Mie	–
Rayleigh	$k_0 a < 1$
Born	$(\varepsilon_r - 1)\ k_0 a \ll 1$
WKB	$(\varepsilon_r - 1)\ k_0 a \gg 1;\ \varepsilon_r - 1 < 1$

The point-matching method is based on the expansion of the fields inside the scatterer and the expansion of the scattered field outside the scatterer. Together with the incident field, these fields must be matched at the surface of the scatterer in such a way that the boundary conditions are satisfied. The expansions are mostly truncated to give approximated results.

The Fredholm integral equation method is based on the integral equation for the scattering of a plane wave. This equation is transformed in order to derive a pair of coupled Fredholm integral equations with a non-singular kernel. These equations are solved by the Schwinger variational method (Singh and Stauffer, 1975) in combination with numerical quadrature.

The point-matching method is mostly used for axially symmetric scatterers (raindrops), although it should be possible to use any smooth scatterer. The method is limited by the Rayleigh hypothesis, which means that the surface of the scatterer must not deviate too much from a sphere, in order to obtain a meaningful solution in terms of spherical expansion functions. For raindrops this means that the calculations are valid up to approximately 60 GHz.

The Fredholm integral equation method is only limited by the available computer power (it is normally quite demanding in that aspect). The method is numerically stable and can handle heavily distorted raindrops.

(b) Point-matching method

The basis of the method is an expansion of the scattered and refracted field (i.e. the field inside the scatterer) in terms of regular vector spherical harmonics (Stratton, 1941).

Outside the raindrop, the total electromagnetic field is the sum of the incident field of the plane wave and the scattered field. The scattered field is expanded in terms of outgoing vector spherical harmonics that satisfy the radiation condition of Sommerfeld. In practice, these functions are Bessel functions of the third kind (Hankel functions of the first kind):

$$\boldsymbol{E}_{\mathrm{s}} = -\sum_{m=-\infty}^{\infty} \sum_{\substack{n \geqslant |m| \\ n \neq 0}} [a_{mn} \boldsymbol{M}_{mn}^{(3)}(k_0) + b_{mn} \boldsymbol{N}_{mn}^{(3)}(k_0)] \tag{4.2.37}$$

$$\boldsymbol{H}_{\mathrm{s}} = \frac{\mathrm{j}k_0}{\omega\mu_0} \sum_{m=-\infty}^{\infty} \sum_{\substack{n \geqslant |m| \\ n \neq 0}} [a_{mn} \boldsymbol{N}_{mn}^{(3)}(k_0) + b_{mn} \boldsymbol{M}_{mn}^{(3)}(k_0)] \tag{4.2.38}$$

where a_{mn} and b_{mn} are the expansion coefficients.

The origin of the coordinate systems is interior to the raindrop. In the origin at $r = 0$ the field must remain finite, and therefore the refracted field is expanded in spherical Bessel functions of the first kind:

$$\boldsymbol{E}_{\mathrm{t}} = -\sum_{m=-\infty}^{\infty} \sum_{\substack{n \geq |m| \\ n \neq 0}} [c_{mn}\boldsymbol{M}_{mn}^{(1)}(k_{\mathrm{s}}) + d_{mn}\boldsymbol{N}_{mn}^{(1)}(k_{\mathrm{s}})] \tag{4.2.39}$$

$$\boldsymbol{H}_{\mathrm{t}} = \frac{\mathrm{j}k_{\mathrm{s}}}{\omega\mu_0} \sum_{m=-\infty}^{\infty} \sum_{\substack{n \geq |m| \\ n \neq 0}} [c_{mn}\boldsymbol{N}_{mn}^{(1)}(k_{\mathrm{s}}) + d_{mn}\boldsymbol{M}_{mn}^{(1)}(k_{\mathrm{s}})] \tag{4.2.40}$$

where c_{mn} and d_{mn} are the expansion coefficients and k_{s} is the wavenumber inside the raindrop; $k_{\mathrm{s}} = nk_0$.

The unknown expansion coefficients a_{mn}, b_{mn}, c_{mn} and d_{mn} must be determined from the boundary conditions. The boundary conditions are that the tangential components of the total electric and magnetic fields must be continuous across the surface of the raindrop.

The drops are chosen to be axially symmetric. In that case the incident plane wave can be expanded in a Fourier series in the azimuthal angle ψ:

$$\boldsymbol{E}_{\mathrm{i}} = \sum_{m=-\infty}^{\infty} \boldsymbol{e}_{\mathrm{m}}(r, \theta)\, \mathrm{e}^{\mathrm{j}m\psi} \tag{4.2.41}$$

$$\boldsymbol{H}_{\mathrm{i}} = \sum_{m=-\infty}^{\infty} \boldsymbol{h}_{m}(r, \theta)\, \mathrm{e}^{\mathrm{j}m\psi}. \tag{4.2.42}$$

In this way, the problem can be decomposed and the boundary conditions satisfied independently for each term of the Fourier series.

The forward-scattering amplitude and the scattering cross-sections can be defined in terms of the coefficients in the expansions of the scattered field.

If the fields are expanded in vector spherical harmonics, a theoretical problem occurs: the validity of the Rayleigh hypothesis (Van den Berg and Fokkema, 1979). The Rayleigh hypothesis concerns the legitimacy of using spherical expansion functions to point-match a non-spherical surface. It asserts that, outside and on the obstacle, the scattered field may be expanded in terms of outward-going wavefunctions. It is analogous to the assumption made by Lord Rayleigh in his treatment of diffraction by a reflecting grating.

In the paper of Van den Berg and Fokkema, the conditions under which the Rayleigh hypothesis holds are estimated for two-dimensional scatterers. They reduced the problem of studying the legitimacy of the Rayleigh hypothesis to one of locating the singularities in the representation of the exterior scattered field. This can be done very easily by taking as a starting point the Rayleigh theory itself.

The investigation of whether the Rayleigh solution is analytical in the domain of interest yields a condition under which the Rayleigh hypothesis can be used. It may be obvious that the Rayleigh hypothesis never holds for surfaces with edges.

Because of the Rayleigh hypothesis, Holt (1982) doubts the correctness of the point-matching method for raindrops above 35 GHz. In practice, calculations show that the point-matching method gives accurate results up to 60 GHz.

To obtain an approximate solution for the expansion coefficients, only a finite number of them is considered. One procedure is to truncate the sums at a given number and then to satisfy the boundary conditions for the same number of points on the surface of the scatterer. This has been done by Oguchi (1973) and it leads to a system of simultaneous linear equations for the coefficients. This procedure, in which the total number of fitting points is equal to the number of unknown coefficients, it called collocation.

Point matching can also be combined with least-squares fitting (Morrison and Cross, 1974). In this case the boundary conditions are satisfied in a least-squares sense at a larger number of points than the number of unknown coefficients in the truncated expansion of the scattered field. In this way a significant improvement can be obtained in the overall fit of the boundary condition. The far-field quantities are not affected as significantly. This is because the higher-order coefficients are more significant on the boundary than in the far field. Another advantage of the least-squares fitting method is that it improves the numerical stability.

Since the fit of the boundary condition for the elliptic cylinder becomes poorer with increasing eccentricity (Rayleigh hypothesis), it is desirable to use least-squares fitting rather than collocation for the raindrop problem.

(c) Fredholm integral equation method

In the equations in this section, the dyadic notation is used (indicated by sans serif symbols). Although this type of notation is uncommon in electromagnetic scattering theory, it is used here to be consistent with the notation of Holt (1980).

The starting point in the Fredholm integral equation method is the volume integral equation for the electromagnetic field $\mathbf{E}(\boldsymbol{r})$ describing the scattering of a plane wave with propagation vector $\boldsymbol{k}_{\mathrm{i}}$ by a scatterer of dielectric constant $\varepsilon(\boldsymbol{r})$ and volume V:

$$\mathbf{E}(\boldsymbol{r}) = \mathbf{J}_{\mathrm{i}}\,\mathrm{e}^{\mathrm{j}\boldsymbol{k}_{\mathrm{i}}\cdot\boldsymbol{r}} + \int_V \mathbf{K}(\boldsymbol{r}')\mathbf{G}(\boldsymbol{r}, \boldsymbol{r}')\cdot\mathbf{E}(\boldsymbol{r}')\,\mathrm{d}\boldsymbol{r}' \qquad (4.2.43)$$

where

$$\mathbf{G}(\boldsymbol{r}, \boldsymbol{r}') = \left(\mathbf{I} + \frac{1}{k_0^2}\nabla\nabla\right)\frac{\mathrm{e}^{\,\mathrm{j}k_0|\boldsymbol{r}-\boldsymbol{r}'|}}{4\pi|\boldsymbol{r} - \boldsymbol{r}'|} \qquad (4.2.44)$$

$$\mathbf{K}(\boldsymbol{r}) = k_0^2[\varepsilon_{\mathrm{r}}(\boldsymbol{r}) - 1] \qquad (4.2.45)$$

where k_0 is the free-space wavenumber, ε_r is the dielectric constant, $\mathbf{I}$ = the identity matrix, $\boldsymbol{r}'$ is another $\boldsymbol{r}$ and

$$\boldsymbol{J}_i = \boldsymbol{I} - \boldsymbol{u}_i \cdot \boldsymbol{u}_i. \tag{4.2.46}$$

In the papers of Holt (e.g. Holt, 1980) reference is made to Newton (latest edition, 1982). Newton uses the Maxwell equations in Gaussian units, but in his papers Holt has transformed the equations to standard SI units.

The scattering amplitude $\boldsymbol{f}(\boldsymbol{k}_s, \boldsymbol{k}_i)$ for the scattering into direction $\boldsymbol{k}_s$ is defined in the standard way:

$$\boldsymbol{E}(\boldsymbol{r}) \overset{r\to\infty}{\sim} \boldsymbol{J}_i \, \mathrm{e}^{\mathrm{j}\boldsymbol{k}_i \cdot \boldsymbol{r}} + \frac{\mathrm{e}^{\mathrm{j}k_0 r}}{r} \boldsymbol{f}(\boldsymbol{k}_s, \boldsymbol{k}_i) + \boldsymbol{O}\left(\frac{1}{r^2}\right). \tag{4.2.47}$$

Writing equation (4.2.43) in its asymptotic form gives

$$\boldsymbol{f}(\boldsymbol{k}_s, \boldsymbol{k}_i) = \frac{1}{4\pi} \boldsymbol{J}_s \cdot \int_V \mathrm{e}^{-\mathrm{j}\boldsymbol{k}_s \cdot \boldsymbol{r}} \boldsymbol{K}(\boldsymbol{r}) \boldsymbol{E}(\boldsymbol{r}) \, \mathrm{d}\boldsymbol{r}. \tag{4.2.48}$$

This equation, which is the generalized version of equation (4.2.4), shows that the scattering is determined from knowledge of the field inside the scatterer only. However, the integral equation (4.2.43) has a singular Green's function. In order to deal with this singularity analytically, equation (4.2.43) is multiplied by $\mathrm{e}^{-\mathrm{j}\boldsymbol{k}_1 \cdot \boldsymbol{r}}$ (which resembles a plane wave) and by $\boldsymbol{K}(\boldsymbol{r})$ and then integrated throughout the volume of the scatterer:

$$\int_V \mathrm{e}^{-\mathrm{j}\boldsymbol{k}_1 \cdot \boldsymbol{r}} \boldsymbol{K}(\boldsymbol{r}) \boldsymbol{E}(\boldsymbol{r}) \, \mathrm{d}\boldsymbol{r}$$

$$= \boldsymbol{J}_i \int_V \mathrm{e}^{-\mathrm{j}(\boldsymbol{k}_1 - \boldsymbol{k}_i) \cdot \boldsymbol{r}} \boldsymbol{K}(\boldsymbol{r}) \, \mathrm{d}\boldsymbol{r} + \int_V \int_V \mathrm{e}^{-\mathrm{j}\boldsymbol{k}_1 \cdot \boldsymbol{r}} \boldsymbol{K}(\boldsymbol{r}) \boldsymbol{G}(\boldsymbol{r}, \boldsymbol{r}') \cdot \boldsymbol{K}(\boldsymbol{r}') \boldsymbol{E}(\boldsymbol{r}') \, \mathrm{d}\boldsymbol{r} \, \mathrm{d}\boldsymbol{r}'. \tag{4.2.49}$$

It is assumed that a solution of equation (4.2.49) is square integrable and it can therefore be expressed as the Fourier transform:

$$\boldsymbol{E}(\boldsymbol{r}) = \int \boldsymbol{C}(\boldsymbol{k}_2) \, \mathrm{e}^{\mathrm{j}\boldsymbol{k}_2 \cdot \boldsymbol{r}} \, \mathrm{d}\boldsymbol{k}_2. \tag{4.2.50}$$

Substituting equation (4.2.50) in equations (4.2.49) and (4.2.48) gives (after some mathematical considerations) a pair of coupled non-singular Fredholm integrals, which determine the scattering amplitude. These can be solved by the Schwinger variational method in combination with numerical quadrature. For the case of homogeneous scatterers, the integrals reduce to two-dimensional integrals.

Equation (4.2.43) is also called the Lippman–Schwinger equation; from this equation the scattering amplitude is defined (equation (4.2.48)). Using the Schwinger variational method the scattering amplitude can be reconstructed (Singh and Stauffer, 1975).

In this method the electric field is replaced by some trial functions containing arbitrary parameters. These parameters are then varied to obtain the stationary value of the scattering amplitude, which is taken to be the approximate value of the scattering amplitude.

Because of computational advantages, the trial functions are often chosen to be linear combinations of some basis functions. Holt (1980) uses functions of the form:

$$\boldsymbol{E}(\boldsymbol{r}) = \sum_{j=1}^{n} \boldsymbol{C}_j \, e^{jk_j \cdot \boldsymbol{r}}. \tag{4.2.51}$$

Imposing some restrictions on the basis functions yields a convergent process to determine the electric field and the scattering amplitude. The error in the stationary value of the scattering amplitude can be shown to be of second order in the error of the basis functions (Singh and Stauffer, 1975).

In the Fredholm integral equation method, convergence tests can be done during the calculations without starting all over again.

(d) Comparison of the two methods (Holt, 1982)

Both methods have the advantage of great flexibility. They can deal with many different scatterer shapes. For the point-matching method (PMM), the scatterers must be smooth and they should not be concave. The Fredholm integral equation method (FIM) has been implemented for bodies with edges.

The FIM requires a separate analysis for each scatterer, leading to different programs for each scatterer and, in practice, only a few model shapes are feasible. The FIM could in principle be implemented for bodies with continuously varying, or even anisotropic, dielectric constants, but no implementations have yet been made. The PMM appears not to be suitable for treating such bodies. The FIM has the advantage that a significant part of the calculation, the integrals, depends only on the size parameter and the axial ratio. Consequently, changing the dielectric constant does not require a complete new calculation. Also, performing convergence checks does not require recalculation of the integrals. Holt (1982) is of the opinion that the

FIM is probably slower for close-to-spherical bodies but fast and more stable for bodies with an axial ratio of 5:1 or more. The FIM method has been applied to other scatterers than axially symmetric ones. However, few calculations have been made, since the problem lacked physical motivation. The PMM can in theory be applied to bodies without an axis of symmetry, but would probably be computationally very costly. Because of the Rayleigh hypothesis, the validity of the PMM calculations for raindrops must be checked for frequencies higher than 60 GHz. The Schwinger variational method assures the stability and convergence of the FIM for any shape and frequency.

4.2.6 Application to millimetre wave propagation

(a) Rayleigh scattering approximation

It is of interest to study the range of validity of Rayleigh scattering in more detail when considering scattering and attenuation of millimetre waves. The general condition $k_0 a < 1$ means that the circumference of the droplet must be smaller than one wavelength. The Kerker condition mentioned in section 4.2.4(c) limits application of the Rayleigh assumption to $k_0 a \leqslant 0.3$. Analysis of the relative error between Rayleigh scattering and (exact) Mie scattering calculations was done for a range of permittivities and sizes. The Rayleigh scattering approximation results in a scattering amplitude function that equals that of a dipole. When particle size increases, this model is no longer accurate. However, for particles that have a scattering cross-section that is small compared with the absorption cross-section, the Rayleigh assumption yields good results, even if the angular distribution of the scattered field deviates from a dipole field.

From the detailed comparison with Mie scattering the following simple condition was found for an accuracy of less than 1% for the Rayleigh approximation:

$$fD|\varepsilon_\mathrm{r}|^{1/2} \leqslant 10 \tag{4.2.52}$$

where f (GHz) is the frequency and D (mm) is the diameter. This is equivalent to

$$k_0 a|\varepsilon_\mathrm{r}|^{1/2} \leqslant 0.1. \tag{4.2.53}$$

Since, for non-conducting particles, the wavenumber inside the particle is given by $k = k_0 \varepsilon_\mathrm{r}^{1/2}$, it appears that the condition for the Rayleigh approximation to be valid is that the size (circumference) of the particle is small compared with the wavelength inside the particle. Whenever the simple rule of equation (4.2.52) is not fulfilled, it is better to use Mie scattering calculations.

de Wolf, Russchenberg and Ligthart (1990) have discussed a Rayleigh scattering model for composite particles as found in the melting layer in stratiform clouds. There also, the condition of validity of a model of small homogeneous particles with a certain effective permittivity was expressed in terms of the 'local' wavelength inside the particles.

(b) Born approximation (Rayleigh–Debye)

In the same manner, an accuracy analysis was carried out for the Born approximation. The general condition, stated in the literature as

$$k_0 a(\varepsilon_r - 1) \ll 1 \tag{4.2.54}$$

was found to be inappropriate for more detailed specifications. The condition for an accuracy of approximately 1% or better can be formulated as follows:

$$\begin{aligned} fD &\leqslant 100 \\ \mathrm{Re}\,(\varepsilon_r) - 1 &\leqslant 0.01 \\ [\mathrm{Im}\,(\varepsilon_r)]^2 fD &\leqslant 0.04. \end{aligned} \tag{4.2.55}$$

For small permittivities the Born approximation is very good, even up to very high frequencies.

(c) WKB interior wavenumber approximation

For spherical particles, the WKB approximation was found to fail in the (sub)millimetre wave range for every practical value of size and relative permittivity. The error was 10% or higher. The reason for this may be that the particles considered are spherical. The approximation uses a plane wave reflection coefficient for normal incidence. This is not a very accurate approximation for a spherical particle, where it can only hold at one point. For large particles with irregular shapes the method may be more useful.

Figure 4.5 shows a logarithmic plot of the scattering efficiency $Q_s = \sigma_s/\pi a^2$ versus frequency for a drop diameter of 1 mm, calculated with the three methods. While condition (4.2.52) limits usage of Rayleigh scattering for accurate calculations of the scattered field to $f < 1.5$ GHz, it is clear that the Rayleigh results are quite close to the exact (Mie) calculations up to some 40 GHz.

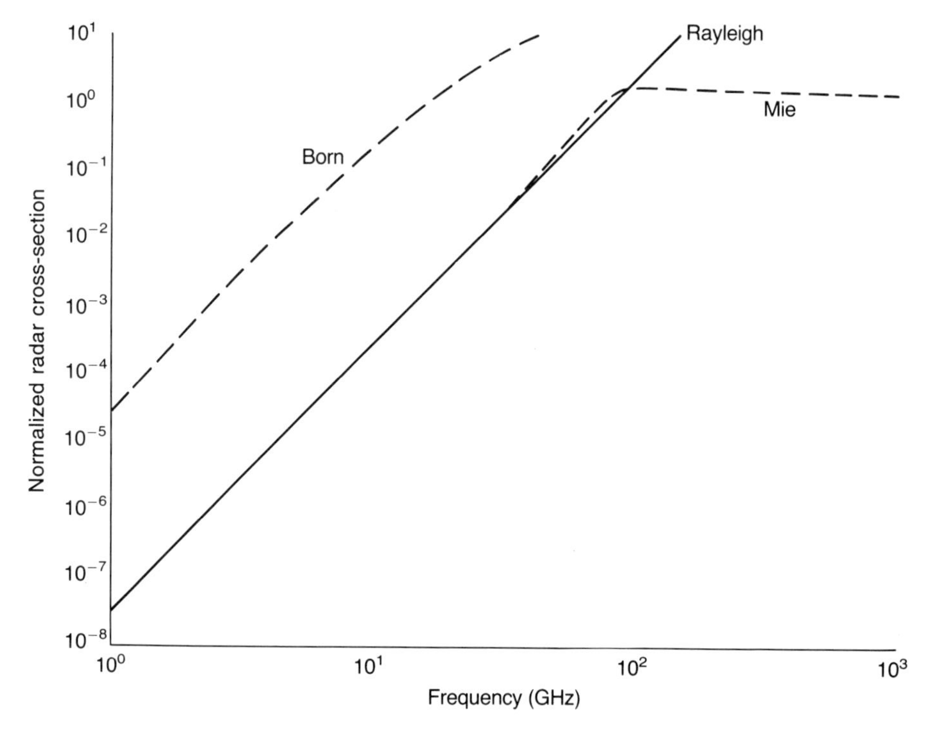

Fig. 4.5 Comparison of different models for spherical raindrop scattering. The Born approximation fails owing to the large value of ε_r for water.

(d) Scattering by water and ice particles

In this section we will examine in more detail the scattering from water and ice particles following Mie theory. For water we have used Manabe's model of permittivity (Chapter 13), taking a temperature of +10 °C. For ice we use Ray's permittivity model, with a temperature of −10 °C.

Scattering by water spheres Figure 4.6 shows the scattering, absorption and total cross-sections as a function of frequency for various drop radii. From Fig. 4.6 it is clear that scattering and absorption cross-sections for raindrops in the millimetre wave range are of comparable magnitudes; hence the scattering albedo σ_s/σ_t is not small. In the millimetre wave range, total extinction is due to both scattering and absorption.

Scattering by ice spheres Calculations of scattering by ice spheres have been performed over the 10–300 GHz frequency range. Figure 4.7 shows the scattering, absorption and total cross-sections as a function of frequency for

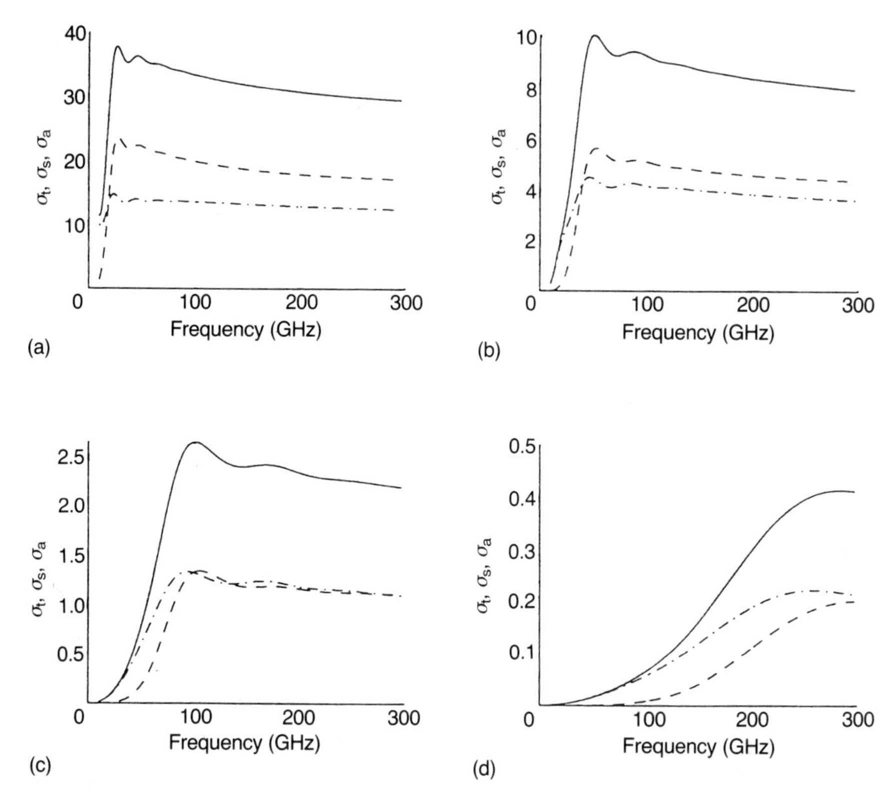

Fig. 4.6 Scattering (– – –), absorption (– · –) and total (——) cross-sections σ_s, σ_a and σ_t for water spheres of various sizes as a function of frequency: (a) $a = 2$ mm; (b) $a = 1$ mm; (c) $a = 0.5$ mm; (d) $a = 0.2$ mm.

various sizes. As expected, the ice absorption cross-sections are much smaller than those for scattering, resulting in almost negligible contribution to total extinction. The occurrence of 'resonant' and 'window' regions for ice can be distinguished. This resonant behaviour occurs in certain frequency regions depending on particle size. For a 2 mm particle the resonant maxima are 50–90 GHz, 145–180 GHz and 250–280 GHz, where ice particles have cross-sections almost 1.5 times those of water drops (Fig. 4.8). Nevertheless in the window regions the ice extinction cross-sections are smaller than the water extinction cross-sections but become comparable as frequency increases. Also we should note the small-scale fluctuations with frequency that appear in scattering for ice. This puzzling effect has attracted scientific interest over recent decades and various explanations have been suggested (this is discussed in the subsequent section on radar cross-sections, where the fluctuations are more pronounced).

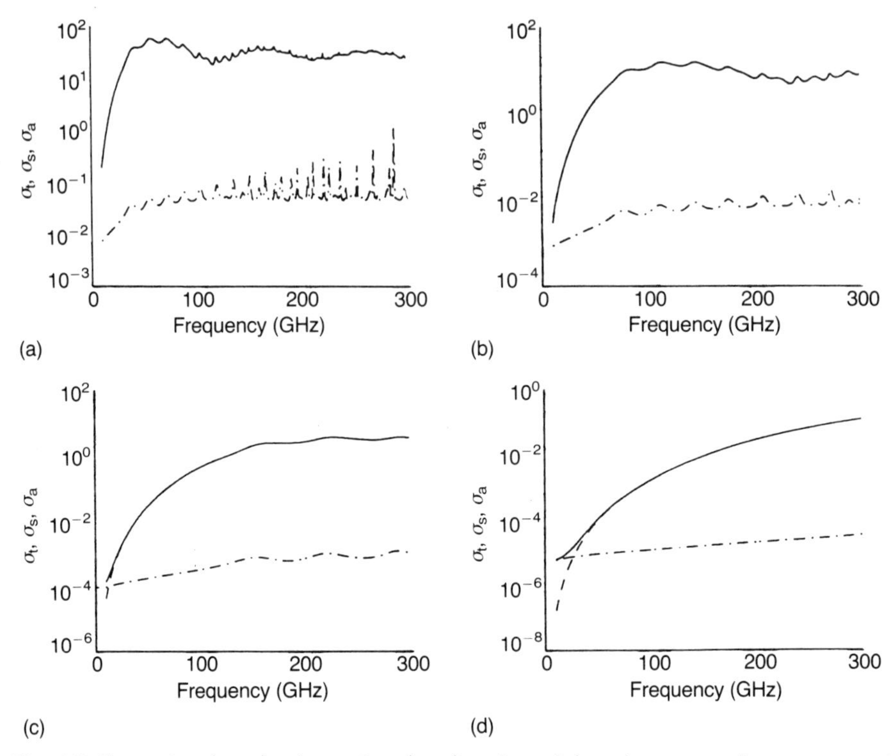

Fig. 4.7 Scattering (– – –), absorption (– · –) and total (——) cross-sections σ_s, σ_a and σ_t for ice spheres of various sizes as a function of frequency: (a) $a = 2$ mm; (b) $a = 1$ mm; (c) $a = 0.5$ mm; (d) $a = 0.2$ mm.

Distinction between scattering and absorption for ice Figure 4.9 shows the scattering albedo as a function of frequency for various radii of ice particles. Looking carefully at Fig. 4.9, we see that small particles absorb a much more significant percentage of the incident energy than the larger ones for frequencies up to 30 GHz. This is seen more easily in Fig. 4.10 where the values of scattering and absorption cross-sections are illustrated for ice spheres ranging from 0.1 mm to 1 mm in radius. For a 0.1 mm radius particle the absorbed energy is greater than the scattered energy for frequencies up to approximately 45 GHz, where the two energies are equal. As the size of ice particles increases, the cross-over point moves rapidly to lower frequencies, e.g. for 0.2 mm particles it is at 25 GHz. Although the proportion of energy absorbed is much greater than that scattered at lower frequencies, the absolute values are of course very small and usually negligible. For frequencies above 60 GHz the resultant attenuation is almost exclusively due to scattering, even for quite small ice particles.

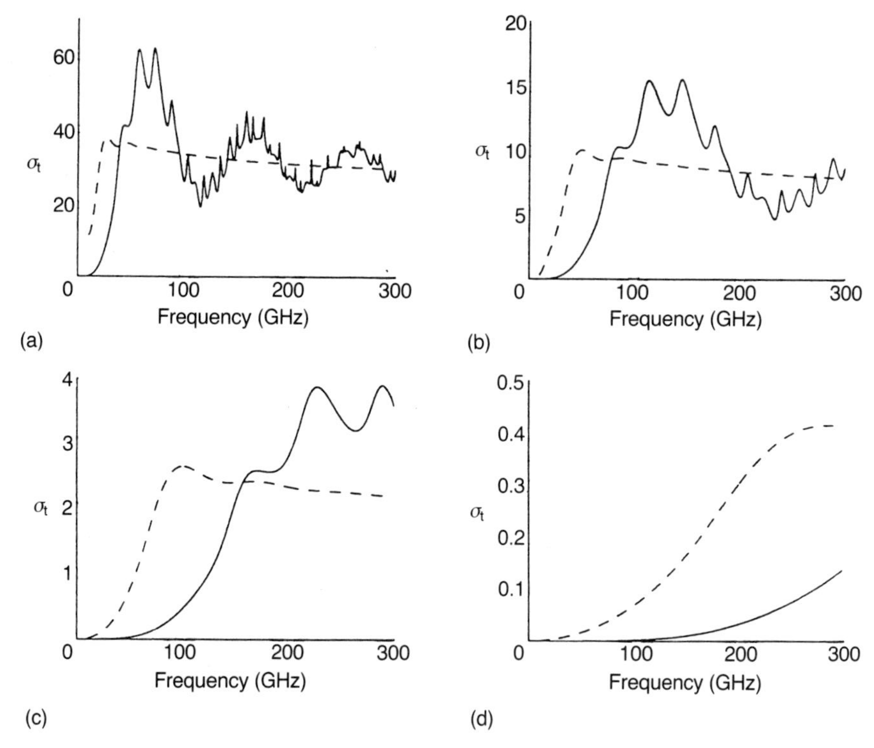

Fig. 4.8 Comparison of extinction cross-sections for ice (——) and water (---) spheres over the microwave and millimetre wave frequency range: (a) $a = 2$ mm; (b) $a = 1$ mm; (c) $a = 0.5$ mm; (d) $a = 0.2$ mm.

Radar cross-sections The normalized radar cross-section $Q = \sigma/\pi a^2$ versus size parameter $2\pi a/\lambda$ is plotted for ice and water spheres in Fig. 4.11 as calculated from the Mie theory for a frequency of 30 GHz. The size parameter increment has been set at 6.28×10^{-4} in order to observe the small-scale fluctuations, which are of the same origin as those in the extinction curve (Fig. 4.8) but much more pronounced. Careful comparison between the data shown in Figs 4.8 and 4.11 reveals that the sharp minima and maxima appearing in the forward scatter and backscatter cases for ice occur at the same values of size parameter. The striking effect of these numerous strong resonances not only emerges in theoretical calculations but has been verified experimentally by Atlas *et al*. (1963). Various explanations have been sought in the past for the existence of these fine-scale resonances, some referring exclusively to forward scatter and others to backscatter, but none appears to be entirely satisfactory. However, recently Papatsoris and Watson (1992) have related the existence of these resonances both for the forward and the

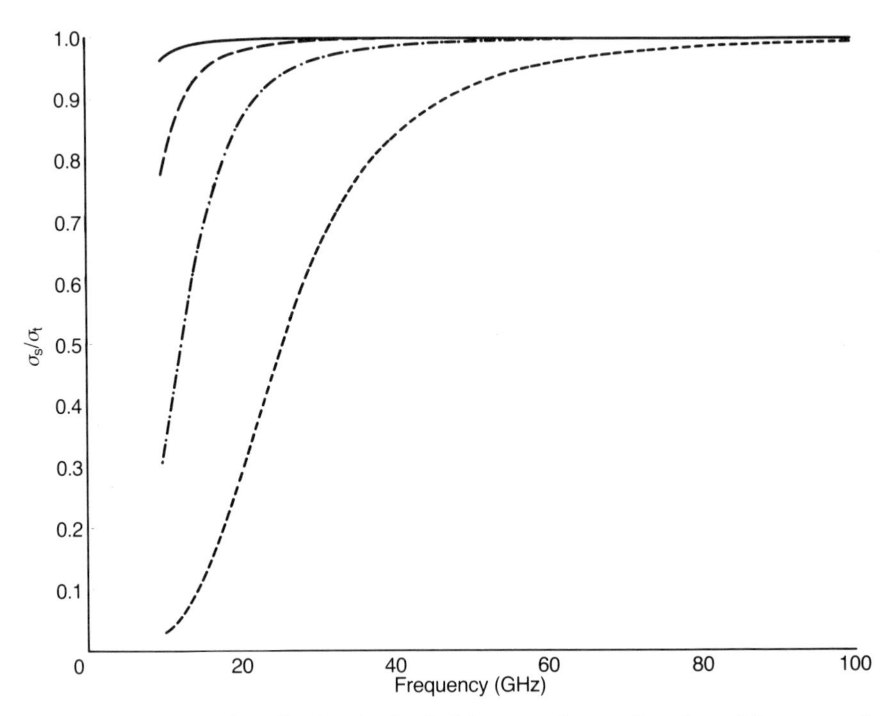

Fig. 4.9 Single-scattering albedo of spherical ice crystals as a function of frequency for various radii: ——, $a = 2.0$ mm; – – –, $a = 1.0$ mm; – · –, $a = 0.5$ mm · · · ·, $a = 0.1$ mm.

backward scattering cases to the excitation of the modes of free oscillation of the ice particles. The explanation may be summarized as follows.

When a periodic external field falls upon a sphere, it gives rise to a forced oscillation of free and bound charges, synchronous with the applied field. This leads to the excitation of a secondary field both inside and outside the sphere, which has to be added vectorially to the primary field in order to determine the total resultant field. The continuity of the tangential components of the fields at the separation surface lead to a transient term arising from the natural modes of oscillation with suitable amplitudes. If the frequency of the applied field lies very close to an eigenfrequency or mode of free oscillation of the material body of the sphere, resonance phenomena will occur. Whether these oscillations will be damped or increased depends on the index of refraction of the material and the scattering direction. Here we must note that even for a dielectric sphere the eigenfrequencies are always complex, whereas the frequency of the applied field is real, so the case of infinite amplitudes does not occur. For ice, which is almost a pure dielectric, the excited oscillations are not significantly damped, giving rise to pronounced fluctu-

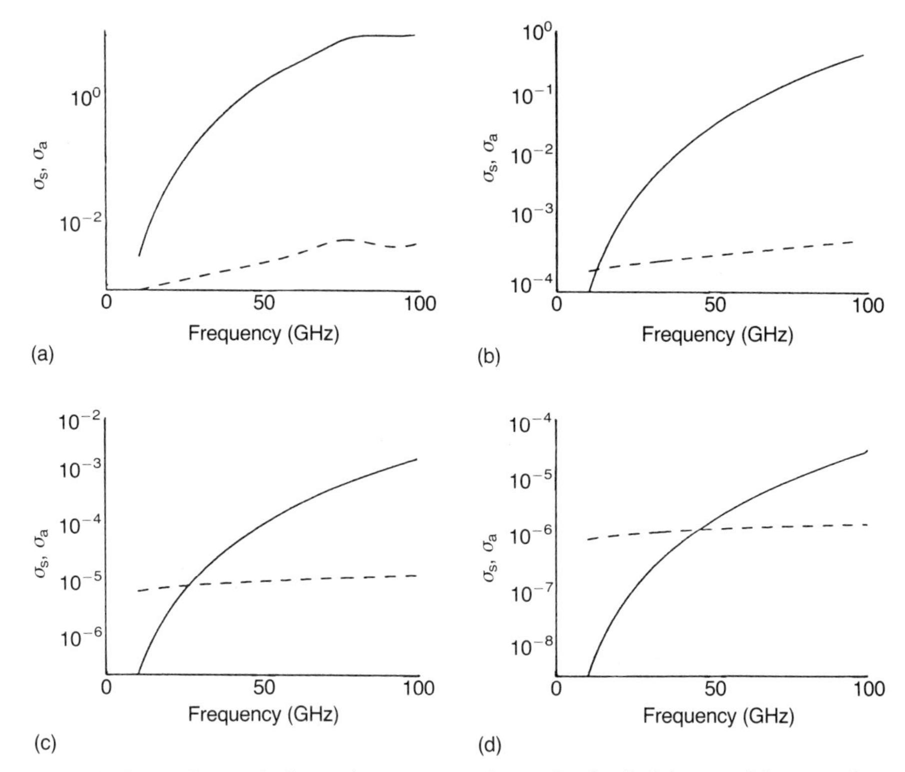

Fig. 4.10 Scattering and absorption cross-sections of spherical ice particles as a function of frequency: (a) $a = 1$ mm; (b) $a = 0.5$ mm; (c) $a = 0.2$ mm; (d) $a = 0.1$ mm.

ations. Such oscillations therefore are responsible for the numerous peaks observed when ice scattering cross-sections versus size parameter are plotted.

4.3 THEORY OF MULTIPLE-PARTICLE SCATTERING

4.3.1 Introduction

In this section the scattering of electromagnetic waves by a random distribution of many particles will be treated. The focus is on attenuation and fluctuation of radiowaves on a line-of-sight path. The discussion, in general, is a plane-wave treatment; the effect of antennas, which are used to obtain narrow-beam transmissions, is not taken into account. The medium is assumed to be non-turbulent.

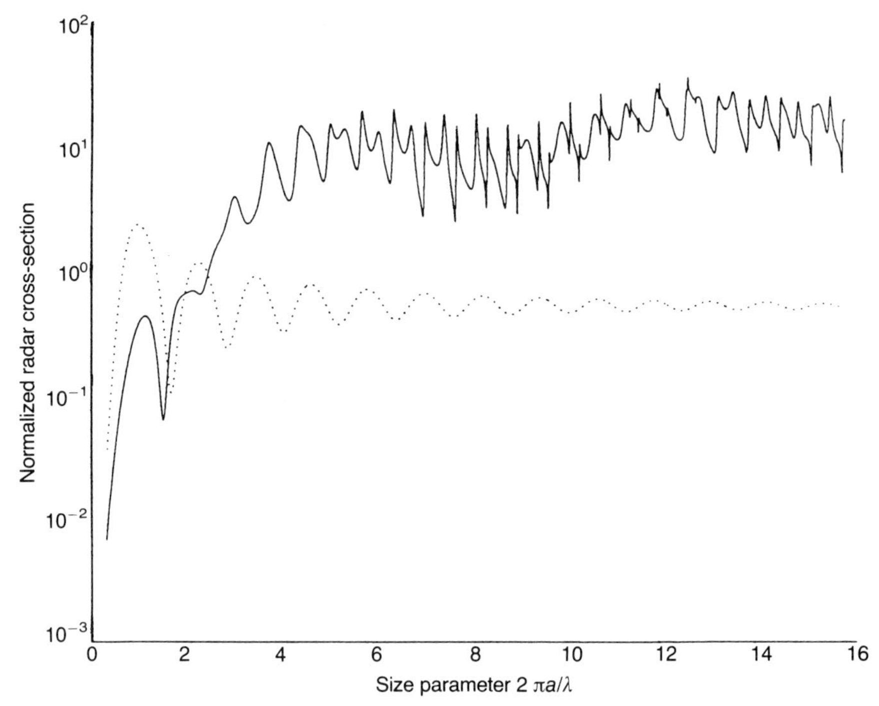

Fig. 4.11 Normalized radar cross-section versus size parameter $2\pi a/\lambda$ for ice (——) and water (· · · ·) spheres at 30 GHz. The temperature for ice is -10 °C and that for water is 10 °C.

4.3.2 Definition of terms

In multiple-particle scattering theory, a few important conceptions are encountered that merit some discussion before presenting the theory of multiple-particle scattering. They are discussed in this section.

(a) Coherent and incoherent field and intensity

Before the different multiple-particle scattering approximations are analysed, two important quantities have to be explained: the coherent and incoherent field (or intensity) (Ishimaru, 1978). Since the particles are randomly distributed and (in general) in constant motion, the scattered field is not constant and its amplitude and phase should fluctuate in a random manner.

The electric or magnetic field $V(\boldsymbol{r}, t)$ is a real function of time t and position $\boldsymbol{r}$. It can be expressed in terms of the complex field $U(\boldsymbol{r}, t)$, the amplitude $A(\boldsymbol{r}, t)$ and the phase $\phi(\boldsymbol{r}, t)$:

$$V(\mathbf{r}, t) = \mathrm{Re}[U(\mathbf{r}, t)\,\mathrm{e}^{-\mathrm{j}\omega_0 t}] \tag{4.3.1}$$

$$U(\mathbf{r}, t) = A(\mathbf{r}, t)\,\mathrm{e}^{\mathrm{j}\phi(\mathbf{r},t)} \tag{4.3.2}$$

where $A(\mathbf{r}, t)$ and $\phi(\mathbf{r}, t)$ are slowly varying functions of time.

The field $U(\mathbf{r}, t)$ is a random function of position $\mathbf{r}$ and time t. Therefore, it can be written as a sum of the average field $\langle U \rangle$ and the fluctuating field U_f:

$$U(\mathbf{r}, t) = \langle U(\mathbf{r}, t) \rangle + U_f(\mathbf{r}, t) \tag{4.3.3}$$

$$\langle U_f(\mathbf{r}, t) \rangle = 0 \tag{4.3.4}$$

where the angle brackets represent the (ensemble) average.

The average field $\langle U \rangle$ is called the coherent field and the fluctuating field U_f is called the incoherent field. The square of the magnitude of the coherent field is called the coherent intensity I_c and the average of the square of the magnitude of the incoherent field is called the incoherent intensity I_i. The sum of I_c and I_i is the average intensity $\langle I \rangle$:

$$\langle I \rangle = \langle |U|^2 \rangle = I_c + I_i \tag{4.3.5}$$

$$I_c = |\langle U \rangle|^2 \tag{4.3.6}$$

$$I_i = \langle |U_f|^2 \rangle. \tag{4.3.7}$$

The correlation function $\boldsymbol{\Gamma}(\mathbf{r}_1, t_1; \mathbf{r}_2, t_2)$ of the field at $\mathbf{r}_1$ and t_1 and the field at $\mathbf{r}_2$ and t_2 is given by

$$\boldsymbol{\Gamma}(\mathbf{r}_1, t_1; \mathbf{r}_2, t_2) = \langle U(\mathbf{r}_1, t_1) U^*(\mathbf{r}_2, t_2) \rangle. \tag{4.3.8}$$

Using equations (4.3.3) and (4.3.4) yields

$$\boldsymbol{\Gamma} = \boldsymbol{\Gamma}_c + \boldsymbol{\Gamma}_f \tag{4.3.9}$$

$$\boldsymbol{\Gamma}_c = \langle U(\mathbf{r}_1, t_1) \rangle \langle U^*(\mathbf{r}_2, t_2) \rangle \tag{4.3.10}$$

$$\boldsymbol{\Gamma}_f = \langle U_f(\mathbf{r}_1, t_1) U_f^*(\mathbf{r}_2, t_2) \rangle \tag{4.3.11}$$

where $\boldsymbol{\Gamma}_f$ is the covariance of the fluctuating field $U_f = U - \langle U \rangle$. The function $\boldsymbol{\Gamma}$ is the mutual coherence function.

In many practical problems, the field $U(\mathbf{r}, t)$ can be considered a stationary random function of time within a limited time period. For example, wave fluctuations in atmospheric turbulence and hydrometeors may be approximately stationary within a few minutes. If the field is stationary, $\boldsymbol{\Gamma}_f$ is a function of $\boldsymbol{r}$ and $\tau = t_2 - t_1$ only:

$$\boldsymbol{\Gamma}_f = \boldsymbol{\Gamma}_f(\boldsymbol{r}, \tau) = \boldsymbol{\Gamma}_f(\boldsymbol{r}, -\tau). \tag{4.3.12}$$

The coherent intensity I_c is a measure of the power flux (power per unit area). The attenuation along a path is therefore expected to decrease with an attenuation constant (attenuation per unit length) which is equal to the number of particles per unit volume multiplied by their total cross-section $\sigma_t = \sigma_s + \sigma_a$:

$$I_c = e^{-M\sigma_t z}. \tag{4.3.13}$$

The intensity scattered out of the propagation path is the incoherent intensity I_i. However, this intensity is also scattered out of all other propagation paths. Therefore the sum $\langle I \rangle$ of I_c and I_i should depend largely on the absorption cross-section of the particles and may be given approximately by

$$\langle I \rangle \approx e^{-M\sigma_a z}. \tag{4.3.14}$$

Figure 4.12 illustrates the behaviour of the fields. Equations (4.3.13) and (4.3.14) do not take into account the effect of backscattering, particle size, receiver characteristics, etc. Therefore, they are only approximate; they serve to give some overall characteristics of the field in the line-of-sight problem.

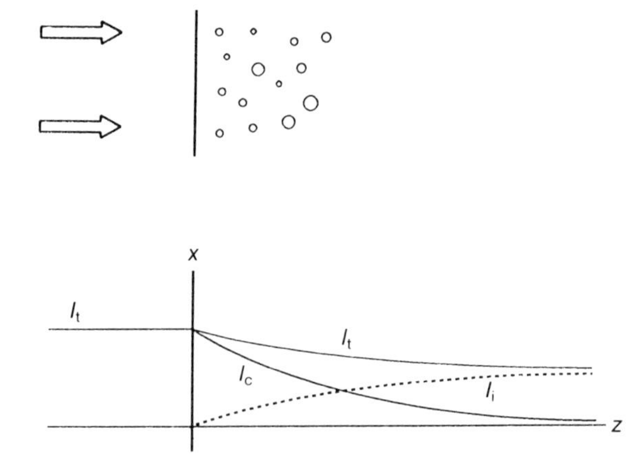

Fig. 4.12 Approximate behaviour of coherent intensity I_c, incoherent intensity I_i and total intensity $I_t = \langle I \rangle$ for a plane wave incident upon a semi-infinite region ($z > 0$) containing particles.

(b) Optical distance

Attenuation of coherent energy transmitted along a path is given by (see also the following sections):

$$I(l) = I(0)\exp\left(-\int_0^l M\sigma_t\,\mathrm{d}z\right) = I(0)\,\mathrm{e}^{-\zeta} \tag{4.3.15}$$

where M is the number density of particles (monodisperse distribution) and σ_t is the total cross-section of one particle. The parameter ζ, introduced here, is called the optical distance (also used in sections 4.3.3(c) and 4.3.3(e)). It is a dimensionless parameter. Stating that the optical distance of a path of length l is ζ is equivalent to stating that the power flux is attenuated, through scattering and absorption, to a factor $\mathrm{e}^{-\zeta}$ of the incident flux.

Assumptions and approximations often are dependent on the total attenuation occurring on a path and may then conveniently be expressed in terms of optical distance. For example, the single-scattering assumption is valid if

$$\zeta \approx M\sigma_t l \ll 1. \tag{4.3.16}$$

Particle size distribution In rain many drops are present with very different sizes. Therefore, for calculations involving the rain medium, the particle size distribution should be taken into account. Other media exist also in which drops with varying sizes are present, such as clouds and fog.

Let $m(D, \boldsymbol{r})\,\mathrm{d}D$ be the number of particles per unit volume located at $\boldsymbol{r}$ having a range of sizes between D and $D + \mathrm{d}D$; then the optical distance ζ may be expressed as

$$\zeta = \int_0^l M\langle\sigma_t\rangle\,\mathrm{d}z \tag{4.3.17}$$

where:

$$M\langle\sigma_t\rangle = \int_0^\infty m(D, \boldsymbol{r})\sigma_t(D)\,\mathrm{d}D \tag{4.3.18}$$

$$M(\boldsymbol{r}) = \int_0^\infty m(D, \boldsymbol{r})\,\mathrm{d}D. \tag{4.3.19}$$

(c) Weak fluctuation region

The weak fluctuation region can be defined for those configurations in which the field is predominantly coherent and the magnitude of the incoherent field is much smaller than that of the coherent field. This corresponds to the case where

$$\zeta \approx M\sigma_t l \ll 1. \tag{4.3.20}$$

It is clear from Fig. 4.12 that when scattering is much smaller than absorption (or the albedo $\sigma_s/\sigma_t \ll 1$) then the incoherent field is smaller than the coherent field, even if the optical distance is comparable with or larger than unity. Therefore, this case can also be regarded as the weak fluctuation case.

Another weak fluctuation case which is often encountered in practice is that in which the receiver has a narrow receiving angle. In this case, the amount of scattered intensity entering the receiver is small compared with the direct coherent intensity and therefore the received field is predominantly coherent. An example of this is microwave transmission through rain, where the optical distance can be large but the fluctuation is weak because of the small receiving angle of the antenna.

In the weak fluctuation approximation, the field $U(\boldsymbol{r}, t)$ at the receiver is given by the sum of the coherent field $\langle U(\boldsymbol{r}, t)\rangle$ and the incoherent field $U_f(\boldsymbol{r}, t)$:

$$U(\boldsymbol{r}, t) = \langle U(\boldsymbol{r}, t)\rangle + U_f(\boldsymbol{r}, t). \tag{4.3.21}$$

The single scattering and the first-order multiple scattering approximations (see below) are valid in the weak fluctuation region. However, these approximations become inadequate as the particle density is increased and the coherent intensity becomes comparable with or less than the incoherent intensity. Then the multiple scattering effects become dominant in determining the fluctuation characteristics of a wave.

4.3.3 Approximate models

(a) Introduction

Figure 4.13 shows different approximations for multiple-particle scattering. If the particle density is tenuous, the single scattering approximation is used. In this approximation the incident wave encounters very few particles. The scattered wave is, on arrival at the receiver, assumed to originate from a scattering by a single particle. All double and multiple scattering is assumed to be negligible. The direct wave reaching the receiver is the incident wave. This approximation is used in weather radar and ocean acoustics experiments. As the particle density increases the assumption that the direct wave is identical to the incident wave no longer holds. The attenuation of the direct wave due to scattering and absorption along its path has to be taken into account. The scattered wave originates from scattering by a single particle, but the wave incident on this particle has likewise been attenuated by scattering and absorption along its path. This approximation takes into

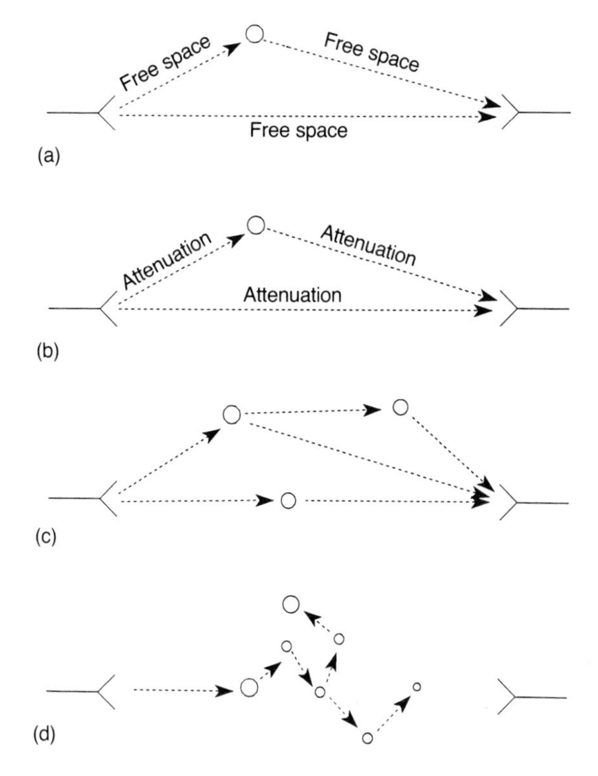

Fig. 4.13 (a) Single scattering, (b) first-order multiple scattering, (c) multiple scattering and (d) diffusion approximation.

account some of the multiple scattering effects and therefore it is called the first-order multiple scattering approximation. This approximation is used in calculations of rain attenuation.

The single scattering and the first-order multiple scattering approximations are applicable to many practical problems. They are useful because of their relative simplicity.

Historically two distinct theories have been developed in dealing with multiple scattering effects in a correct way: the analytical theory, or as it is mostly called the multiple scattering theory, and the transport theory.

The multiple scattering theory starts with basic differential equations such as the Maxwell equations or the wave equation. Then the scattering and absorption characteristics of particles are introduced and appropriate differential or integral equations are obtained for the statistical quantities such as variances and correlation functions. In principle all the multiple scattering, diffraction and interference effects can be included. In practice it is impossible to obtain a formulation which completely includes all these effects. Therefore the various theories which yield useful solutions are all approximate, each being useful in a specific range of parameters. An important example for radio wave transmission is the Twersky theory, which gives rise to the Foldy–Twersky integral equation.

Transport theory deals directly with the transport of energy through a medium containing particles. The development of the theory is heuristic and it lacks the mathematical rigour of the multiple scattering theory. Even though diffraction and interference effects are included in the description of the scattering and absorption characteristics of a single particle, transport theory itself does not include diffraction effects. It is assumed in transport theory that there is no correlation between fields, and therefore, the addition of power rather than the addition of fields holds. For the transport theory the basic differential equation is called the equation of transfer, which is equivalent to Boltzmann's equation used in the kinetic theory of gases and in neutron transport theory.

Even though the starting points of the multiple scattering theory and the transport theory are different, there exists a fundamental relationship between them. The specific intensity used in transport theory and the mutual coherence function used in multiple scattering theory are related through a Fourier transform. This means that even though transport theory was developed on the basis of the addition of powers, it contains information about the correlation of the fields.

For densely distributed particles the so-called diffusion approximation can be used. This approximation is based on the transport theory. The solution of this approximation is the so-called equation of radiative transfer. Between the two extremes of the tenuous and dense media, a range of particle densities exists for which multiple scattering effects are important. These effects can be described by the multiple scattering approximation.

The distinction among the different approximations must be made on the basis of optical distance (section 4.3.2(b)) and albedo (section 4.3.3(e)), which depend on wavelength, particle size and scattering characteristics as well as the transmitter and receiver characteristics.

To gain an impression of the range of validity of the models, the volume of the particles' density can be used. The volume density is the ratio of the volume occupied by particles to the total volume of the medium. If the volume density is considerably smaller than 0.1% the single scattering and first-order solutions are applicable. If the volume density is much larger than 1% the transport theory can be approximated by the diffusion approximation. For a volume density of about 1% neither the first-order solution nor the diffuse approximation may be valid and the complete equation of transfer must be solved or multiple scattering theory must be used.

If the scatterers are in motion, the fields become functions of time. In that case the correlation of the fields in time as well as in space has to be considered.

(b) Single scattering

The volume V contains a random distribution of particles. Consider a small volume $\mathrm{d}V$ containing $M\,\mathrm{d}V$ particles where M, the number density, is the number of particles per unit volume. It is assumed that this volume is located far from both the transmitter and the receiver. Let $\boldsymbol{u}_i$ and $\boldsymbol{u}_s$ be the unit vectors in the direction of the incident wave to and the scattered wave from the volume $\mathrm{d}V$ (Fig. 4.14). Let r_1 and r_2 be the distances from the volume $\mathrm{d}V$ to the transmitter and receiver respectively, and let S_i be the power density of the incident plane wave. At the receiver the scattered power density S_r due to a single particle is then given by

$$S_r = \frac{|\boldsymbol{f}(\boldsymbol{u}_s, \boldsymbol{u}_i)|^2}{r_2^2} S_i. \tag{4.3.22}$$

Because of the randomness of particle distribution, all interference between waves scattered by different particles is neglected and the power in the different waves scattered once from a particle is added. Thus the scattered power density S_r at the receiver due to the particles in $\mathrm{d}V$ is

$$S_r = \frac{M|\boldsymbol{f}(\boldsymbol{u}_s, \boldsymbol{u}_i)|^2}{r_2^2} S_i\,\mathrm{d}V. \tag{4.3.23}$$

From this equation, the total received power S_r can be calculated using system characteristics.

Applications of the single scattering approximation are mono- and bistatic radar equations, when optical distance (attenuation) is small.

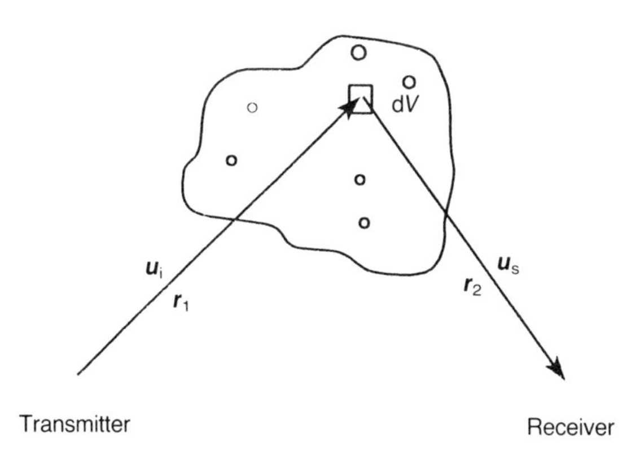

Fig. 4.14 Geometry for single scattering approximation.

(c) First-order multiple scattering

For the first-order multiple scattering approximation the incident power density S_i at dV in Fig. 4.14 should include attenuation due to the total cross-section. Therefore

$$S_i' = S_i e^{-\zeta_1} \tag{4.3.24}$$

where ζ_1 is the optical distance from the transmitter to dV along path r_1 (section 4.3.2(b)):

$$\zeta_1 = \int_0^{R_1} M\sigma_t \, dz. \tag{4.3.25}$$

Similarly, the scattered power density S_r should be multiplied by $e^{-\zeta_2}$ where ζ_2 is the optical distance from dV to the receiver:

$$\zeta_2 = \int_0^{R_2} M\sigma_t \, dz. \tag{4.3.26}$$

Using equation (4.3.19) yields

$$S_r = \frac{M|\boldsymbol{f}(\boldsymbol{u}_s, \boldsymbol{u}_i)|^2}{r_2^2} e^{-\zeta_1-\zeta_2} S_i \, dV. \tag{4.3.27}$$

(d) Rytov solution

The Rytov solution for amplitude and phase fluctuation uses an alternative representation of the field in terms of the fluctuation of the 'complex phase' $\varphi(\boldsymbol{r}, t)$:

$$U(\boldsymbol{r}, t) = U_0(\boldsymbol{r}, t) e^{\varphi(\boldsymbol{r}, t)} \tag{4.3.28}$$

where $U_0(\boldsymbol{r}, t)$ is a reference field with definite amplitude and phase. This field can be chosen to be the field in free space when the random medium is removed or to be the average field $\langle U(\boldsymbol{r}, t)\rangle$. Here U_0 is chosen to be equal to $\langle U \rangle$.

The function $\varphi(\boldsymbol{r}, t)$ is in general complex:

$$\varphi(\boldsymbol{r}, t) = \varphi'(\boldsymbol{r}, t) + j\varphi''(\boldsymbol{r}, t). \tag{4.3.29}$$

U and U_0 can be expressed in terms of the amplitudes A and A_0 and the phases ϕ and ϕ_0:

$$U(\boldsymbol{r}, t) = A(\boldsymbol{r}, t) e^{j\phi(\boldsymbol{r}, t)} \tag{4.3.30}$$

$$U_0(\boldsymbol{r}, t) = A_0(\boldsymbol{r}, t)\,\mathrm{e}^{\mathrm{j}\phi_0(\boldsymbol{r}, t)}. \tag{4.3.31}$$

Now the functions φ' and φ'' in equation (4.3.29) become

$$\varphi'(\boldsymbol{r}, t) = \ln\left[A(\boldsymbol{r}, t)/A_0(\boldsymbol{r}, t)\right] \tag{4.3.32}$$

$$\varphi''(\boldsymbol{r}, t) = \phi(\boldsymbol{r}, t) - \phi_0(\boldsymbol{r}, t). \tag{4.3.33}$$

The real part φ' of φ represents the fluctuation of the logarithm of the amplitude A and therefore it is called the log-amplitude fluctuation. The imaginary part φ'' of φ represents the fluctuation of phase.

Expression (4.3.28) is more suitable for line-of-sight propagation, because we may model the field as being modified by successive layers of the medium. The output field is the product of individual contributions, each successive layer having the output of the previous layer as its input (Fig. 4.15(a)). In that way, the exponent φ represents the sum of all fluctuations along the propagation path and it is more convenient to express the field in the exponential form.

For the scattering problem, expression (4.3.3) is more suitable since it is more convenient to view the scattered field as a sum of many contributions from different parts of the random medium (Fig. 4.15(b)).

The solution of the scattering problem using the Rytov method can be obtained by transforming the wave equation for U into a non-linear Riccati differential equation for φ and then solving it by an iterative procedure. The first iteration solution is

$$U(\boldsymbol{r}, t) = \langle U(\boldsymbol{r}, t)\rangle\,\mathrm{e}^{U_{\mathrm{f}}(\boldsymbol{r}, t)/\langle U(\boldsymbol{r}, t)\rangle} \tag{4.3.34}$$

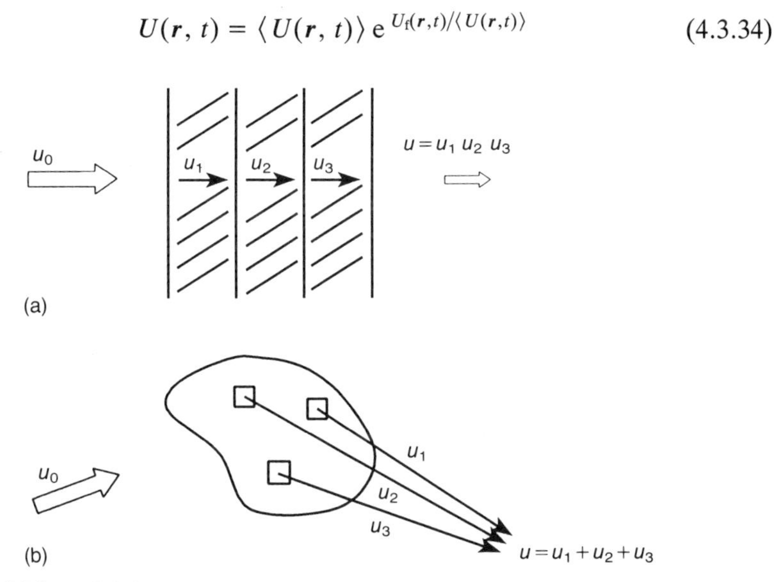

Fig. 4.15 (a) Line-of-sight description and (b) scattering problem description.

where $U_f(\boldsymbol{r}, t)$ is the first-order multiple scattering solution. If the magnitude of $U_f/\langle U \rangle$ is small compared with unity, the exponent can be expanded in a series:

$$U = \langle U \rangle (1 + U_f/\langle U \rangle + \ldots). \tag{4.3.35}$$

The first two terms of this series are identical to the approximation in the weak fluctuation region (section 4.3.2(c)). Further iterations of the equation result in a solution which is accurate over a wider range of particle densities.

(e) Transport theory

Transport theory does not start with the wave equation but deals exclusively with the transport of energy through a medium containing scatterers.

If we know the power flux density and its variation with direction we can find for a given location and for a given direction the average power flux density per unit of solid angle per unit bandwidth. This quantity is called specific intensity, measured in watts per square metre per steradian per hertz. Specific intensity may best be described as the statistical average of the Poynting vector within a given solid angle. This illustrates the fact that transport theory is applicable to randomly varying, uncorrelated fields.

For a description of scattering and absorption by a medium containing particles, we consider a specific intensity $I(\boldsymbol{r}, \boldsymbol{u}_s)$ incident upon a cylindrical elementary volume with unit cross-section and length $\mathrm{d}l$ (Fig. 4.16). The volume $\mathrm{d}l$ contains $M\,\mathrm{d}l$ particles, where M is the number of particles in a unit volume and is called the number density. Each particle absorbs a power

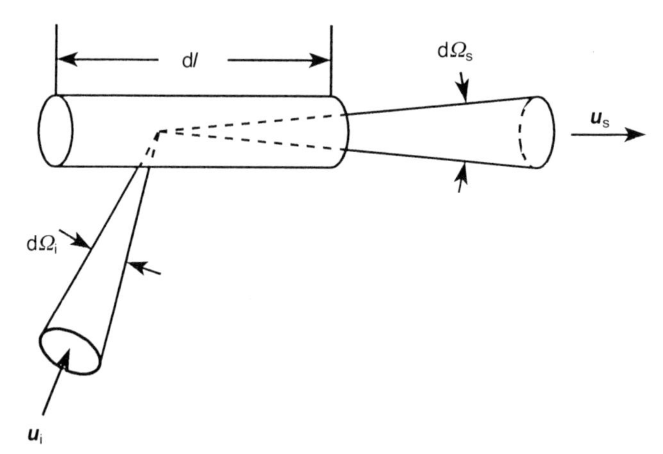

Fig. 4.16 Scattering of specific intensity incident upon the volume $\mathrm{d}l$ from the direction $\boldsymbol{u}_i$ into the direction $\boldsymbol{u}_s$.

$\sigma_a I$ and scatters a power $\sigma_s I$ and therefore the decrease of the specific intensity $dI(\boldsymbol{r}, \boldsymbol{u}_s)$ for the volume dl is expressed as

$$dI(\boldsymbol{r}, \boldsymbol{u}_s) = -M\, dl(\sigma_a + \sigma_s)I = -M\, dl\, \sigma_t I. \tag{4.3.36}$$

At the same time, the specific intensity increases because a portion of the specific intensity $I(\boldsymbol{r}, \boldsymbol{u}_s)$ incident on this volume from other directions $\boldsymbol{u}_i$ is scattered into the direction $\boldsymbol{u}_s$ and is added to the intensity $I(\boldsymbol{r}, \boldsymbol{u}_s)$.

The incident flux density through a solid angle $d\Omega_i$ is given by $S_i = I(\boldsymbol{r}, \boldsymbol{u}_i)\, d\Omega_i$. This flux is incident on the particles in volume dl. The power flux density S_r of the wave scattered by a single particle in the direction $\boldsymbol{u}_s$ at a distance r from the particle is then given by

$$S_r = \frac{|\boldsymbol{f}(\boldsymbol{u}_s, \boldsymbol{u}_i)|^2}{r^2} S_i \tag{4.3.37}$$

where $\boldsymbol{f}(\boldsymbol{u}_s, \boldsymbol{u}_i)$ is the scattering amplitude defined in section 4.2.2. The scattered specific intensity in the direction $\boldsymbol{u}_s$ due to S_i is therefore

$$S_r r^2 = |\boldsymbol{f}(\boldsymbol{u}_s, \boldsymbol{u}_i)|^2 S_i = |\boldsymbol{f}(\boldsymbol{u}_s, \boldsymbol{u}_i)|^2 I(\boldsymbol{r}, \boldsymbol{u}_i)\, d\Omega_i. \tag{4.3.38}$$

Adding the incident flux from all directions $\boldsymbol{u}_i$, the specific intensity scattered into the direction $\boldsymbol{u}_s$ by $M\, dl$ particles in the volume dl is given by

$$\int_{4\pi} M\, dl\, |\boldsymbol{f}(\boldsymbol{u}_s, \boldsymbol{u}_i)|^2 I(\boldsymbol{r}, \boldsymbol{u}_i)\, d\Omega_i \tag{4.3.39}$$

where the integration over all Ω_i is taken to include the contributions from all directions $\boldsymbol{u}_i$.

Using

$$\Phi(\boldsymbol{u}_s, \boldsymbol{u}_i) = \frac{4\pi}{\sigma_t} |\boldsymbol{f}(\boldsymbol{u}_s, \boldsymbol{u}_i)|^2 \tag{4.3.40}$$

yields

$$\frac{1}{4\pi} \int_{4\pi} \Phi(\boldsymbol{u}_s, \boldsymbol{u}_i)\, d\Omega_s = \Phi_0 = \frac{\sigma_s}{\sigma_t} \tag{4.3.41}$$

where Φ is called the phase function (which has no relation to the phase of a wave, but originates in analogy to lunar phases) and Φ_0 is the albedo of a single particle.

The specific intensity may also increase as a result of emission from within the volume dl. Denoting by $Y(\boldsymbol{r}, \boldsymbol{u}_s)$ the power radiation per unit volume per unit solid angle in the direction $\boldsymbol{u}_s$, the increase of the specific intensity is given by

$$\mathrm{d}l\, Y(\boldsymbol{r}, \boldsymbol{u}_s). \tag{4.3.42}$$

Adding the contributions (4.3.36), (4.3.39) and (4.3.42), the equation of transfer becomes

$$\frac{\mathrm{d}I(\boldsymbol{r}, \boldsymbol{u}_s)}{\mathrm{d}s} = -M\sigma_t I(\boldsymbol{r}, \boldsymbol{u}_s) + \frac{M\sigma_t}{4\pi}\int_{4\pi} \Phi(\boldsymbol{u}_s, \boldsymbol{u}_i) I(\boldsymbol{r}, \boldsymbol{u}_i)\,\mathrm{d}\Omega_i + Y(\boldsymbol{r}, \boldsymbol{u}_s). \tag{4.3.43}$$

In this equation, particle density and size can be different at different locations and therefore $M\sigma_t$ and Φ can be functions of $\boldsymbol{r}$. Therefore, it is sometimes convenient to measure the distance in terms of the dimensionless optical distance ζ (section 4.3.2(b)):

$$\zeta = \int_0^l M\sigma_t\,\mathrm{d}z.$$

Equation (4.3.43) now becomes

$$\frac{\mathrm{d}I(\boldsymbol{\zeta}, \boldsymbol{u}_s)}{\mathrm{d}\zeta} = -I(\boldsymbol{\zeta}, \boldsymbol{u}_s) + \frac{1}{4\pi}\int_{4\pi} \Phi(\boldsymbol{u}_s, \boldsymbol{u}_i) I(\boldsymbol{\zeta}, \boldsymbol{u}_i)\,\mathrm{d}\Omega_i + J(\boldsymbol{\zeta}, \boldsymbol{u}_s) \tag{4.3.44}$$

where the vector $\boldsymbol{\zeta}$ has length ζ and direction of $\boldsymbol{r}$ and $J(\boldsymbol{r}, \boldsymbol{u}_s) = Y(\boldsymbol{r}, \boldsymbol{u}_s)/M\sigma_t$ is called the source function. It can be shown that the power is conserved if the medium is lossless ($\sigma_a = 0$) and if there is no source ($Y(\boldsymbol{r}, \boldsymbol{u}_s) = 0$).

Adding to the differential equation (4.3.44) an appropriate boundary condition (e.g. for a plane-parallel atmosphere), a solution for the specific intensity can be found. It is possible to write the differential equation together with the boundary conditions in the form of an integral equation (Ishimaru, 1978).

The transport theory of wave propagation in the presence of random particles can be applied to only a limited number of cases to obtain exact solutions. In the majority of practical cases it is necessary to use approximate or numerical solutions.

Transport theory can also be applied to a tenuous medium. In that case the solutions obtained are equivalent to the single-particle scattering and first-order multiple scattering approximations. Transport theory, however, gives the solution in terms of specific intensity, which can be convenient in some applications.

(f) Multiple scattering theory

The multiple scattering theory, also called the analytical theory, starts with fundamental differential equations governing field quantities and then it introduces statistical considerations.

Twersky (1962) obtained in this way consistent sets of integral equations. His theory gives a clear physical picture of various processes of multiple scattering. Many authors have contributed to multiple scattering theory and developed solutions for particular problems (Ishimaru, 1978). Only Twersky's theory will be presented here, because of its usefulness and clearness for propagation problems.

Consider a random distribution of M particles located at $\boldsymbol{r}_1, \boldsymbol{r}_2, \ldots, \boldsymbol{r}_n$ in a volume V. The particles need not necessarily be identical in shape and size. The scalar field Ψ_{a} at $\boldsymbol{r}_{\mathrm{a}}$, a point in space between the scatterers, satisfies the wave equation

$$(\nabla^2 + k_0^2)\Psi_{\mathrm{a}} = 0 \tag{4.3.45}$$

where $k_0 = 2\pi/\lambda$ is the wavenumber of the medium surrounding the particles.

$\Psi_{\mathrm{i}}^{\mathrm{a}}$ is the incident wave in the absence of any particles at $\boldsymbol{r}^{\mathrm{a}}$ (the superscript of a field quantity denotes the location of observation, the subscript denotes the origin of the field). The field Ψ^{a} at $\boldsymbol{r}^{\mathrm{a}}$ is the sum of the incident wave $\Psi_{\mathrm{i}}^{\mathrm{a}}$ and the contributions Ψ_b^{a} from all M particles located at $\boldsymbol{r}^b$, $b = 1, 2, \ldots, M$ (Fig. 4.17):

$$\Psi^{\mathrm{a}} = \Psi_{\mathrm{i}}^{\mathrm{a}} + \sum_{b=1}^{M} \Psi_b^{\mathrm{a}}. \tag{4.3.46}$$

Ψ_b^{a} is the wave at $\boldsymbol{r}^{\mathrm{a}}$ scattered from the scatterers located at $\boldsymbol{r}^b$ and can be expressed in terms of the wave Ψ^b incident upon the scatterer at $\boldsymbol{r}^b$ and the scattering characteristics u_{b}^{a} of the particle located at $\boldsymbol{r}^b$ as observed at $\boldsymbol{r}^{\mathrm{a}}$ (Fig. 4.18):

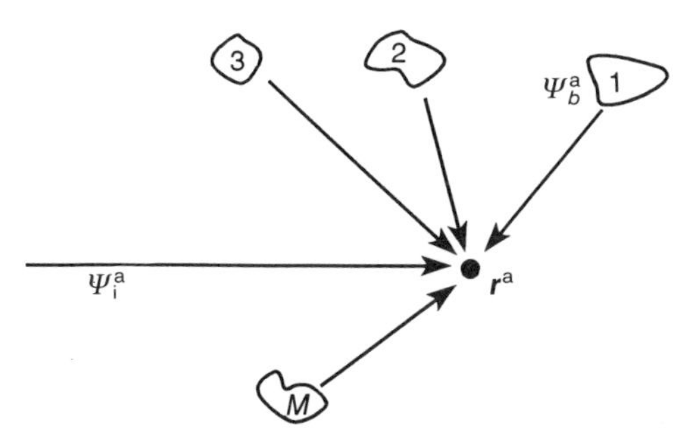

Fig. 4.17 Summing the field at $\boldsymbol{r}^{\mathrm{a}}$.

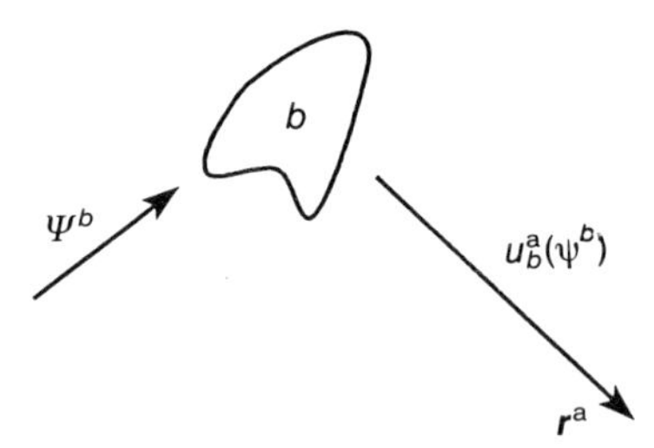

Fig. 4.18 The contribution from particle b.

$$\Psi_b^{\mathrm{a}} = u_b^{\mathrm{a}}(\Psi^b). \tag{4.3.47}$$

In general, $u_b^{\mathrm{a}}(\Psi^b)$ does not mean the product of u_b^{a} and Ψ^b, but a functional relationship to indicate the field at $\boldsymbol{r}^{\mathrm{a}}$ due to the scatterer at $\boldsymbol{r}^b$.

Ψ^b consists of the wave Ψ_{i}^b, incident on the medium, and the wave scattered from all the particles, except the one at $\boldsymbol{r}^b$ (Fig. 4.19):

$$\Psi^b = \Psi_{\mathrm{i}}^b + \sum_{c=1,c\neq b}^{M} \Psi_c^b = \Psi_{\mathrm{i}}^b + \sum_{c=1,c\neq b}^{M} u_c^b(\Psi^c). \tag{4.3.48}$$

From equations (4.3.47) and (4.3.48), Ψ^b can be eliminated by iteration and then the solution Ψ^{a} for a given incident wave Ψ_{i} is obtained:

$$\begin{aligned}\Psi^{\mathrm{a}} &= \Psi_{\mathrm{i}}^{\mathrm{a}} + \sum_{b=1}^{M} u_b^{\mathrm{a}}\left[\Psi_{\mathrm{i}}^b + \sum_{c=1,c\neq b}^{M} u_c^b(\Psi^c)\right] \\ &= \Psi_{\mathrm{i}}^{\mathrm{a}} + \sum_{b=1}^{M} u_b^{\mathrm{a}}(\Psi_{\mathrm{i}}^b) + \sum_{b=1}^{M} \sum_{c=1,c\neq b}^{M} u_b^{\mathrm{a}}[u_c^b(\Psi_{\mathrm{i}}^c)] \\ &\quad + \sum_{b=1}^{M} \sum_{c=1,c\neq b}^{M} \sum_{d=1,d\neq c}^{M} u_b^{\mathrm{a}}\{u_c^b[u_d^c(\Psi_{\mathrm{i}}^d)]\} + \ldots.\end{aligned} \tag{4.3.49}$$

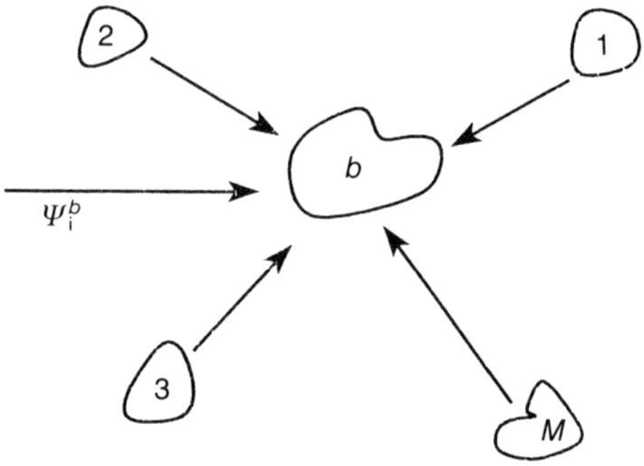

Fig. 4.19 The effective field on particle b.

(Here, use is being made of the linearity of the Maxwell equations and the medium considered.) The first term is the incident wave and the next term represents all the single scattering (Fig. 4.20(a)). The next summation represents all the double scattering (Fig. 4.20(b)). The third summation is triple, but the terms $c = b$ and $d = c$ are excluded, while the term $b = d$ is not excluded. Therefore it is written so as to include the terms containing separate b, c and d plus the terms corresponding to $b = d$:

$$\sum_{b=1}^{M} \sum_{c=1,c\neq b}^{M} \sum_{d=1,d\neq c}^{M} u_b^{\mathrm{a}}\{u_c^b[u_d^c(\Psi_{\mathrm{i}}^d)]\}$$

$$= \sum_{b=1}^{M} \sum_{c=1,c\neq b}^{M} \sum_{d=1,d\neq c,d\neq b}^{M} u_b^{\mathrm{a}}\{u_c^b[u_d^c(\Psi_{\mathrm{i}}^d)]\} + \sum_{b=1}^{M} \sum_{c=1,c\neq b}^{M} u_b^{\mathrm{a}}\{u_c^b[u_b^c(\Psi_{\mathrm{i}}^b)]\}. \tag{4.3.50}$$

The first triple summation is pictured in Fig. 4.20(c). The second summation involves only two scatterers at $\boldsymbol{r}^b$ and $\boldsymbol{r}^c$ and is pictured in Fig. 4.20(d). Using this observation the summations can be divided into two groups: terms involving chains of successive scattering going through different scatterers – the first summation in equation (4.3.50) (Fig. 4.21(a)); and terms containing

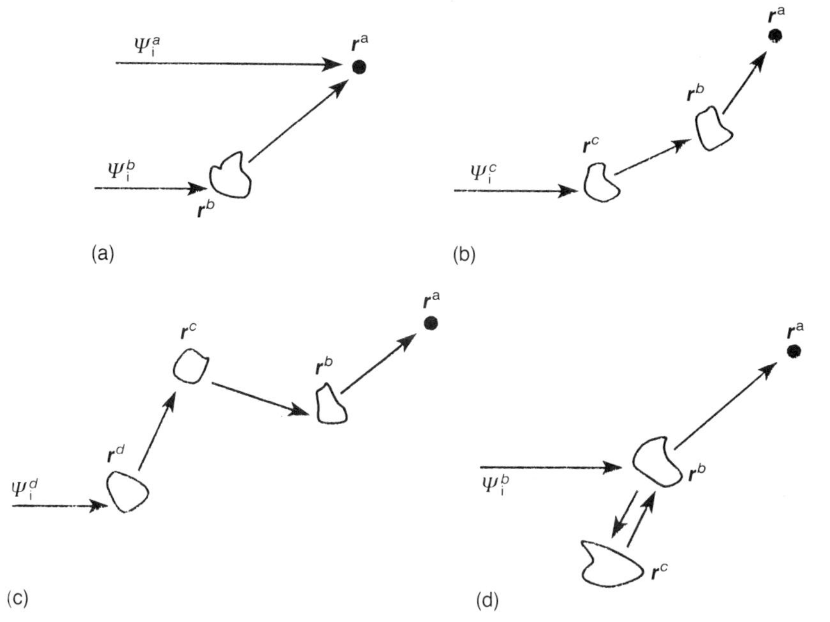

Fig. 4.20 Various scattering terms: (a) single scattering; (b) double scattering; (c) triple scattering through different particles; (d) triple scattering when the propagation path goes through the same particle more than once.

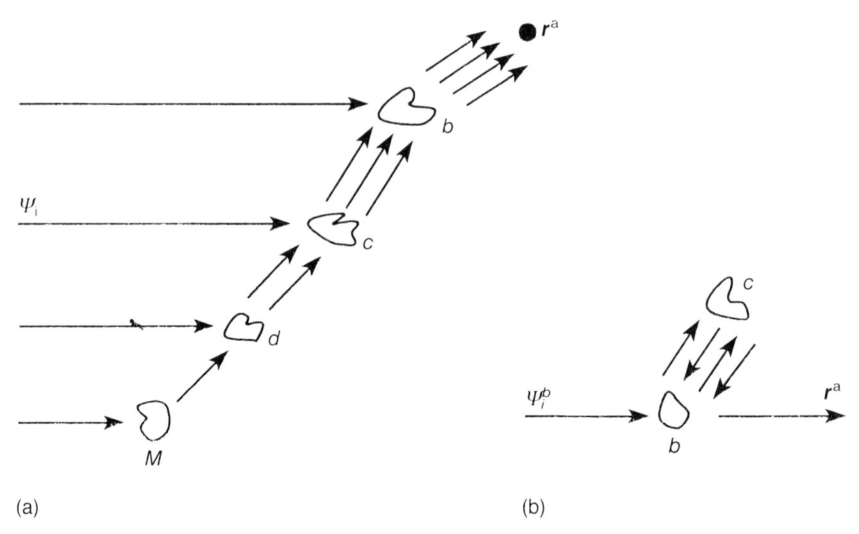

Fig. 4.21 (a) Chains of successive scattering paths going through different scatterers and (b) chains of scattering paths which go through the same scatterer more than once.

paths which go through a scatterer more than once – the second summation in equation (4.3.50) (Fig. 4.21(b)). Twersky's theory includes all terms belonging to the first group but neglects the terms corresponding to the second group. Therefore, this theory should only give excellent results when the backscattering is insignificant compared with the scattering in other directions.

Thus Twersky's theory is based on the following form of the field:

$$\Psi^a = \Psi_i^a + \sum_{b=1}^{M} u_b^a(\Psi_i^b) + \sum_{b=1}^{M} \sum_{c=1, c\neq b}^{M} u_b^a[u_c^b(\Psi_i^c)]$$

$$+ \sum_{b=1}^{M} \sum_{c=1, c\neq b}^{M} \sum_{d=1, d\neq c, d\neq b}^{M} u_b^a\{u_c^b[u_d^c(\Psi_i^d)]\} + \ldots \quad (4.3.51)$$

This equation, which is called the expanded representation by Twersky, is useful for understanding the scattering processes involved, but is impractical for the evaluation of important quantities. Therefore Foldy (1945) and Twersky developed consistent integral equations.

In order to arrive at these integral equations, some assumptions must be made with respect to statistics.

- It is assumed that the particle density is low and that the particle size is much smaller than the separation between particles.

- In this case the finite size of particles can be neglected and the location and characteristics of each scatterer can be assumed to be independent of the location and characteristics of other scatterers. This also means that all particles are considered as point particles and the effect of size appears only in the scattering characteristics.
- Finally, it is also assumed that all scatterers have the same statistical characteristics.

With these assumptions, the coherent field $\langle \Psi^{\mathrm{a}} \rangle$ using Twersky's theory becomes

$$\langle \Psi^{\mathrm{a}} \rangle = \Psi_{\mathrm{i}}^{\mathrm{a}} + \int u_b^{\mathrm{a}}(\Psi_{\mathrm{i}}^{b}) M(\boldsymbol{r}^b)\,\mathrm{d}\boldsymbol{r}^b + \iint u_b^{\mathrm{a}}[u_c^{b}(\Psi_{\mathrm{i}}^{c})] M(\boldsymbol{r}^b) M(\boldsymbol{r}^c)\,\mathrm{d}\boldsymbol{r}^b\,\mathrm{d}\boldsymbol{r}^c$$
$$+ \iiint u_b^{\mathrm{a}}\{u_c^{b}[u_d^{c}(\Psi_{\mathrm{i}}^{d})]\} M(\boldsymbol{r}^b) M(\boldsymbol{r}^c) M(\boldsymbol{r}^d)\,\mathrm{d}\boldsymbol{r}^b\,\mathrm{d}\boldsymbol{r}^c\,\mathrm{d}\boldsymbol{r}^d + \ldots \tag{4.3.52}$$

where $M(\boldsymbol{r})$ (m^{-3}) is the number density of particles at location $\boldsymbol{r}$. This can also be written as the Foldy–Twersky integral equation:

$$\langle \Psi^{\mathrm{a}} \rangle = \Psi_{\mathrm{i}}^{\mathrm{a}} + \int u_b^{\mathrm{a}} \langle \Psi^{b} \rangle M(\boldsymbol{r}^b)\,\mathrm{d}\boldsymbol{r}^b. \tag{4.3.53}$$

Now a plane wave incident normally upon a slab of thickness l is considered (Fig. 4.22). The incident wave with unit amplitude is given by

$$\Psi_{\mathrm{i}}(\boldsymbol{r}) = \mathrm{e}^{\,\mathrm{j}k_0 z}. \tag{4.3.54}$$

The coherent field $\langle \Psi \rangle$ inside the slab is given by equation (4.3.53). Since

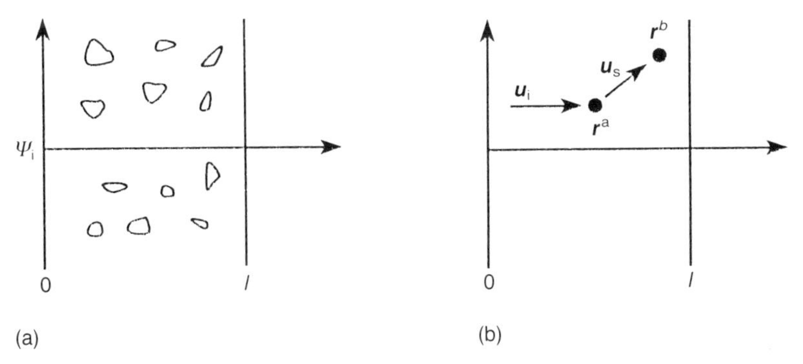

Fig. 4.22 (a) Plane wave incident on a slab of thickness l and (b) location of $\boldsymbol{r}^{\mathrm{a}}$ and $\boldsymbol{r}^{b}$.

the incident wave and the geometry of the slab are independent of the x and y coordinates, the coherent field $\langle \Psi \rangle$ must also be independent of x and y and thus $\langle \Psi \rangle$ should behave like a plane wave propagating in the $+z$ direction.

If $\boldsymbol{r}^{\mathrm{a}}$ is in the far-field zone of the particle at $\boldsymbol{r}^{b}$, $u_b^{\mathrm{a}}(\langle \Psi^b \rangle)$ becomes

$$u_b^{\mathrm{a}}(\langle \Psi^b \rangle) = f(\boldsymbol{u}_{\mathrm{s}}, \boldsymbol{u}_{\mathrm{i}}) \frac{\mathrm{e}^{\mathrm{j}k_0|\boldsymbol{r}^{\mathrm{a}} - \boldsymbol{r}^{b}|}}{|\boldsymbol{r}^{\mathrm{a}} - \boldsymbol{r}^{b}|} \langle \Psi^b \rangle \tag{4.3.55}$$

where $\boldsymbol{u}_{\mathrm{i}}$ is a unit vector in the direction of propagation of $\langle \Psi^b \rangle$ ($\boldsymbol{u}_{\mathrm{i}} = \boldsymbol{u}_z$) and $\boldsymbol{u}_{\mathrm{s}}$ is a unit vector in the direction of $\boldsymbol{r}^{a} - \boldsymbol{r}^{b}$. Using the method of stationary phase, equation (4.3.53) can be written as

$$\begin{aligned} \langle \Psi(z) \rangle = \mathrm{e}^{\mathrm{j}kz} &+ \int_0^z \frac{2\pi\mathrm{j}}{k} \mathrm{e}^{\mathrm{j}k_0(z - z_{\mathrm{s}})} f(\boldsymbol{u}_{\mathrm{i}}, \boldsymbol{u}_{\mathrm{i}}) M \langle \Psi(z_{\mathrm{s}}) \rangle \, \mathrm{d}z_{\mathrm{s}} \\ &+ \int_z^l \frac{2\pi\mathrm{j}}{k} \mathrm{e}^{\mathrm{j}k_0(z_{\mathrm{s}} - z)} f(-\boldsymbol{u}_{\mathrm{i}}, \boldsymbol{u}_{\mathrm{i}}) M \langle \Psi(z_{\mathrm{s}}) \rangle \, \mathrm{d}z_{\mathrm{s}}. \end{aligned} \tag{4.3.56}$$

Assuming that, in general, the ratio of forward scattering to backscattering is sufficiently high, the second integral in equation (4.3.56) is usually neglected. However, this assumption still remains to be checked for the millimetre wave region and for the atmospheric propagation medium (rain, clouds).

Assuming that the density M is constant, equation (4.3.56) becomes

$$\langle \Psi(z) \rangle = \mathrm{e}^{\mathrm{j}k_0 z} \left[1 + \frac{2\pi\mathrm{j}}{k_0} f(\boldsymbol{u}_{\mathrm{i}}, \boldsymbol{u}_{\mathrm{i}}) M \int_0^z \mathrm{e}^{-\mathrm{j}k_0 z_{\mathrm{s}}} \langle \Psi(z_{\mathrm{s}}) \rangle \, \mathrm{d}z_{\mathrm{s}} \right]. \tag{4.3.57}$$

The solution of this equation is:

$$\langle \Psi(z) \rangle = A\, \mathrm{e}^{\mathrm{j}k_{\mathrm{s}} z} \tag{4.3.58}$$

$$A = 1, \quad k_{\mathrm{s}} = k_0 + \frac{2\pi f(\boldsymbol{u}_{\mathrm{i}}, \boldsymbol{u}_{\mathrm{i}})}{k_0} M. \tag{4.3.59}$$

This means that the average field $\langle \Psi \rangle$ for a plane wave propagates in a slab with propagation constant k_{s}. More generally, the average field $\langle \Psi \rangle$ for any wave in a medium containing random particles can be represented by assuming that it satisfies the wave equation

$$(\nabla^2 + k_{\mathrm{s}}^2) \langle \Psi(\boldsymbol{r}) \rangle = 0. \tag{4.3.60}$$

$f(\boldsymbol{u}_{\mathrm{i}}, \boldsymbol{u}_{\mathrm{i}})$ is in general complex even for a lossless scatterer and so the coherent field $\langle \Psi(\boldsymbol{r}) \rangle$ attenuates as it propagates through particles. This is of course due to scattering and is related to scattering cross-section. To investigate this point, consider the coherent intensity for an incident plane wave:

$$|\langle \Psi(z) \rangle|^2 = \exp\left[\frac{-4\pi M}{k} \operatorname{Im} f(\boldsymbol{u}_i, \boldsymbol{u}_i)\, z\right]. \tag{4.3.61}$$

According to the forward scattering theorem (section 4.2.3), which states

$$\frac{4\pi}{k} \operatorname{Im} f(\boldsymbol{u}_i, \boldsymbol{u}_i) = \sigma_s + \sigma_a, \tag{4.3.62}$$

equation (4.3.61) becomes

$$|\langle \Psi(z) \rangle|^2 = e^{-M(\sigma_s+\sigma_a)z}, \quad 0 < z < l. \tag{4.3.63}$$

This shows that the coherent intensity attenuates exponentially and that the attenuation constant is proportional to the density M and to the total cross-section $\sigma_t = \sigma_s + \sigma_a$.

Although the analysis is devoted to the problem of a plane wave incident on a slab, its generalization given by equation (4.3.60) is considered to be a valid approximation in many practical situations.

Consistent with the Foldy–Twersky integral equation (4.3.53), Twersky obtained an integral equation for the correlation function $\langle \Psi^a \Psi^b \rangle$ from which total field intensity can be calculated. Using this equation it can be shown that, inside a slab of scatterers, the total intensity decays as

$$\langle |\Psi^a|^2 \rangle = e^{-M\sigma_a z}. \tag{4.3.64}$$

Likewise, outside the slab it follows that the total field received by the receiver aligned with the incident direction $\boldsymbol{u}_i$ is given by

$$\langle |\Psi^a|^2 \rangle = e^{-M\sigma_a l}[e^{-M\sigma_s l} + q(1 - e^{-M\sigma_s l})] \tag{4.3.65}$$

where

$$q = \frac{\int_{\Omega_r} |f(\boldsymbol{u}_i, \boldsymbol{u}_s)|^2 \, d\Omega_s}{\int_{4\pi} |f(\boldsymbol{u}_i, \boldsymbol{u}_s)|^2 \, d\Omega_s}$$

is the fraction of the total scattered power that is scattered in the receiving angle Ω_r of the receiver. This result has been widely used, derived in several ways, in millimetre wave propagation studies.

It can be shown that the treatment by transport theory is closely related to the Twersky multiple scattering theory. In fact, the results shown above may be derived in an equivalent manner from transport theory and the specific intensity in transport theory is closely related to the correlation function used in multiple scattering theory (Ishimaru, 1978).

4.4 PLANE WAVE TREATMENT OF MICROWAVE SCATTERING

Another way to describe scattering effects, apart from the classical formulations given above, is the plane wave expansion method (Haworth, 1980). This approach uses the plane wave theory to deduce expressions for the coherent and incoherent effects due to propagation of a plane wave through a thin slab of hydrometeors. Haworth (1980) uses a tensor description of the scattering process, which has been simplified here, consistent with equation (4.2.3).

4.4.1 Plane wave formulation

The plane wave approach makes it possible to derive an integral from which the power transmitted between antennas of any size can be calculated. Approximate methods of evaluating this integral can then be used to estimate the effect of operating within the near field of antennas. Supposing the existence of a matched load across the receive antenna terminals, it is found that the received amplitude of the electric field is given by

$$T = -\frac{1}{2Z_0}\int \boldsymbol{F}_{\mathrm{r}}(-\boldsymbol{u}_{\mathrm{s}}) \cdot \boldsymbol{F}_{\mathrm{t}}(\boldsymbol{u}_{\mathrm{s}})\, \mathrm{e}^{\,\mathrm{j}k_0\boldsymbol{u}_{\mathrm{s}}\cdot\boldsymbol{r}}\, \mathrm{d}\Omega \tag{4.4.1}$$

where T is the transmission coefficient, $|T|^2$ is the power dissipated by a matched load attached to the receiver, Z_0 is the matched load of the receive antenna, $\boldsymbol{F}_{\mathrm{r}}$ is the plane wave component for the receive antenna alone, behaving as a transmitter and transmitting a unit total power, $\boldsymbol{F}_{\mathrm{t}}$ is the plane wave component for the transmitting antenna, transmitting a unit total power, $\boldsymbol{r}$ is the location of the transmitter relative to the receiver, and $\mathrm{d}\Omega$ is the elementary solid angle around the direction vector $\boldsymbol{u}_{\mathrm{s}}$. We can convert from the full wave expansion (4.4.1) to an asymptotic representation of the antenna interaction by using the Weyl angular decomposition formula:

$$\frac{\mathrm{e}^{\,\mathrm{j}k_0 r}}{r} = \frac{k_0}{2\pi\mathrm{j}}\int \mathrm{e}^{\,\mathrm{j}k_0\boldsymbol{u}_{\mathrm{s}}\cdot\boldsymbol{r}}\, \mathrm{d}\Omega \tag{4.4.2}$$

where $r = |\boldsymbol{r}|$.

Now the interaction of a receive antenna with the scattered signal from a hydrometeor illuminated by a plane wave may be derived. A hydrometeor is illuminated by a plane wave having a field $\boldsymbol{E}_{\mathrm{i}}$ at the phase centre $\boldsymbol{r}$ of the hydrometeor, as measured from the receive antenna, and propagating in a direction $\boldsymbol{u}_\mathrm{i}$ (Fig. 4.23). The scattered field at the receive antenna is, using the definition of equation (4.2.3) in section 4.2.2,

$$\boldsymbol{E}_\mathrm{s} \sim \boldsymbol{f}(\boldsymbol{u}_\mathrm{s}, \boldsymbol{u}_\mathrm{i}) E_\mathrm{i} \frac{\mathrm{e}^{\,\mathrm{j}k_0 r}}{r} \text{ with } \boldsymbol{E}_\mathrm{i} = E_\mathrm{i} \boldsymbol{u}_\mathrm{e}. \tag{4.4.3}$$

From the angular decomposition formula (4.4.2) we can deduce that equation (4.4.3) is an asymptotic representation of

$$\boldsymbol{E}_\mathrm{s} = \frac{k_0}{2\pi\mathrm{j}} \int \boldsymbol{f}(\boldsymbol{u}_\mathrm{s}, \boldsymbol{u}_\mathrm{i}) E_\mathrm{i} \, \mathrm{e}^{\,\mathrm{j}k_0 \boldsymbol{u}_\mathrm{s} \cdot \boldsymbol{r}} \, \mathrm{d}\Omega \tag{4.4.4}$$

and it can be shown that the interaction with the receive antenna gives a transmission coefficient

$$T_\mathrm{s} = -\frac{1}{2Z_0} \int \boldsymbol{F}_\mathrm{r}(-\boldsymbol{u}_\mathrm{s}) \cdot \boldsymbol{f}(\boldsymbol{u}_\mathrm{s}, \boldsymbol{u}_\mathrm{i}) E_\mathrm{i} \, \mathrm{e}^{\,\mathrm{j}k_0 \boldsymbol{u}_\mathrm{s} \cdot \boldsymbol{r}} \, \mathrm{d}\Omega. \tag{4.4.5}$$

Now the transmission coefficient for an unscattered plane wave may be derived as follows. It can be shown that the angular representation of the plane wave is

$$\boldsymbol{F}_\mathrm{t}(\boldsymbol{u}_\mathrm{s}) = \frac{2\pi\mathrm{j}}{k_0} \boldsymbol{E}_\mathrm{i} \delta(\boldsymbol{u}_\mathrm{s} - \boldsymbol{u}_\mathrm{i}) \tag{4.4.6}$$

where δ is the Dirac function for vectors. Substitution into equation (4.4.1) gives for plane wave reception

$$T_\mathrm{d} = \frac{\pi}{\mathrm{j}k_0 Z_0} \boldsymbol{F}_\mathrm{r}(-\boldsymbol{u}_\mathrm{i}) \cdot \boldsymbol{E}_\mathrm{i} \, \mathrm{e}^{\,\mathrm{j}k_0 \boldsymbol{u}_\mathrm{i} \cdot \boldsymbol{r}}. \tag{4.4.7}$$

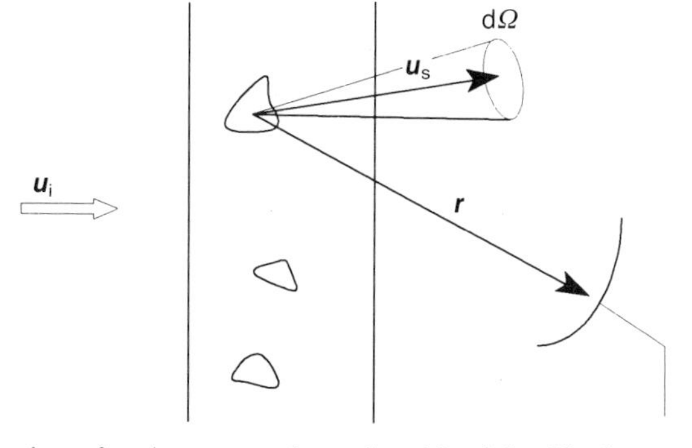

Fig. 4.23 Propagation of a plane wave through a thin slab of hydrometeors.

So, equations (4.4.5) and (4.4.7) define the transmission coefficients at the receive antenna due to an incident plane wave and to scattering from a single hydrometeor, respectively. The total transmission coefficient is the sum of T_s and T_d.

4.4.2 Analysis of coherent and incoherent effects

(a) Coherent case

A thin infinite slab of scatterers with a plane wave incident normally on to the slab is considered. The transmission coefficient T, in the single scattering approximation, is a linear superposition of a directly received signal component and the contributions from all scatterers:

$$T = \frac{\pi}{\mathrm{j}k_0 Z_0} \boldsymbol{F}_\mathrm{r}(-\boldsymbol{u}_\mathrm{i}) \cdot \boldsymbol{E}_\mathrm{i} + \sum_k - \frac{1}{2Z_0} \int \boldsymbol{F}_\mathrm{r}(-\boldsymbol{u}_\mathrm{s}) \cdot \boldsymbol{f}(\boldsymbol{u}_\mathrm{s}, \boldsymbol{u}_\mathrm{i}) E_\mathrm{i}\, \mathrm{e}^{\mathrm{j}k_0(\boldsymbol{u}_\mathrm{s}-\boldsymbol{u}_\mathrm{i})\cdot \boldsymbol{r}_k}\, \mathrm{d}\Omega \tag{4.4.8}$$

where the phase has been normalized to the directly received signal phase. To evaluate the coherent received signal component, we consider each hydrometeor to have a random position within the slab and to move independently of the positions of the other hydrometeors. We also assume, for simplicity, that the hydrometeors have the same physical shape, size and orientation. Using the number density M per unit volume, we find

$$T = \frac{\pi}{\mathrm{j}k_0 Z_0} \boldsymbol{F}_\mathrm{r}(-\boldsymbol{u}_\mathrm{i}) \cdot \boldsymbol{E}_\mathrm{i} - \frac{M}{2Z_0} \int\int \boldsymbol{F}_\mathrm{r}(-\boldsymbol{u}_\mathrm{s}) \cdot \boldsymbol{f}(\boldsymbol{u}_\mathrm{s}, \boldsymbol{u}_\mathrm{i}) E_\mathrm{i}\, \mathrm{e}^{\mathrm{j}k_0(\boldsymbol{u}_\mathrm{s}-\boldsymbol{u}_\mathrm{i})\cdot \boldsymbol{r}}\, \mathrm{d}V\, \mathrm{d}\Omega \tag{4.4.9}$$

where $\mathrm{d}V$ is an elementary volume within the slab. The volume integral can be partially evaluated by integrating with respect to the components of $\boldsymbol{r} = (x, y, z)$ and since the only function depending on x and y is the exponential

$$T = \frac{\pi}{\mathrm{j}k_0 Z_0} \boldsymbol{F}_\mathrm{r}(-\boldsymbol{u}_\mathrm{i}) \cdot E_\mathrm{i} - \frac{M}{2Z_0} \int \boldsymbol{F}_\mathrm{r}(-\boldsymbol{u}_\mathrm{s}) \cdot \boldsymbol{f}(\boldsymbol{u}_\mathrm{s}, \boldsymbol{u}_\mathrm{i}) E_\mathrm{i} \left(\frac{2\pi}{k_0}\right)^2 \delta(\boldsymbol{u}_\mathrm{s} - \boldsymbol{u}_\mathrm{i})\, l\, \mathrm{d}\Omega \tag{4.4.10}$$

where l is the thickness of the slab (in the direction of z) and where the identity

$$\int_{-\infty}^{\infty} \mathrm{e}^{\mathrm{j}ax}\,\mathrm{d}a = 2\pi\delta(x) \tag{4.4.11}$$

has been used twice. The angular integration in equation (4.4.10) now results in

$$T = \frac{\pi}{\mathrm{j}k_0 Z_0} \boldsymbol{F}_\mathrm{r}(-\boldsymbol{u}_\mathrm{i}) \cdot \left[\boldsymbol{E}_\mathrm{i} + \frac{2\pi}{\mathrm{j}k_0} Ml\boldsymbol{f}(\boldsymbol{u}_\mathrm{s}, \boldsymbol{u}_\mathrm{i}) E_\mathrm{i} \right]. \tag{4.4.12}$$

As pointed out by Van de Hulst (1981), the bracketed term in equation (4.4.12) can be recognized as the first two terms of an exponential describing a medium having an equivalent refractive index n with

$$n = 1 + \frac{2\pi}{k_0^2} Mf(\boldsymbol{u}_\mathrm{i}, \boldsymbol{u}_\mathrm{i}). \tag{4.4.13}$$

The result is independent of whether the slab is located in the near or the far field of the receive antenna; this was the main point of the plane wave expansion as presented by Haworth, McEwan and Watson (1978) and Haworth (1980).

(b) Incoherent case

For the same thin slab of hydrometeors, the incoherent signal power can also be determined using the plane wave spectrum technique. Because of the discreteness and random location of the hydrometeors their instantaneous contribution to the received signal will fluctuate about its mean value and, in the case of a uniformly random distribution of hydrometeors, the fluctuation will have an independent contribution from each scatterer which is then termed incoherent.

The incoherent signal is contributed entirely by the hydrometeors and can thus be viewed as a scattered signal. The signal average power is denoted $\overline{T_\mathrm{s} T_\mathrm{s}^{*\mathrm{i}}}$ where the subscript ‘s’ denotes a signal scattered from the hydrometeors and the superscript ‘i’ indicates that the signal is incoherent. Under the assumption of an independent location for each scatterer within the slab, the total incoherent scattered signal power becomes

$$\overline{T_\mathrm{s} T_\mathrm{s}^{*\mathrm{i}}} = \frac{1}{4Z_0^2} \sum_a \overline{\int\int U(\boldsymbol{u}_\mathrm{s}) U^*(\boldsymbol{u}'_\mathrm{s})\, \mathrm{e}^{\mathrm{j}k_0(\boldsymbol{u}-\boldsymbol{u}')\cdot \boldsymbol{r}_\mathrm{a}}\, \mathrm{d}\Omega\, \mathrm{d}\Omega'} \tag{4.4.14}$$

where $U(\boldsymbol{u}_\mathrm{s}) = \boldsymbol{F}_\mathrm{r}(-\boldsymbol{u}_\mathrm{s}) \cdot \boldsymbol{f}(\boldsymbol{u}_\mathrm{s}, \boldsymbol{u}_\mathrm{i}) E_\mathrm{i}$.

For a uniform number density within the slab and a single scatterer size, equation (4.4.14) becomes

$$\overline{T_s T_s^{*i}} = \left(\frac{2\pi}{k_0}\right)^2 \frac{Ml}{4Z_0^2} \iint U(\boldsymbol{u}_s) U^*(\boldsymbol{u}_s') \delta(\boldsymbol{u}_s - \boldsymbol{u}_s') \, \mathrm{d}\Omega \, \mathrm{d}\Omega'$$

$$= \left(\frac{2\pi}{k_0}\right)^2 \frac{Ml}{4Z_0^2} \int UU^* \, \mathrm{d}\Omega. \qquad (4.4.15)$$

Equation (4.4.15) shows that there are independent contributions to the received signal power from every elementary angular sector $\mathrm{d}\Omega$. This is fundamental to the radiative transfer formulation and has again been derived independently of the distance of the hydrometeor slab from the receive antenna. It is equivalent to the result of equation (4.3.65) in section 4.3.3(f).

4.4.3 Conclusions

The plane wave spectrum technique may be applied for propagation through a thin slab of hydrometeors. Fundamental equations for the coherent and incoherent signals at the receive antenna can be successfully derived independently of the distance of the slab from the receive antenna. This indicates that the equivalent refractive index description of the propagation of the coherent field and the radiative transfer description of the propagation of the incoherent field are, in fact, also applicable to propagation through hydrometeors in the near field of an antenna.

A basic assumption in the derivations using the plane wave spectrum technique was that the transverse dimensions of the slab are infinite. So for practical applications for both coherent and incoherent propagation, the dimensions of the slab must be greater than the illumination region of the antenna.

4.5 SINGLE-PARTICLE SCATTERING PROPERTIES

4.5.1 Characterization

Methods for calculation of the scattering properties for single particles are chosen according to frequency range and particle type. For certain types of particles (e.g. rain and smooth ice particles of low eccentricity) very accurate characterization is possible over all frequency ranges, whereas for other types of particles (e.g. ice–water mixtures and large, highly eccentric ice particles) approximations must be used with caution. Table 4.2 gives a summary of methods available and their regimes of operation.

As it happens, the size range of atmospheric particles is such that over microwave and millimetre wavelengths we are concerned with resonance (or Mie) scattering. This makes calculations more difficult. For a population of naturally occurring particles, resonance fluctuations are smoothed out, so that over this range of frequencies attenuation cross-sections for a population generally increase with frequency. Also, the relative contribution of the absorption and scattering processes to attenuation varies with frequency and particle type. Dry ice particles may be modelled as pure dielectric particles, with negligible contribution from absorption, whereas for rain, absorption and scattering contributions to attenuation both are important, with absorption dominant up to about 20 GHz.

4.5.2 Problem areas

For frequencies in the range up to 20 GHz where high availability satellite systems are planned, we are concerned mainly with attenuation from liquid particles in heavy rain. In this region our methods for calculation of scattering properties are very accurate and more uncertainty resides in the characterization of the particles (composition, number density, dielectric properties, shape, etc.) than in the analytical calculations.

In heavy rain the melting process gives a relatively low contribution to attenuation in the frequency range 20–40 GHz when compared with 10–20 GHz. However, if we switch our interest to low availability systems, then we will need to place more emphasis on modelling light, widespread rain with a significant and identifiable melting band. It may therefore be of interest to have a better representation of melting particle scattering properties, than hitherto available. The 'T-matrix' and 'extended Rayleigh' methods have been used so far, but further developments may be required. This may also be valuable for the prediction of interference (bistatic scatter) conditions in the higher bands.

In the millimetre wave bands, the smaller raindrops make a greater contribution to attenuation as frequency increases (Fig. 4.24). This makes scattering calculations simpler (since the smaller drops can be approximated by spheres) but leaves again a major uncertainty in population density and size distributions in fog, cloud and light rain.

Scattering from dry ice particles becomes more important with increasing frequency in contributing to attenuation and to interference. Up to the lower millimetre wave bands, we usually resort to Rayleigh scattering calculations as the most tractable for needle-like particles. This approximation will not hold at higher frequencies. Calculations on irregular ice particles (especially non-smooth particles) are difficult in all frequency ranges, but especially so when their dimensions are comparable with the wavelength.

Table 4.2 Methods of calculating single-particle scattering

Method of calculation	Particular advantages	Constraints	Computational requirements	References
Rayleigh	Simple analytical method	Small particles only, in particular for $0 < k_0an < 0.3$	Trivial	Numerous
Mie–Stratton	Covers whole frequency range	Homogeneous spherical scatterers only	Modest	Numerous
Extended Mie for multilayer spheres	Covers whole frequency range	Spherical scatterers only	Modest	Aden and Kerker (1951)
Perturbation	Modest computer requirements	Small deformations from sphere; homogeneous particles only	Modest	Oguchi (1973), Morrison and Chu (1973), Watson and Arbabi (1973), Morrison and Cross (1974)
Spheroidal function expansion	Fast and accurate for spheroidal particles, stable even for large deformations; covers whole frequency range	Spheroidal homogeneous particles; k_0a should be < 17 for $a/b = 2$ for prolates	Significant storage requirements increasing with size and deformation	Oguchi (1973), Asano (1979)
Collocation (point matching)	Irregular shapes can be used (but quicker for symmetrical particles)	Convergence tricky, especially for large deformities; large numbers of modes and points necessary for accuracy; homogeneous particles only	Significant memory and precision requirements	Oguchi (1973), Dissanayake and Watson (1979)

T-matrix	Highly deformed irregular and multi-layer particles can be used	Convergence poor at high frequencies and for large particles	Powerful processing and significant storage requirements	Waterman (1965), Peterson and Strom (1975), Bringi and Seliga (1977)
Fredholm integral	Convergence guaranteed; most stable method for heavily distorted scatterers	Many numerical integrations for non-symmetrical scatterers; homogeneous particles only (or continuously varying dielectric properties)	Slower than other powerful techniques for near-to spheres but faster and more stable for high axial ratios	Holt and Santoso (1972), Holt, Uzunoglu and Evans (1978)
Unimoment method	Arbitrary shapes and inhomogeneity	Up to $k_0a = 60$	Increases as the fifth power of size parameter, k_0a; inefficient for large axial ratios	Fang and Lee (1978)
Extended Rayleigh method	Simple; wider validity range than Rayleigh; applicable to melting band	Small particles	Modest	de Wolf, Russchenberg and Ligthart (1990)

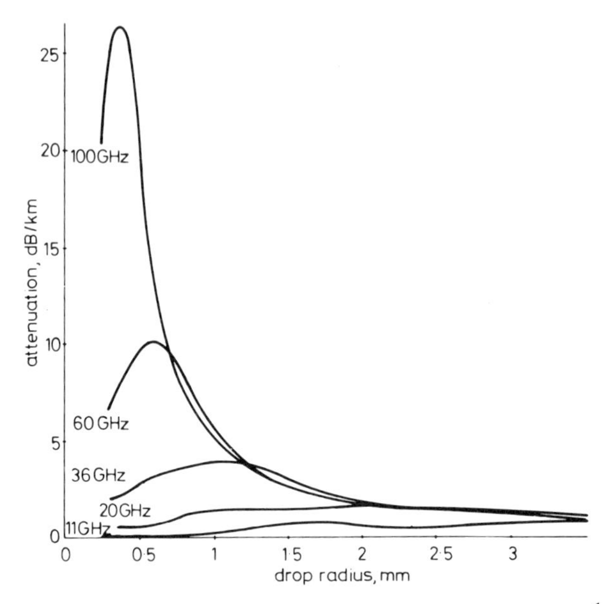

Fig. 4.24 Attenuation for a fixed ground rainfall intensity (12.5 mm h^{-1}) for monodispersed rain as a function of drop size (Humpleman and Watson, 1978).

4.6 TRANSMISSION EQUATIONS FOR PROPAGATION IN A POPULATION OF HYDROMETEORS

4.6.1 Transmission equations for the coherent field

The theory as outlined above is often simplified in the form of transmission equations. This presentation and some qualitative explanations found generally in the literature are discussed in this section. This may clarify conceptually the treatment of the previous sections and at the same time allow introduction of polarization properties of the waves.

Transmission through a slab of scatterers is conveniently expressed in the form of a general transmission matrix in two polarizations (for coherent signals) or in terms of Stokes vectors or a coherency matrix for incoherent signals. The relevant equations are developed below using the notations of the previous sections, where possible. The distinction between coherent and incoherent fields for propagation in a population of hydrometeors and representation of the incoherent field in terms of Stokes vectors are described.

For a single scatterer we may write the incident field $\boldsymbol{E}_{\mathrm{i}}$ and scattered field $\boldsymbol{E}_{\mathrm{s}}$ as

$$\begin{bmatrix} E_{\mathrm{s}\perp} \\ E_{\mathrm{s}\parallel} \end{bmatrix} = \frac{\mathrm{e}^{\,\mathrm{j}k_0 r}}{r} \begin{bmatrix} f_{11} & f_{12} \\ f_{21} & f_{22} \end{bmatrix} \begin{bmatrix} E_{\mathrm{i}\perp} \\ E_{\mathrm{i}\parallel} \end{bmatrix} \tag{4.6.1}$$

where f_{mn}, the complex scattering amplitudes, are functions of the direction of the scattered waves. $\perp$ and $\parallel$ denote hard and soft polarizations respectively. The complex forward scatter amplitudes may be denoted by $f_{mn}(0)$.

For any arbitrary transmission medium, a dual-polarization transmission matrix may be written with complete generality for the coherent (received) field, i.e.

$$\begin{bmatrix} E_{r_2} \\ E_{r_1} \end{bmatrix} = \begin{bmatrix} T_{11} & T_{12} \\ T_{21} & T_{22} \end{bmatrix} \begin{bmatrix} E_{i_1} \\ E_{i_2} \end{bmatrix} \tag{4.6.2}$$

where the basis vectors may be chosen for orthogonal linear, circular or elliptical polarizations. For the coherent field it can be shown that for any medium there are two incident polarizations (characteristic polarizations) which are transmitted without change of polarization (Beckman, 1968; Born and Wolf, 1986).

For attenuation, we simply take

$$A_1 = -20 \log |T_{11}| \tag{4.6.3a}$$

and

$$A_2 = -20 \log |T_{22}|. \tag{4.6.3b}$$

Evidently, the relationship between the $\boldsymbol{T}$ elements and the single-particle scattering amplitudes $f_{mn}(0)$ for a cloud of particles follows from the theory outlined in the previous sections. This relationship for rain depends on two basic modelling assumptions:

- use of the coherent field approximation;
- raindrop shape and orientation models.

The raindrop shape and orientation models are based on the physical descriptions elaborated in more detail in Chapter 13.

4.6.2 Distinction between coherent and incoherent fields

For particles along a line of sight between two antennas, contained within a volume defined by the first Fresnel zone, the field strengths can be regarded as adding with relative phase angle distributed around zero (i.e. in phase), and small changes in particle position will not give rise to rapidly fluctuating fields. This situation describes, in qualitative terms, the coherent field.

Again in qualitative terms, for particles outside of the first Fresnel zone for a line-of-sight path, or for energy scattered significantly off axis from line of sight, the scattering results in the incoherent field, with a range of received signal phase angles distributed over 2π and therefore with a zero-average field strength (but non-zero power).

It should be noted that, for all practical purposes, the coherent received signal increases with antenna beamwidth, whereas the incoherent energy is integrated over the beam, as shown in sections 4.3 and 4.4. Consequently, the ratio of incoherent to coherent signal power is dependent on antenna beamwidth.

Multiple scattering can make a contribution to either the coherent or incoherent fields. This point is examined later.

4.6.3 Calculation of the coherent field

Ignoring (for simplicity) polarization dependences, the coherent transmission coefficient may be written as

$$T = \mathrm{e}^{-(\gamma - \mathrm{j}k_0)z} = \mathrm{e}^{-\mathrm{j}(k-k_0)z} \tag{4.6.4}$$

where γ is the bulk propagation coefficient of the medium and $k_0 = f/(\varepsilon_0\mu_0)^{1/2}$ is the wavenumber for vacuum. Following the arguments of the previous section, the classical form for k for a monodispersive medium is

$$k = k_0 + \frac{2\pi}{k_0} Mf(\boldsymbol{u}_\mathrm{i}, \boldsymbol{u}_\mathrm{i}) \tag{4.6.5}$$

where M is the number density of the (spherical) scatterers and $f(\boldsymbol{u}_\mathrm{i}, \boldsymbol{u}_\mathrm{i})$ is the forward scattering amplitude (equation (4.2.4)). Also, for a medium with a distribution of sizes $M(r)$,

$$k = k_0 + \frac{2\pi}{k_0} \int f(\boldsymbol{u}_\mathrm{i}, \boldsymbol{u}_\mathrm{i}, r) M(r)\,\mathrm{d}r. \tag{4.6.6}$$

Van de Hulst (1981), in elaborating his formulation for thin slabs of media, invokes a stationary phase approximation. However, it should be clear from the discussion in the previous sections that the Van de Hulst refractive index formula for a population of particles in a slab can be derived in more general circumstances (including the near field of an antenna) without the use of a stationary phase approximation (Capsoni and Paraboni, 1978). Also, Rogers and Olsen (1976) have pointed out that the Van de Hulst approach accounts for all forward multiple scattering paths that have components in the forward direction. This also emerges from the plane wave spectrum treatment. However, it is important to note that a basic assumption in the derivations using the plane wave spectrum treatment (which also underlies the Van de Hulst treatment) is that the transverse dimensions of the slab are infinite or, in practice, are large compared with the illumination region of the antenna.

4.6.4 Incoherent scattering

Incoherent scattering has been addressed by Crane (1971), DeLogne *et al.* (1975), Capsoni and Paraboni (1978), Ishimaru, Lesselier and Yeh (1984), and Oguchi (1986). Early workers solved the radiative transfer equation using iterative procedures or discrete ordinate methods. Oguchi (1980) has reported a numerical solution of the radiative transfer equation using an extended spherical harmonics method and more recently Oguchi (1986) has solved the transfer equation analytically using generalized spherical functions (sets of ordinary differential equations, which are solved numerically).

In order to take into account polarization effects, Stokes vectors are normally used and, since the equations are simpler for circularly polarized basis vectors than for linear polarizations, the circular polarization representation is usually taken. The Stokes vector can then be defined as

$$\boldsymbol{J}_\mathrm{c} = \begin{bmatrix} \overline{E_+ E_-^*} \\ \overline{E_+ E_+^*} \\ \overline{E_- E_-^*} \\ \overline{E_- E_+^*} \end{bmatrix} \tag{4.6.7}$$

where E_+ and E_- are the right- and left-handed components of the circularly polarized wave, respectively, * denotes the complex conjugate and $\overline{}$ denotes the time average. In the case of spherical scatterers, the radiative transfer equation for Stokes vector $\boldsymbol{J}_\mathrm{c}$ is written as

$$\frac{\mathrm{d}\boldsymbol{J}_\mathrm{c}}{\mathrm{d}r} = -\sigma \boldsymbol{J}_\mathrm{c} + \int \boldsymbol{\Sigma}_\mathrm{c}(\boldsymbol{u}_\mathrm{s}, \boldsymbol{u}_\mathrm{i}) \boldsymbol{J}_\mathrm{c}(\boldsymbol{u}_\mathrm{i}) \, \mathrm{d}\Omega_\mathrm{i} \tag{4.6.8}$$

where r is the distance from the antenna, $\boldsymbol{u}_\mathrm{s}$ and $\boldsymbol{u}_\mathrm{i}$ are direction vectors, with $\boldsymbol{u}_\mathrm{i}$ pointing to solid angle element $\mathrm{d}\Omega_\mathrm{i}$, σ is the total cross-section (scalar) per unit volume and $\boldsymbol{\Sigma}_\mathrm{c}(\boldsymbol{u}_\mathrm{s}, \boldsymbol{u}_\mathrm{i})$ is a fourth-order square matrix, representing the scattering cross-section per unit solid angle and per unit volume of an assembly of hydrometeors, when a stream of radiation is incident on the scatterers in the direction $\boldsymbol{u}_\mathrm{i}$ and the scattered radiation is observed in direction $\boldsymbol{u}_\mathrm{s}$. The Stokes vector may be divided into two parts, representing coherent and incoherent terms respectively:

$$\boldsymbol{J}_\mathrm{c} = \boldsymbol{J}_\mathrm{cc} \delta(\boldsymbol{u}_\mathrm{s} - \boldsymbol{u}_\mathrm{c}) + \boldsymbol{J}_\mathrm{ci}(\boldsymbol{u}_\mathrm{s}) \tag{4.6.9}$$

where $\boldsymbol{u}_\mathrm{c}$ is the propagation direction of the incident coherent radiation.

The behaviour of the coherent and incoherent terms may then be expressed separately as

$$\frac{\mathrm{d}\boldsymbol{J}_{\mathrm{cc}}}{\mathrm{d}r} = -\sigma \boldsymbol{J}_{\mathrm{cc}} \tag{4.6.10}$$

and

$$\frac{\mathrm{d}\boldsymbol{J}_{\mathrm{ci}}}{\mathrm{d}r} = -\sigma \boldsymbol{J}_{\mathrm{ci}} + \int \boldsymbol{\Sigma}_{\mathrm{c}}(\boldsymbol{u}_{\mathrm{s}}, \boldsymbol{u}_{\mathrm{i}}) \boldsymbol{J}_{\mathrm{ci}}(\boldsymbol{u}_{\mathrm{i}})\, \mathrm{d}\Omega_{\mathrm{i}} + \boldsymbol{\Sigma}_{\mathrm{c}}(\boldsymbol{u}_{\mathrm{s}}, \boldsymbol{u}_{\mathrm{c}}) \boldsymbol{J}_{\mathrm{cc}}. \tag{4.6.11}$$

4.6.5 Discussion of results

Oguchi has written $\boldsymbol{\Sigma}_{\mathrm{c}}(\boldsymbol{u}_{\mathrm{s}}, \boldsymbol{u}_{\mathrm{i}})$ in terms of generalized spherical functions, assuming the hydrometeors to be spherical. This yields a set of ordinary differential equations which are then solved numerically. Solutions are presented for circularly and linearly polarized plane waves. Calculations were made at 34.8, 82, 140 and 245.5 GHz and one optical wavelength (0.6328 μm). Table 4.3 illustrates some of the results. The incoherent power at a receiver was calculated for a Laws and Parsons drop size distribution up to a rain intensity of 150 mm h^{-1}. The following conclusions were reached.

- The directional patterns of incoherent waves in a semi-infinite rain medium tend to become sharper as the propagation distance increases. This is explained by the fact that the multiple scattered waves, which contribute to angular spreading, are more strongly absorbed than the waves scattered in the forward direction, because of their longer electrical path through the attenuating medium.
- For a practical transmitter–receiver distance and in the millimeter wave region, the coherent power is still larger than the incoherent power, even at 245.5 GHz and for 150 mm h^{-1} of rain, unless the beamwidth is too large.
- For optical wave propagation in fog, the incoherent power is larger than the coherent power even for a half-power beamwidth of 0.1°.

No measurements of incoherent versus coherent scattering at high millimetre wavelengths have been reported, but the calculations of Oguchi and others suggest that generally the incoherent terms in forward scatter are relatively small for most practical radio links up to at least 200 GHz. However, in the case of low-gain antennas, the incoherent to coherent ratio (and the consequent scintillation) can be relatively high (of the order of 1 dB) (Paraboni, personal communication).

Incoherent scattering is, of course, important for bistatic or interference geometries. This is described in Chapter 11.

The effect of incoherent scattering on channel transfer characteristics and pulse shape has been evaluated by Ishimaru and Lin (1972) and Hong and Ishimaru (1976). At a rain rate of 2.5 mm h^{-1} for a 5 km transmission path pulses of a carrier at 100 GHz show the principal part of the pulse transmitted

Table 4.3 Received incoherent power as a function of antenna beamwidth (ϑ) for various frequencies and optical slab thicknesses ζ_s (the incident wave is LHC polarized)

$R = 150\ \text{mm h}^{-1}$; coherent signal level, −66.59 dB; $\zeta_s = 15.33$

		Received incoherent power		
f (GHz)	Polarization	$\vartheta = 1°$	$\vartheta = 5°$	$\vartheta = 10°$
34.8	LHC	−89.9	−75.9	−69.8
	RHC	−94.3	−80.3	−74.3
82.0	LHC	−82.6	−68.6	−62.6
	RHC	−86.7	−72.7	−66.6

$R = 150\ \text{mm h}^{-1}$; coherent signal level, −59.35 dB; $\zeta_s = 13.66$

		Received incoherent power		
f (GHz)	Polarization	$\vartheta = 1°$	$\vartheta = 5°$	$\vartheta = 10°$
140.0	LHC	−72.1	−58.1	−52.0
	RHC	−77.9	−63.9	−57.9
245.5	LHC	−67.2	−53.2	−47.2
	RHC	−75.0	−61.1	−55.0

Monodispersion of water droplets (radius, 1 μm); water content, $0.015\ \text{g m}^{-3}$; coherent signal level, −65.44 dB; $\zeta_s = 15$

		Received incoherent power		
λ (μm)	Polarization	$\vartheta = 0.1°$	$\vartheta = 0.3°$	$\vartheta = 0.5°$
0.6328	LHC	−55.4	−45.8	−41.4
	RHC	−55.7	−46.2	−41.8

unchanged but followed by a tail of small amplitude. With a rain rate of $25\ \text{mm h}^{-1}$, distortion was predicted to be significant, with the pulse spread over 7 μs. However, in the latter instance, attenuation was 60 dB.

Attempts to measure bandwidth fluctuation from incoherent effects to date have all been at relatively low frequencies (30 GHz or less) and have all proved to be negative (Cox, Arnold and Leck, 1980).

4.6.6 Multiple scattering effects

For the incoherent field these effects are included in the radiative transfer treatments of Oguchi and others. For the coherent field Rogers and Olsen

(1976) have shown (from the work of Twersky, described in section 4.3.3(f)) that all forward multiple scattering processes are accounted for in the Van de Hulst formula. Van de Hulst, in effect, applied the single scattering form to successive thin layers of the medium, resulting in the microscopic forward multiple scattering processes being reinstated in the theory, through the macroscopic forward multiple scattering between the thin layers.

The effect of backward multiple scattering processes was also investigated by Rogers and Olsen up to 1000 GHz with very high rain rates and found to be negligible.

4.7 SPECIFIC EXPERIMENTS FOR VERIFICATION OF THE THEORY

4.7.1 Apparent discrepancies

An early review paper (Medhurst, 1965) pointed to apparent discrepancies between the theory and measurement. As a result a number of model-oriented experiments were established (described in the following). The majority of the discrepancies highlighted in Medhurst's paper were later seen to be from instrumentation or of trivial origin (Watson, 1976).

4.7.2 Single-particle measurements

Measurements on droplets supported in wind tunnels (Mathews, Goodall and Watson, 1976) and ice particles supported by thin cord (Aucterlonie and Bryant, 1981) have been reported. Results are in agreement with theory within the range of measurement accuracy.

4.7.3 Rain range experiment

Crane (1974) has reported an experiment in which water was sprayed from a gantry so as to create artificial rain. Measurements were undertaken at 9.4 GHz and 35 GHz over a short radio link. This experiment was set up in view of the apparent discrepancies mentioned earlier. The results of the experiment are puzzlingly ambiguous. At 9.4 GHz good agreement between theory and measurements is reported but at 35 GHz a near-field correction was made in order to gain agreement. The need for this correction has been questioned by Haworth, McEwan and Watson (1978) and in the plane wave spectrum treatment of Haworth (1980) is seen not to be required.

4.7.4 Densely instrumented short links

Perhaps the best substantiation of theoretical predictions is given by this type of measurement, on very short radio links with closely spaced rain gauges. The measurements of Norbury and White (1972), Humpleman and Watson (1978), Keizer, Sneider and de Haan (1978) and Gibbins *et al*. (1987) all show excellent agreement with theory within the range of experimental error.

4.7.5 Conclusions from model-oriented experiments

All of the experiments set up to test the accuracy of the scattering theory have succeeded in doing so (with the exception of that of Crane which requires further explanation). We conclude that the single-particle scattering theory combined with the coherent scattering model gives accurate predictions of rain attenuation well into the millimetre wave range.

REFERENCES

Aden, A.L. and Kerker, M. (1951) Scattering of electromagnetic waves by two concentric spheres. *J. Appl. Phys.*, **22**, 1242–6.

Asano, S. (1979) Light scattering properties of spheroidal particles. *Appl. Opt.*, **18**(5), 712–23.

Atlas, D., Battan, L.J., Harper, W.G., Herman, B.M., Kerker, M. and Matijevic, E. (1963) Backscatter by dielectric spheres (refractive index 1.6). *IEEE Trans. Antennas Propag.*, **11**, 68–72.

Aucterlonie, L.J. and Bryant, D.L. (1981) Experimental study of millimetre-wave scattering from simulated hailstones in an open resonator. *Proc. IEE, Part H*, **128**, 236–42.

Beckman, P. (1968) *The Depolarization of Electromagnetic Waves*, Golem, Boulder, CO.

Born, M. and Wolf, E. (1986) *Principles of Optics*, Pergamon, Oxford.

Bringi, V.N. and Seliga, T.A. (1977) Scattering from axisymmetric dielectrics or perfect conductors imbedded in an axisymmetric dielectric. *IEEE Trans. Antennas Propag.*, **25**(4), 575–81.

Capsoni, C. and Paraboni, A. (1978) Properties of the forward-scattered incoherent radiation through intense precipitation. *IEEE Trans. Antennas Propag.*, **26**(6), 804–9.

Cox, D.C., Arnold, H.W. and Leck, R.P. (1980) Phase and amplitude dispersion for earth-satellite propagation in the 20- to 30-GHz frequency range. *IEEE Trans. Antennas Propag.*, **28**, 359–66.

Crane, R.K. (1971) Propagation phenomena affecting satellite communication systems operating in the centimetre and millimetre wavelength bands. *Proc. IEEE,* **59**(2), 173–88.

Crane, R.K. (1974) The rain range experiment – propagation through a simulated rain environment. *IEEE Trans. Antennas Propag.*, **22**, 321–8.
DeLogne, P., Osvath, P., Sobieski, P. and Van Vye, J. (1975) Theory and experimental study of microwave depolarisation due to rain. URSI, Lima.
Dissanayake, A.W. and Watson, P.A. (1979) Forward scattering and cross-polarisation from spheroidal ice particles, *Electron. Lett.*, **13**, 140–2.
Fang, D.J. and Lee, F.J. (1978) Tabulations of raindrop induced forward and backscattering amplitudes. *COMSAT Tech. Rev.*, **8**, 455–86.
Foldy, L.O. (1945) The multiple scattering of waves. *Phys. Rev.*, **67**(3, 4), 107–19.
Gibbins, C.J., Carter, D.G., Egget, P.A., Lidiard, K.A., Pike, M.G., Tracey, M.A., White, E.H., Woodroffe, J.M. and Yilmaz, U.M. (1987) A 500 m experimental range for propagation studies at millimetre, infra-red and optical wavelengths. *J. IERE*, **57**, 227–34.
Haworth, D.P. (1980) Plane wave spectrum treatment of microwave scattering by hydrometeors on an earth-satellite link. NATO/AGARD Conf. on Propagation Effects in Space/Earth Paths, Conf. Preprints, 284, Paper 4, pp. 1–11.
Haworth, D.P., McEwan, N.J. and Watson, P.A. (1978) Effect of rain in the near field of an antenna. *Electron. Lett.*, **14**, 94–6.
Holt, A.R. (1980) The Fredholm integral equation method and comparison with the T-matrix approach, in *Acoustic, Electromagnetic and Elastic Wave Scattering – Focus on the T-Matrix Approach* (eds V.K. Varadan and V.V. Varadan), Pergamon, New York, pp. 255–68.
Holt, A.R. (1982) The scattering of electromagnetic waves by single hydrometeors. *Radio Sci.*, **17**(5), 929–45.
Holt, A.R. and Santoso, B. (1972) Fredholm integral equation method for scattering phase shifts. *J. Phys. B*, **5**(3), 497.
Holt, A.R., Uzunoglu, N.K. and Evans, B.G. (1978) An integral equation solution to the scattering of electromagnetic radiation by dielectric spheroids and ellipsoids. *IEEE Trans. Antennas Propag.*, **26**(5), 706–12.
Hong, S.T. and Ishimaru, A. (1976) Two-frequency mutual coherence function, coherence bandwidth and coherence time of millimeter and optical waves in rain, fog, and turbulence. *Radio Sci.*, **11**(6), 551–9.
Humpleman, R.J. and Watson, P.A. (1978) Investigation of attenuation by rainfall at 60 GHz. *Proc. IEE*, **125**, 85–91.
Ishimaru, A. (1978) *Wave Propagation and Scattering in Random Media*, Academic Press, New York.
Ishimaru, A. and Lin, J.C. (1972) Multiple scattering effects on wave propagation through rain, in Proc. Conf. on Telecommunications Aspects of Frequencies between 10 and 100 GHz, NATO, Norway, pp. 1.1–1.13.
Ishimaru, A., Lesselier, D. and Yeh, C. (1984) Multiple scattering calculations for non-spherical particles based on the vector radiative transfer theory. *Radio Sci.*, **19**, 1356–66.
Keizer, W.P.M.N., Snieder, J. and de Haan, C.D.L. (1978) Propagation measurements at 94 GHz and comparison of experimental rain attenuation with theoretical results derived from actually measured raindrop size distributions, IEE Int. Conf. on Antennas and Propagation, ICAP-78, IEEE Conf. Publ. 169, pp. 72–6.
Kerker, M. (1969) *The Scattering of Light*, Academic Press, New York.
Mathews, N.A., Goodall, F. and Watson, P.A. (1976) Amplitude and phase measure-

ments of forward-scattered microwave radiation from water drops. *Electron. Lett.*, **12**, 157–8.

Medhurst, R.G. (1965) Rainfall attenuation of centimetre waves, comparison of theory and measurement. *IEEE Trans. Antennas Propag.*, **13**, 550–64.

Mie, G. (1908) *Ann. Phys.*, **25**(4), 377.

Morrison, J.A. and Chu, T.S. (1973) Perturbation calculations of rain induced differential attenuation and differential phase shift at microwave frequencies. *Bell Syst. Tech. J.*, **52**(10), 1907–13.

Morrison, J.A. and Cross, M.J. (1974) Scattering of a plane electromagnetic wave by axisymmetric raindrops. *Bell Syst. Tech. J.*, **53**, 955–1019.

Newton, R.G. (1982) *Scattering Theory of Waves and Particles*, 2nd edn, Springer, New York.

Norbury, J.R. and White, W.J.K. (1972) Microwave attenuation at 35.8 GHz due to rainfall. *Electron. Lett.*, **8**(4), 91–2.

Oguchi, T. (1973) Attenuation and phase rotation of radio waves due to rain; calculations at 19.3 and 34.8 GHz. *Radio Sci.*, **8**, 31–8.

Oguchi, T. (1980) Effect of incoherent scattering on attenuation and cross polarisation of millimetre waves due to rain. Preliminary calculations at 34.8 and 82 GHz for spherical drops. *J. Radio Res. Lab. Jpn.*, **27**, 1–51.

Oguchi, T. (1986) Effects of incoherent scattering on attenuation and depolarisation of millimetre and optical waves due to hydrometeors. *Radio Sci.*, **21**(4), 717–30.

Papatsoris, A.D. and Watson, P.A. (1992) A rigorous explanation of the resonances observed in the scattering from ice particles. Proc. URSI Commission F Int. Symp., Ravenscar, pp. 5.6.1–5.6.5.

Paraboni, A. (1991) Personal communication.

Peterson, B. and Strom, S. (1975) Matrix formulation of acoustic scattering from multilayered scatterers. *J. Acoust. Soc.*, **57**(1), 2–13.

Pruppacher, H.R. and Pitter, R.L. (1971) A semi-empirical determination of the shape of cloud and raindrops. *J. Atmos. Sci.*, **28**, 86–94.

Rogers, D.V. and Olsen, R.L. (1976) Calculation of radiowave attenuation due to rain at frequencies up to 1000 GHz. *CRC Rep. 1299*, Communications Research Centre, Ottawa.

Saxon, D.S. (1955) Lectures on the scattering of light. *Rep. 9*, UCLA Department of Meteorological Science.

Singh, S.R. and Stauffer, A.D. (1975) A convergence proof for the Schwinger variational method for the scattering amplitude. *J. Phys. A*, **8**(9), 1379–83.

Stratton, J. (1941) *Electromagnetic Theory*, McGraw-Hill, New York.

Twersky, V. (1962) On scattering of waves by random distributions. *J. Math. Phys.*, **3**(4), 700–34.

Van de Hulst, H.C. (1981) *Light Scattering by Small Particles*. Dover Publications, New York.

Van den Berg, P.M. and Fokkema, J.T. (1979) The Rayleigh hypothesis in the theory of diffraction by a cylindrical obstacle. *IEEE Trans. Antennas Propag.*, **27**(5), 577–83.

Waterman, P.C. (1965) Matrix formulation of electromagnetic scattering. *Proc. IEEE*, **53**, 805–12.

Watson, P.A. (1976) Survey of measurements of attenuation by rain and other hydrometeors. *Proc. IEE*, **123**, 863–71.

Watson, P.A. and Arbabi, M. (1973) Rainfall cross polarisation at microwave frequencies. *Proc. IEE*, **120**, 413–18.

de Wolf, D.A., Russchenberg, H.W.J. and Ligthart, L.P. (1990) Effective permittivity of and scattering from wet snow and ice droplets at weather radar wavelengths. *IEEE Trans. Antennas Propag.*, **38**(9), 1317–25.

Yeh, C. (1964) Perturbation approach to the diffraction of electromagnetic waves by arbitrary shaped dielectric obstacles. *Phys. Rev.*, **135**, A1193–A1201.

Yeh, C. (1969) Scattering by liquid-coated prolate spheroids, *JASA*, **46**, 797–801.

5

Thermal radiation from hydrometeors and atmospheric gases

5.1 THE EMISSION PROCESS

5.1.1 Thermodynamic equilibrium (Brussaard, 1985)

The steady state of any closed system is governed by the two fundamental laws of thermodynamics, i.e. conservation of energy and maximization of entropy. The latter has as a consequence that energy is distributed equally over all degrees of freedom. Therefore, in thermodynamic equilibrium, energy is distributed over both possible orthogonal polarizations.

From the law of quantization of energy it then follows that the average energy density W per unit bandwidth in an isothermal enclosure is given by Planck's function:

$$W\ (\mathrm{J\,m^{-3}\,Hz^{-1}}) = \frac{8\pi h f^3}{c^3} \frac{1}{\mathrm{e}^{hf/kT} - 1} \tag{5.1.1}$$

where $h = 6.6262 \times 10^{-34}$ J Hz^{-1} is Planck's constant, f (Hz) is the frequency, $k = 1.380\,62 \times 10^{-23}$ J K^{-1} is Boltzmann's constant, T (K) is the thermodynamic temperature and $c = 2.997\,924\,6 \times 10^8$ m s^{-1} is the velocity of light. In thermodynamic equilibrium, energy flow is independent of direction. Therefore, the brightness or specific intensity in thermodynamic equilibrium is given by

$$B\ (\mathrm{W\,m^{-2}\,sr^{-1}\,Hz^{-1}}) = \frac{2hf^3}{c^2} \frac{1}{\mathrm{e}^{hf/kT} - 1} \tag{5.1.2}$$

and for $hf \ll kT$ this reduces to

$$B = \frac{2kT}{\lambda^2} \tag{5.1.3}$$

where λ (m) is the wavelength. This is known as the Rayleigh–Jeans law. The total power per unit bandwidth delivered to the matched load of a lossless antenna placed inside the medium is

$$P_r\ (\mathrm{W\,Hz^{-1}}) = 4\pi B \frac{\lambda^2}{4\pi} = 2kT. \tag{5.1.4}$$

This power is equally distributed over the two polarizations received. Each antenna port therefore receives kT.

The energy flowing in all directions is absorbed by the medium, while this medium also radiates energy spontaneously. If the intensity is to remain constant, the ratio of the mass emission and absorption coefficients must be a universal constant which depends only on temperature and not on the properties of the medium. This is known as Kirchhoff's law for thermodynamic equilibrium.

5.1.2 Extension and limitations of the concept of thermodynamic equilibrium

When thermal noise is observed in the atmosphere, the conditions for strict thermodynamic equilibrium are not fulfilled, as the atmosphere is a non-isothermal, finite medium without a black enclosure. Hence, strictly speaking, Kirchhoff's law does not apply. Nevertheless, an equilibrium may exist, such that any volume element at a certain local temperature $T(\boldsymbol{r})$ has the same absorption and emission characteristics as it would have in the state of thermodynamic equilibrium at that temperature. This is the case if the probability of excitation and decay of molecular and atomic energy levels is controlled primarily by collision effects, and if the influence of radiation intensity can be neglected. This condition of so-called 'local thermodynamic equilibrium' is the basis for the study of radiative transfer through rain at microwave frequencies. For this condition to exist, neither an isothermal medium nor an isotropic radiation intensity is required

If, on the other hand, the effect of collisions between atoms and molecules can be neglected, the equilibrium state is controlled by the radiation intensity and by the spontaneous decay of the energy levels excited by the incident radiation. This is the case of monochromatic radiation equilibrium. This condition occurs when the medium has low density, e.g. in the upper part of the atmosphere.

These equilibrium conditions are extreme cases of a general equilibrium state, where both radiative and collision effects play a role. For the treatment of emission properties of the higher atmosphere, therefore, different relations between absorption, emission, temperature and density exist. For the calculation of total radiance and absorption of the atmosphere, this generally is neglected. It is taken into account, however, in the theory relating to the remote sensing of trace gases in the middle and higher atmosphere (above 20 km) by limb sounding.

The simplification $hf \ll kT$ (Rayleigh–Jeans) needs some attention. For brightness temperatures in the order of a few hundreds of kelvin or lower, the condition is not fulfilled for frequencies of a few hundreds of gigahertz. For example, for $f = 300$ GHz and $T = 300$ K,

$$hf \approx 2 \times 10^{-22} \quad \text{and} \quad kT \approx 4 \times 10^{-21}.$$

This leads to a relative error of 3% when using the Rayleigh–Jeans law.

5.1.3 Atmospheric gases

At microwave and millimetre wave frequencies, atmospheric gases do not cause any scatter. Emissivity and attenuation are directly related by Kirchhoff's law. The shape of absorption lines and the dependence of absorption on atmospheric parameters is discussed in Chapter 2.

For considerations of total attenuation, at frequencies below 300 GHz only water vapour and oxygen produce significant absorption. The absorption lines of trace gases (such as CO, O_3, H_2O_2) are too weak to be detected in attenuation measurements. Even with radiometric measurements through the atmosphere (either up- or downlooking) no reliable detection is possible, since the total emission is dominated by water vapour and oxygen. Limb sounding, in which radiometric measurements are made through the middle and upper atmosphere only from an aircraft, balloons or a spacecraft, has the ability to detect trace gases around 205 GHz and to quantify their density (Waters, 1976, Poynter and Pickett, 1984; Waters *et al.*, 1988).

5.1.4 Hydrometeors

Absorption and scattering by hydrometeors has been treated in Chapter 4.

Emission of thermal radiation is again governed by Kirchhoff's law on the assumption of local thermodynamic equilibrium. This is equivalent to assuming that no higher-power radiation is present which might influence the energy state of the water molecules. This condition might be violated in the case of nuclear explosions and for power satellite systems employing microwave beams for power transfer.

Observations have been made of enhanced radiation associated with so-called 'wet' scintillations (Vanhoenacker and Vandervorst, 1985). It is not clear, however, what the exact nature is of the process involved. Most probably this is caused by variations in water vapour and/or liquid water at the edges of cumulus cloud structures, where much turbulence exists.

5.1.5 Other particles

Particles such as dust, smoke and sand consist of solid matter that in general has low absorption (Chapter 6). Since signal extinction by these particles is largely due to scatter, their emissivity is very low and no literature has been found on detection of solid particles by passive sensing at microwave frequencies.

5.2 ELECTROMAGNETIC MODELLING

5.2.1 Radiative transfer theory

The theory of radiative transfer (section 4.3.3(e)) considers the transfer of energy as it is governed by the basic laws of thermodynamics. It takes into account the scattering and absorption characteristics of particles, under the assumption that there is no correlation in the fields generated by individual particles, so that power addition may be applied.

Signal extinction along the propagation path considered is expressed as the power $\mathrm{d}P$ removed from the incident radiation. For a polydispersion of particles, such as rain, cloud drops and cloud ice, this takes the form

$$\mathrm{d}P = -M\sigma_\mathrm{t} P_\mathrm{i}\,\mathrm{d}z \tag{5.2.1}$$

where M (m^{-3}) is the volumetric number density of particles, σ_t (m^2) is the extinction cross-section and P_i (W) is the incident power. A further assumption in the treatment of scattering in radiative transfer theory is that the particles are in each other's far field. Hence, clouds and rain are generally classified as 'tenuous polydispersions of particles'.

Extinction is the result of scattering and absorption. Both being linear processes, we may write

$$\sigma_\mathrm{t} = \sigma_\mathrm{s} + \sigma_\mathrm{a} \qquad \text{(scattering and absorption cross-sections).}$$

The bistatic cross-section is defined by

$$\sigma_{\mathrm{bi}} = 4\pi |S(\boldsymbol{u}_\mathrm{s}, \boldsymbol{u}_\mathrm{i})|^2 \tag{5.2.2}$$

where $S(\boldsymbol{u}_\mathrm{s}, \boldsymbol{u}_\mathrm{i})$ is the complex scattering function for incident direction $\boldsymbol{u}_\mathrm{i}$ and scattering direction $\boldsymbol{u}_\mathrm{s}$, for which

$$\int_{4\pi} |S(\boldsymbol{u}_\mathrm{s}, \boldsymbol{u}_\mathrm{i})|^2 \,\mathrm{d}\Omega_\mathrm{s} = \sigma_\mathrm{s} \tag{5.2.3}$$

where the scattering direction vector $\boldsymbol{u}_\mathrm{s}$ points to solid angle element $\mathrm{d}\Omega_\mathrm{s}$.

Defining the phase function

$$\Phi(\boldsymbol{u}_s, \boldsymbol{u}_i) = \sigma_{bi}(\boldsymbol{u}_s, \boldsymbol{u}_i)/\sigma_t \tag{5.2.4}$$

we find, for non-uniform scattering,

$$\sigma_s = \frac{\sigma_t}{4\pi}\int_{4\pi} \Phi(\boldsymbol{u}_s, \boldsymbol{u}_i)\,\mathrm{d}\Omega_s. \tag{5.2.5}$$

For uniform scattering the phase function is constant:

$$\Phi_0 = \frac{\sigma_s}{\sigma_t} \qquad \text{(scattering albedo).} \tag{5.2.6}$$

In its general form the radiative transfer equation describes the variation of intensity along the propagation path as a differential equation:

$$\frac{1}{M\sigma_t}\frac{\mathrm{d}I}{\mathrm{d}z} = -I(z) + J(z). \tag{5.2.7}$$

Defining the optical distance ζ between two locations z_1 and z_2 as

$$\zeta(z_1, z_2) = \int_{z_1}^{z_2} M\sigma_t\,\mathrm{d}z \tag{5.2.8}$$

we write the radiative transfer equation as

$$\frac{\mathrm{d}I}{\mathrm{d}\zeta} = -I + J \tag{5.2.9}$$

and its formal solution is

$$I(z) = I(0)\,\mathrm{e}^{-\zeta(0,z)} + \int_0^z M(z')\sigma_t(z')J(z')\,\mathrm{e}^{-\zeta(z',z)}\,\mathrm{d}z'. \tag{5.2.10}$$

This expression can be interpreted as follows. The radiation at a certain point is the sum of

- the attenuated incident radiation and
- the integral of the contributions from all previous points, attenuated by the medium between.

The source function $J(z)$ describes the generation of radiative energy in the propagation direction by thermal emission and scattering of energy which is incident from other directions. In general, for a monodispersion of particles this source function is

$$J(z, \boldsymbol{u}_\mathrm{s}) = \frac{1}{4\pi}\int_{4\pi} \Phi(\boldsymbol{u}_\mathrm{s}, \boldsymbol{u}'_\mathrm{s}) I(z, \boldsymbol{u}'_\mathrm{s})\,\mathrm{d}\Omega'_\mathrm{s} + \frac{\varepsilon(z, \boldsymbol{u}_\mathrm{s})}{M(z)\sigma_\mathrm{t}(z, \boldsymbol{u}_\mathrm{s})} \tag{5.2.11}$$

where ε is the emission coefficient, which for local thermodynamic equilibrium is $M\sigma_\mathrm{a}B$.

5.2.2 Assumptions and limitations

Combining equations (5.2.10) and (5.2.11) leads to an integral equation which may be solved for particular configurations, often combined with several simplifying assumptions. The most common simplification consists of reducing the three-dimensional integration (5.2.11) to a one-dimensional integration by assuming a layered medium (stratified atmosphere). Assuming isotropic scattering, equation (5.2.11) then reduces to

$$J(z) = \frac{\sigma_\mathrm{s}}{\sigma_\mathrm{t}}U(z) + \left(1 - \frac{\sigma_\mathrm{s}}{\sigma_\mathrm{t}}\right)B \tag{5.2.12}$$

where

$$U = \frac{1}{4\pi}\int I\,\mathrm{d}\Omega$$

is the average intensity.

It should be noted that the theory itself takes multiple scattering into account completely. Simplifications occurring in the literature are always due to a need to reduce computational complexity.

5.3 PREDICTION MODELS

5.3.1 Clear sky noise temperature

The new ITU-R Recommendation PN 526 (Radio Noise) (ITU, 1993), superseding Recommendation 677, now provides the reference formulation for the calculation of sky brightness temperatures for system performance calculations. It uses the formulation of equations (5.2.10) and (5.2.11) for a non-scattering atmosphere, i.e. $\Phi(\boldsymbol{u}_\mathrm{s}, \boldsymbol{u}'_\mathrm{s}) = 0$ in equation (5.2.11), which, for a pure absorber, leads to an equation of the form

$$T_\mathrm{b}(z) = T_\mathrm{b}(0)\,\mathrm{e}^{-\zeta(0,z)} + \int_0^z M(z')\sigma_\mathrm{t}(z')T(z')\,\mathrm{e}^{-\zeta(z',z)}\,\mathrm{d}z', \tag{5.3.1}$$

where T_b is the brightness temperature and T is the physical temperature. Here, $T_b(0)$ is the background radiation and $T_b(z)$ is the emerging radiation after passage through the atmosphere. Using this formulation brightness temperatures are calculated for US Standard Atmospheres for various conditions of temperature, humidity and latitude. Figure 5.1 shows an example.

Galactic noise is neglected in the ITU-R method for frequencies above 2 GHz except for the sun and a few isolated sources. For the strongest radio sources, values in the range 85 MHz–10.7 GHz have been tabulated by Wielebinski (1976) (Fig. 5.2).

The above formulation is the basis for all models for the analysis of remote sensing applications. For uplooking radiometry, $T_b(0)$ is the background galactic and cosmic noise and

$$z = (h_a - h) \sec \epsilon \tag{5.3.2}$$

where h_a is the height of the atmosphere, h is the height above ground and ϵ is the elevation angle (> 5°).

For an exponential atmosphere, the 'weighting function' $M\sigma_t e^{-\zeta(z',z)}$ is

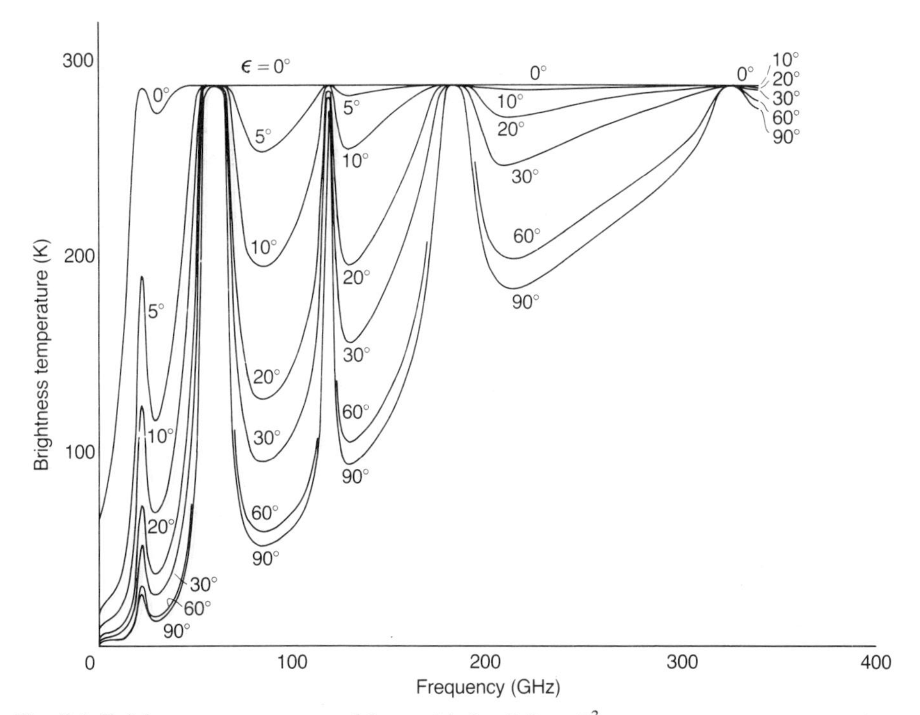

Fig. 5.1 Brightness temperature (clean air) for 7.5 g m^{-3} water vapour concentration (surface temperature, 15 °C; surface pressure, 1023 mbar) where ϵ is the elevation angle (ITU, 1993).

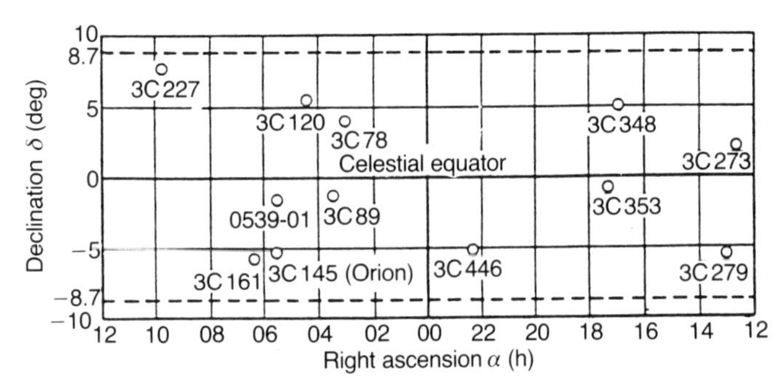

Fig. 5.2 Locations of strongest radio sources (o) for a band of ±8.7° about the celestial equator (Wielebinski, 1976). The numbers refer to catalogue designations, e.g. 3C indicates the third Cambridge Catalogue.

exponential with a rate of decay which is frequency dependent. From multiple-frequency observations, therefore, profiles of atmospheric parameters may be obtained through inversion (Künzi, 1984). A well-known example is the extraction of integrated water vapour content and temperature profile from inversion of radiometric observations at 20 and 30 GHz and a series of channels in the 50 GHz band.

For downlooking (air- or spaceborne) radiometry $z = h \sec \epsilon$ and the combination of the increasing sensitivity ($e^{-\zeta(z',z)}$) and decreasing emission with height results in a weighting function which has a distinct maximum at a height which is determined by the frequency chosen. A combination of frequencies on the flank of an absorption line therefore is a very powerful method to obtain profiles of temperature and pressure (50 GHz and 110 GHz) or water vapour (180 GHz).

Precise knowledge of pressure broadening of absorption lines was combined with experimental and measured atmospheric profiles to obtain an optimum set of frequencies for water vapour radiometry in the 20–30 GHz bands. Optimization was done in terms of the sensitivity of the inversion process to the profile of water vapour content, the objective being the determination of integrated water vapour content. As a result, a frequency slightly below or above the resonance frequency for water vapour (22.2 GHz) is chosen for water vapour radiometry.

5.3.2 Emission by clouds and rain

Modelling of radiation temperature of clouds is generally based on the following assumptions:

- spherical drops;
- Rayleigh scattering;
- plane parallel medium.

Tsang *et al.* (1977) give a model for emission by a layer of cloud or rain, including the added influence of gaseous absorption and emission by the earth's surface which uses a cloud drop size distribution from Deirmendjian (1969), water permittivity from Saxton and Lane (1952) and an integration of the integral transfer equation by Gaussian quadrature from Chandrasekhar (1960). Scattering functions were calculated using Mie theory, making the model applicable both to rain and to clouds. Brightness temperatures were analysed in terms of the scattering contribution, the influence of viewing angle and the effect of polarization. Polarization effects are the result of the asymmetry in the geometric configuration, giving rise to a partially polarized scattering contribution in off-nadir brightness observations.

A comparable analysis is given by Ishimaru and Cheung (1980), who analysed the accuracy of the estimation of attenuation when neglecting or including the effect of scattering in this estimation.

A model capable of calculating the effect of scattering in thermal emissions from rain cells of limited size is given by Brussaard (1985). In this model the effect of limited cell size is analysed using isotropic scattering. The radiative transfer equation is solved using eigenvalue decomposition of the radiation field. An important feature of this model is that it shows that the brightness temperature of a homogeneous cell is no longer uniform when taking particle scattering into account. The brightness decreases when moving from the centre of the cell towards the edge. Thus the error made by neglecting scatter in radiative equations is dependent on the location within the cell.

For practical calculations of noise temperature, induced by rain in a receiver front end of a satellite–earth station, use is generally made of the concept of equivalent absorber temperature, or effective medium temperature. This is the thermodynamic temperature of a hypothetical absorber which gives the same relation between attenuation and emissivity (noise temperature).

For remote sensing applications where rain rate is to be determined from passive multiple-frequency observations the approach mentioned above (Tsang *et al.*, 1977) is generally used. A comprehensive treatment of the requirements for remote sensing of rain is given by Atlas and Thiele (1981) and by Staelin and Rosenkrantz (1978). A study of the effect of rain in degrading the quality of imaging microwave radiometer (IMR) observations of the sea surface is given by Guissard (1980). Practical system evaluations–simulations often make use of simplifications of the calculation of the scattering integral. Such simplifications must be checked very carefully, regarding their accuracy.

REFERENCES

Atlas, D. and Thiele, O.W. (eds) (1981) Precipitation measurements from space. *NASA Workshop Rep.*, GSFC, Greenbelt, MD.

Brussaard, G. (1985) Radiometry – a useful prediction tool? *ESA SP-1071*, ESA, Paris.

Chandrasekhar, S. (1960) *Radiative Transfer*, Dover Publications, New York.

Deirmendjian, D. (1969) *Electromagnetic Scattering on Spherical Polydispersions*, Elsevier, New York.

Guissard, A. (1980) Study on the influence of the atmosphere on the performance of an imaging microwave radiometer. *ESA Contract Rep. 4124/79/NL/DG (SC)*.

Ishimaru, A. and Cheung, R.L.-T. (1980) Multiple-scattering effects on radiometric determination of rain attenuation at millimeter wavelengths. *Radio Sci.*, **15**, 507.

ITU (1993) ITU-R PN Series of Recommendations (propagation in non-ionized media).

Künzi, K.F. (1984) Passive microwave remote sensing in meteorology and atmospheric physics. *J. Quant. Spectrosc. Radiat. Transfer*, **32**(5/6), 435–8.

Poynter, R.L. and Pickett, H.M. (1984) Submillimeter, millimeter and microwave spectral line catalogue. *JPL Publ. 80–23*, Rev. 2, Caltech–JPL.

Saxton, J.A. and Lane J.A. (1952) Electrical properties of sea water. *Wireless Eng.*, **29**, 269.

Staelin, Dh. and Rosenkrantz, P. (1978) High resolution passive microwave satellites. Contract NAS 5-23677.

Tsang, L., Kong, J.A., Njoku, E., Staelin, D.H. and Waters, J.W. (1977) Theory of microwave emission from a layer of cloud or rain. *IEEE Trans. Antennas Propag.*, **25**(5), 650–7.

Vanhoenacker, D. and Vandervorst, A. (1985) Atmospheric fluctuation spectra and radio system implications, 4th Int. Conf. on Antennas and Propagation ICAP. *IEE Conf. Publ.*, **248**, 67–71.

Waters, J.W. (1976) Absorption and emission by atmospheric gases, in *Methods of Experimental Physics* (ed. M.L. Meeks), Vol. 12, Astrophysics, Part B, Radio Telescopes, Academic Press, New York, Section 2.3.

Waters, J.W., Stachnik, R.A., Hardy, J.C. and Jarnot, R.F. (1988) ClO and O_3 stratospheric profiles: balloon-borne measurements. *Geophys. Res. Lett.*, **15**(8), 708–83.

Wielebinski, R. (1976) Antenna calibration, in *Methods of Experimental Physics* (ed. M.L. Meeks), Vol. 12, Astrophysics, Part B, Radio Telescopes, Academic Press, New York, Section 1.5.

6

Scattering and absorption in sand and dust particle populations

6.1 PHYSICAL MODELLING

6.1.1 Model of the scattering process

The general model for scattering in sand and dust particle populations is essentially the same as that for a population of hydrometeors as described in section 4.6. In view of the size, shape and dielectric properties of sand and dust particles, certain assumptions and approximations can usefully be made. McEwan *et al*. (1985) have outlined an approach to modelling and calculating the scattering properties of such particle populations in the frequency range 3–37 GHz. We will first describe this approach and then highlight problems in extending the frequency range above 37 GHz.

6.1.2 Dielectric properties of dust particles: permittivity

The real part of the relative permittivity ε_r' of dust particles is an important factor in determining depolarization, and the imaginary part ε_r'' is a critical factor in the attenuation caused by the energy absorption. In turn, the moisture uptake is significant for the real part and critical for the imaginary part. Extensive permittivity measurements have been performed on several real dust samples at a variety of moisture regains. (The regain is defined as the mass of absorbed water as a fraction of the sample's mass when oven dried.) Table 6.1 displays such measurements as reported by McEwan *et al*. (1985) for 10 GHz. For full clarification of sample references in the tables, refer to McEwan's report.

6.1.3 Size distribution and density

In a dust storm, all particles have a radius of less than 100 μm. Microwave impairments arising directly from these suspended dust particles are propor-

Table 6.1 Relative permittivity of dust-and-water systems at 10 GHz, measured dry and predicted by Looyenga's (1965) formula at other moisture states

Sample (number/site/year)	ε_r'	ε_r''	ε_r'	ε_r''	ε_r'	ε_r''
	0% regain		2% regain		4% regain	
Balcony	4.09	−0.020				
1/KH/82	5.09	−0.102	6.35	−0.440	7.76	−0.860
2/KH/82	4.81	−0.072	6.03	−0.395	7.40	−0.800
3/KH/82	4.82	−0.061	6.05	−0.384	7.42	−0.787
4/KH/82	4.78	−0.084	6.00	−0.409	7.36	−0.814
1/AT/83	5.28	−0.166	6.57	−0.520	8.00	−0.957
SWAN	4.52	−0.045	5.70	−0.354	7.03	−0.742
Clay G	6.05	−0.178	7.43	−0.561	8.79	−1.03
	6% regain		8% regain		10% regain	
1/KH/82	9.31	−1.36				
2/KH/82	8.91	−1.29				
3/KH/82	8.93	−1.27				
4/KH/82	8.87	−1.30	10.52	−1.88		
1/AT/83	9.58	−1.48				
SWAN	8.49	−1.21	10.11	−1.77		
Clay G	10.65	−1.58	12.48	−2.22	14.45	−2.95

tional to the suspended mass density. It is necessary to find a mean relationship between this quantity and the optical visibility. Since the mass is a measure of total particle volume, and optical extinction relates to total particle geometric area, larger particles have more mass for the same visibility.

The adopted 'best estimate' of the mean relation between the mass density ρ (kg m^{-3}) and visibility z_v (km) given by McEwan *et al.* (1985) is

$$\rho = \frac{56 \times 10^{-9}}{z_v^{1.25}} \tag{6.1.1}$$

where z_v is defined as $15/\alpha_0$ (km), with α_0 the optical attenuation coefficient.

Here the density and the visibility are defined at a reference height of 2 m. The decay of dust density with height is a critical factor in the microwave effects, even for a terrestrial link, and for the elevated slant paths imposes a limit on the integrated effect, or effective path length. A relation in which ρ has a power law decay with height up to a ceiling of 2 km has been considered ($\rho \propto h^{-x}$). The power law index x is approximated as a deterministic function of visibility. Table 6.2 displays the adopted parameters for the calculation of microwave effects.

Table 6.2 Adopted parameters for calculation of microwave effects

z_v (m)	2	10	20	100	1000	5000
$\rho(h = 2$ m) (kg m^{-3})	1.3×10^{-4}	1.7×10^{-5}	7.4×10^{-6}	1.0×10^{-7}	5.6×10^{-8}	7.5×10^{-9}
Effective radius a (μm)	38.3	25.7	21.6	14.5	8.14	5.44
Size distribution	1/KH/75	Inter-polated	1/KH/82	Inter-polated	Inter-polated	Patterson
Assumed wind sheer velocity (m s^{-1})	1.5	1.5	1.5	0.8	0.6	0.4
x (ρ–h index)	0.77	0.57	0.51	0.42	0.33	0.29

6.1.4 Shape

The shape of the particles carried in dust storms is crucial to the investigation of their depolarizing effect, for two distinct reasons. The particles can only cause depolarization if they both are non-spherical and have some degree of systematic orientation as they fall. Particle shape is a factor in determining a variety of aerodynamic and other forces which tend to create or destroy systematic orientation of the following particles. Hence the shape is both a direct and an indirect factor in the effective microwave anisotropy of the dust clouds.

For calculating the microwave effects of a dust particle, it is sufficient to replace it with a best-fit ellipsoid. This is also true for calculating all of the forces affecting particle orientation, except for that which requires a measure of the asymmetry of the particle. Various dust storm samples have been studied and it has been found that the shapes have a weak systematic dependence on particle size, tending to become more spherical for smaller particles. It has also been found that different samples gave broadly similar results. For these reasons, McEwan *et al.* (1985) adopted a ground average 'standard' particle shape for all subsequent calculations. This standard model is size independent and is an ellipsoid with the axis ratios $a_2/a_1 = 0.71$ and $a_3/a_1 = 0.57$, derived from all the aggregated data.

In using this model, the simple but justifiable approximation is being made that for any calculation, whether of microwave scatter or orientation torques, all real particles can be replaced by equivolumic particles of the standard shape. This is a more extreme simplification than just using an ellipsoid model for each particle individually, and it is justified on the assumption that, in any real situation, particles more eccentric than the mean will roughly balance the less eccentric particles.

6.1.5 Orientation

Depolarization will only occur if the particles are both non-spherical and non-uniformly oriented. Turbulence tends to destroy alignment of the particles, while an 'inertial torque' produced by air flow round a particle tends to align it so that its shortest axis is vertical. Electrostatic fields tend to align it so that its longest axis lies along the field. If the particle is more asymmetric than an ellipsoid, an asymmetry torque is produced, tending to make its long axis vertical. A detailed theoretical study of the preceding alignment mechanisms is given by McEwan *et al*. (1985). Figure 6.1 depicts all factors influencing dust particle alignment.

The alignment theory leads to the conclusion that the larger particles will be aligned with their shortest axes vertical. Then there remains the question of whether the azimuths of the longest axes, lying in a horizontal plane, would be random or would also be systematically aligned by either wind shear or

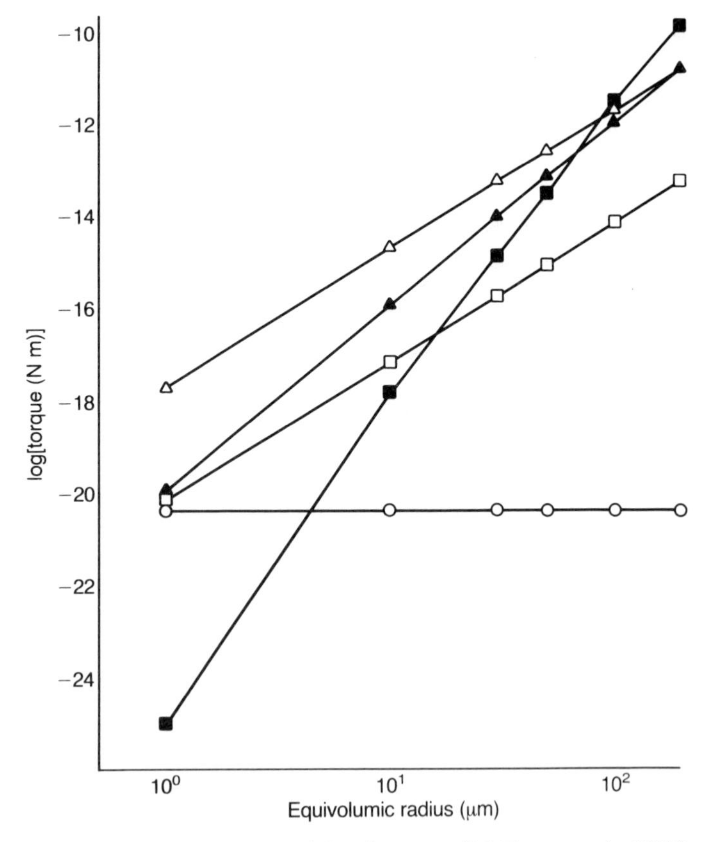

Fig. 6.1 Torques influencing dust particle alignment (McEwan *et al.*, 1985): +, asymmetry; ○, Brownian; ▲, electrostatic; ×, inertial; □, turbulence.

electrostatic forces. The theory of azimuth alignment is of a higher order of complexity and uncertainty than that used to predict the basic 'short axis vertical' state. Nevertheless, it was concluded that the systematic alignment of azimuths was more probable than random alignment for the relevant particle sizes. With this assumption, the preferred azimuth could probably be in any direction at a given time, although some directions might be favoured owing to a prevailing wind sheer.

Synthesizing the alignment theory and experiments, final calculations use a model in which

- all particles with radius $a > 50\ \mu m$ are horizontally aligned,
- 15% of particles between 30 and 50 μm are aligned vertically and the rest horizontally, and
- smaller particles are randomly aligned.

For propagation calculations, we assume that the vertical particles neutralize the anisotropy of an equal fraction of horizontal particles and define an equivalent mass fraction q_m of horizontal alignment with a mass fraction $1 - q_m$ randomly aligned. q_m is approximated as a fixed function of visibility, equal to 50% at $z_v = 1$ km and 71% at $z_v = 10$ m. Table 6.3 shows how q_m has been defined for the various samples which have been studied.

6.1.6 Electromagnetic scattering model

For calculations up to 30 GHz, the scattering properties of dust particles may be calculated with sufficient accuracy by using the Rayleigh approximation. This approximation requires that the scatterer satisfies both

$$\frac{2\pi a}{\lambda} \ll 1 \text{ and } \frac{2\pi a|\varepsilon_r - 1|}{\lambda} \ll 1 \tag{6.1.2}$$

where a is its largest semi-axis. The first condition allows the simplification that the particle may be considered to be placed in a uniform electric field $\boldsymbol{E}$, while the second condition ensures that the phase of the internal fields is not very different from that of the external field. The particle then radiates like a simple oscillating dipole whose dipole moment is proportional to the applied field: $\boldsymbol{p} = \mathcal{A}\mathrm{E}$, where the tensor $\mathcal{A}$ is obtained by an electrostatic calculation with the same applied $\boldsymbol{E}$ field as that in the incident wave.

At 30 GHz, a particle with $a = 0.1$ mm and $\varepsilon_r \approx 9$ (an extreme case for dust) has

$$\frac{2\pi a(\varepsilon_r - 1)}{\lambda} = 0.5$$

Table 6.3 Values of q_m (mass fraction of particles with horizontal alignment)

Sample Mass (%) with	1/KH/82	2/KH/82	3/KH/82	4/KH/75	1/KH/83	1/AT/83	SWAN	Ghobrial	Patterson
$a < 30$ μm	18.5	4.2	6.1	2.2	7.4	3.0	3.0	0.44	50.0
30 μm $< a <$ 50 μm	49.0	9.3	10.5	5.0	26.6	8.6	10.3	1.56	8.0
$a > 50$ μm	32.5	86.5	83.4	92.8	66.0	88.4	86.7	98.0	42.0
q_m (%)	64.9	92.6	90.3	96.0	83.6	94.0	93.4	99.0	47.2

showing that the approximation may approach its limits of validity in the extreme cases of size and permittivity around 30 GHz. However, comparisons of Rayleigh theory with an exact theory have shown it to be quite acceptable up to 37 GHz by Ansari and Evans (1982) and Bashir (1984). Details of the exact calculations need not be given here.

For a fixed shape, permittivity and orientation, the dipole moment induced in a particle by an electrostatic field is obviously proportional to its volume. Thus the Rayleigh approximation applied to dust storms leads to a very convenient simplification: if the shape, orientation and permittivity of the particles are size independent, the specific attenuation or phase shift produced by a storm is dependent only on the total fractional volume of dust in the air, regardless of how it is distributed between particle sizes. Furthermore, if the permittivity is frequency independent, the attenuation and phase shift will vary in direct proportion to frequency.

It should be noted, however, that these statements on frequency and volume dependence of attenuation are based on the assumption that the attenuation is dominated by absorption rather than by scatter. It is quite simple to calculate a scatter contribution to attenuation in the Rayleigh approximation by calculating the total power radiated from the assumed oscillating dipole. If in fact the particle material has negligible loss tangent, there is no absorption and the attenuation is then entirely due to scatter. A check on this point was made by Bashir (1984), who showed that, even for the least lossy dust materials that have been found, the scatter contribution to attenuation is quite negligible up to 30 GHz.

To make any progress with the dust storm calculation, it is necessary to assume that the scattering properties of a dust particle can be approximated as those of an ellipsoid which is a best fit to the actual particle shape. Though no rigorous proof can be given, this seems well justified by physical intuition, where particle size is small compared to wavelength. It also appears likely that the errors will tend to average out over many particles. So now the problem of a homogeneous dielectric ellipsoid placed in an electrostatic field has to be considered. According to Stratton (1941) and Van de Hulst (1957, section 5.32), the dipole moment is given by

$$\boldsymbol{p} = b_1 E_{01} \boldsymbol{u}_x + b_2 E_{02} \boldsymbol{u}_y + b_3 E_{03} \boldsymbol{u}_z \tag{6.1.3}$$

where the coordinate axes have been chosen so that they coincide with the a, b, c axes of the ellipsoid, and

$$b_i = \frac{V\varepsilon_0}{u_i + 1/(\varepsilon_r - 1)}, \; i = 1, 2, 3. \tag{6.1.4}$$

Here V denotes the ellipsoid volume, and u_i is a shape-dependent factor defined as

$$u_1 = \frac{a_1 a_2 a_3}{2} \int_0^{\infty} \frac{dx}{(x + a_1^2) R_s}, \quad R_s = [(x + a_1^2)(x + a_2^2)(x + a_3^2)]^{1/2}, \quad (6.1.5)$$

while the factors u_2 and u_3 are defined in a similar way. It is interesting that

$$u_1 + u_2 + u_3 = 1$$

Van de Hulst states that, in order to determine u_{1-3}, use of equation (6.1.5) is not always necessary. Approximate values can be obtained from

$$u_1 + u_2 + u_3 = 1$$

$$u_1 : u_2 : u_3 = 1/a_1 : 1/a_2 : 1/a_3.$$

6.1.7 Model for a population of particles

If a plane wave of wavelength λ propagates through an assembly of particles which all have the same shape, permittivity and orientation, and the incident electric field lies along the ith axis of the particle, it is then easily shown from the method of Van de Hulst (1957) that the specific attenuation and phase shift are given by

$$\alpha_i \text{ (dB km}^{-1}) = -8.686 \times 10^5 \, \pi \frac{q_v}{\lambda} \, \text{Im} \left[\frac{1}{u_i + 1/(\varepsilon_r - 1)} \right] \quad (6.1.6)$$

and

$$\beta_i \text{ (deg km}^{-1}) = \frac{1.8 \times 10^7}{\pi} \frac{q_v}{\lambda} \, \text{Re} \left[\frac{1}{u_i + 1/(\varepsilon_r - 1)} \right] \quad (6.1.7)$$

where q_v is the fractional volume of space occupied by the particles and λ (cm) is the wavelength. The first expression includes only attenuation due to absorption, the attenuation due to scattering being neglected as already noted. Where the population of particles is not homogeneous in alignment, shape or permittivity, the above equation may of course be re-expressed as sums of contributions from various subgroups of particles.

A further approximation can be made in these calculations, assuming that every particle in a real storm may be replaced by an equivolumic particle whose shape is the mean shape for that size range, i.e. its axis ratios are the mean ratios, derived in the particle shape measurements. This amounts to assuming that

$$\frac{1}{u_i(a_2/a_1, a_3/a_1) + 1/(\varepsilon_r - 1)} = \frac{1}{u_i(\overline{a_2/a_1}, \overline{a_3/a_1}) + 1/(\varepsilon_r - 1)}. \qquad (6.1.8)$$

Although an exact calculation, using individual measured particle shapes, is easy in principle, it leads to much cumbersome computation. Moreover, the approximation is believed to be a good one. Since little dependence of particle shape on size was found, it is also convenient simply to calculate the u_i for the overall 'standard shape' particle which has a size-independent shape. For the standard particle we have $u_1 = 0.22$, $u_2 = 0.34$ and $u_3 = 0.44$.

It is important to note one other approximation that has been made: the particles are modelled as homogeneous, with the permittivity ε_{rh} that was obtained by volume scaling from permittivity measurements on bulk samples. Although this leads to conceptual difficulties when the particles contain both water and minerals, it is not expected to lead to gross errors. In fact all authors on microwave dust effects have adopted this assumption without question.

It is now useful to examine some specimen values of the parameters

$$C_i = \mathrm{Re}\left[\frac{1}{u_i + 1/(\varepsilon_{rh} - 1)}\right] \qquad (6.1.9)$$

and

$$D_i = \mathrm{Im}\left[\frac{1}{u_i + 1/(\varepsilon_{rh} - 1)}\right] \qquad (6.1.10)$$

to be tabulated for various samples. Values of ε_{rh} may be taken from Table 6.4. The sample 3/KH/82 has been selected as representative. This sample has moisture regains of 2%, 4% and 6% at relative humidities of 15%, 60% and 90% respectively. For example, if attenuation has to be calculated, the parameter D_1 is to be used in equation (6.1.6) for an incident wave polarized with its electric field along the longest (a) axis of the particles and D_3 for polarization along the shortest (c) axis.

Table 6.4 Specimen parameters for propagation calculation

Conditions	Typical ε_{rh}	C_1	C_2	C_3	D_1	D_2	D_3
Dry	4.82 – j0.061	2.08	1.66	1.43	0.018	0.011	0.008
15% RH	6.05 – j0.384	2.40	1.86	1.56	0.086	0.052	0.036
60% RH	7.42 – j0.787	2.67	2.02	1.68	0.13	0.077	0.053
90% RH	8.93 – j1.27	2.91	2.16	1.77	0.17	0.092	0.062

6.2 PREDICTION METHODS: TRANSMISSION PARAMETERS

So far, it has been seen that, using the Rayleigh scattering approximation, the microwave attenuation and phase shift per unit distance can be determined from a knowledge of the suspended mass density ρ, but the total effect is proportional to $\int \rho \, dz$.

6.2.1 Effective path lengths through dust storms

The effective length over which a dust storm acts on a given microwave link is a vital parameter in assessing the impairments but is very hard to determine from available data. For high elevation slant paths, the path length is mainly determined by the decay of dust density with height. For low elevation or terrestrial paths, both the height decay and the horizontal structure of the storm are critical. Unfortunately there is almost no hard information on the horizontal extent of the high density dust regions in storms. So a 'dust cell' model has to be used, consisting of a vertical cylinder 10 km in diameter. Density is negligible outside the cylinder. Inside the cylinder it is horizontally uniform but has the power law delay with height up to a ceiling of 2 km. The diameter of the cell is justified by taking the product of typical wind speeds, with typical statements about the duration of very low visibility episodes within dust storm periods. For a given link geometry the effective path length L_e is defined as that length which, multiplied by the dust density ρ_{2m} at a reference height of 2 m within the cell, gives the actual path integral of suspended mass density:

$$\int_{\text{path}} \rho \, dz = L_e \, \rho_{2m}(z_v). \tag{6.2.1}$$

ρ_{2m} is approximated as a deterministic function of z_v using a best-fit relation. L_e clearly depends on the height decay index, which is also approximated as a function of z_v. Table 6.5 displays the worst case effective path lengths through model dust storm cells.

6.2.2 Specific attenuation and phase shift in dust

Specific attenuations and phase shifts at 14 GHz in dust are summarized in Tables 6.6 and 6.7 for reference, as functions of visibility and humidity. They are given for the three cases where the incident electric field is parallel to the longest, shortest and intermediate axes of aligned particles. The differential

Table 6.5 Worst-case effective path lengths L_e (m) through model dust storm cells

Visibility (m)		5000	1000	100	20	10	2	–	–	–
Height decay index α		0.29	0.33	0.42	0.51	0.57	0.77	1.0	1.5	2.0
Slant Paths										
Elevation	Antenna height (m)									
90°	20	366	291	176	108	78	28	9	1	0.2
	2	377	302	186	116	86	34	14	4	2
60°	20	422	337	204	125	96	32	11	1	0.2
	2	436	349	215	134	99	39	16	4	2
45°	20	517	412	249	153	111	39	13	2	0.3
	2	533	428	263	165	122	48	20	5	3
30°	20	731	583	353	216	156	56	18	2	0.4
	2	754	605	372	233	172	68	28	8	4
20°	20	1069	852	516	315	229	81	27	3	0.6
	2	1103	884	544	340	252	99	40	11	6
10°	20	1934	1548	945	583	425	154	52	7	1
	2	1990	1600	995	629	468	188	78	22	12
5°	20	2297	1885	1213	787	593	237	87	12	2
	2	2400	1983	1308	876	677	304	140	44	23
3°	20	2605	2173	1451	975	751	321	126	20	4
	2	2760	2330	1603	1120	887	432	213	74	38
Terrestrial link[a]										
$k = 0.7$[b]		5620	5190	4340	3628	3220	2164	1370	507	187
$k = 4/3$[c]		4330	3856	2974	2293	1928	1082	557	131	31

[a]Length, 40 km; both antenna heights 60 m.
[b]Ray path height at link centre, 14.6 m.
[c]Ray path height at link centre, 35.9 m.

Table 6.6 Attenuation coefficient α (dB km^{-1}) at 14 GHz in dust storms

Visibility (m)		5000	1000	100	20	10	2
Standard dust mass density ρ_2 (gm cm^{-3})		7.5×10^{-9}	5.6×10^{-8}	1.0×10^{-6}	7.4×10^{-6}	1.7×10^{-5}	1.3×10^{-4}
Fraction of mass in aligned particles		46%	50%	59%	66%	71%	84%
Multiplier for sample 'Ghobrial'		11.2	7.5	4.2	2.8	2.4	1.6
Multiplier for sample 1/AT/83		8.9	6.0	3.4	2.2	1.9	1.3
Humidity	Polarization						
0%	*a*	5.4×10^{-5}	4.1×10^{-4}	7.6×10^{-3}	0.057	0.13	1.07
	b	4.2×10^{-5}	3.1×10^{-4}	5.6×10^{-3}	0.041	0.093	0.70
	c	3.7×10^{-5}	2.7×10^{-4}	4.7×10^{-3}	0.034	0.076	0.54
	Differential	1.1×10^{-5}	0.0×10^{-5}	1.0×10^{-3}	0.015	0.038	0.34
15%	*a*	2.6×10^{-4}	1.9×10^{-3}	3.6×10^{-2}	0.27	0.64	5.1
	b	2.0×10^{-4}	1.5×10^{-3}	2.6×10^{-2}	0.19	0.44	3.3
	c	1.7×10^{-4}	1.3×10^{-3}	2.2×10^{-2}	0.16	0.35	2.5
	Differential	5.5×10^{-5}	4.4×10^{-4}	9.4×10^{-3}	0.078	0.19	1.7
60%	*a*	3.9×10^{-4}	2.9×10^{-3}	5.4×10^{-2}	0.41	0.96	7.7
	b	3.0×10^{-4}	2.2×10^{-3}	3.9×10^{-2}	0.29	0.65	4.9
	c	2.6×10^{-4}	1.9×10^{-3}	3.2×10^{-2}	0.23	0.51	3.7
	Differential	8.4×10^{-5}	6.8×10^{-4}	1.4×10^{-2}	0.12	0.29	2.7
90%	*a*	4.9×10^{-4}	3.8×10^{-3}	7.0×10^{-2}	0.53	1.24	10.0
	b	3.6×10^{-4}	2.7×10^{-3}	4.8×10^{-2}	0.35	0.79	5.9
	c	3.1×10^{-4}	2.3×10^{-3}	3.9×10^{-2}	0.28	0.62	4.3
	Differential	1.1×10^{-4}	9.3×10^{-4}	2.0×10^{-2}	0.16	0.40	3.6

Table 6.7 Phase shift coefficient β (deg km^{-1}) at 14 GHz in dust storms

Visibility (m)		5000	1000	100	20	10	2
Standard dust mass density ρ_2 (gm cm^{-3})		7.5×10^{-9}	5.6×10^{-8}	1.0×10^{-6}	7.4×10^{-6}	1.7×10^{-5}	1.3×10^{-4}
Fraction of mass in aligned particles		46%	50%	59%	66%	71%	84%
Multiplier for sample 'Ghobrial'		11.2	7.5	4.2	2.8	2.4	1.6
Multiplier for sample 1/AT/83		8.9	6.0	3.4	2.2	1.9	1.3
Humidity	Polarization						
0%	*a*	4.5×10^{-2}	0.34	6.2	46.1	106.9	837
	b	4.0×10^{-2}	0.30	5.4	39.6	90.8	691
	c	3.8×10^{-2}	0.28	4.9	36.0	82.0	611
	Differential	4.8×10^{-3}	3.9×10^{-2}	0.82	6.8	16.9	153
15%	*a*	5.1×10^{-2}	0.39	7.0	52.8	122.6	962
	b	4.5×10^{-2}	0.34	6.0	44.4	101.9	775
	c	4.2×10^{-2}	0.31	5.5	39.8	90.3	670
	Differential	6.2×10^{-3}	5.2×10^{-2}	1.1	8.8	21.9	128
60%	*a*	5.7×10^{-2}	0.43	7.8	58.5	135.8	1068
	b	5.0×10^{-2}	0.37	6.6	48.4	110.9	842
	c	4.6×10^{-2}	0.34	5.9	43.1	97.8	724
	Differential	7.3×10^{-3}	5.9×10^{-2}	1.2	10.4	25.6	231
90%	*a*	6.1×10^{-2}	0.46	8.4	63.5	147.5	1161
	b	5.3×10^{-2}	0.40	7.0	51.8	118.7	901
	c	4.9×10^{-2}	0.36	6.3	45.8	103.7	766
	Differential	8.4×10^{-3}	6.8×10^{-2}	1.4	11.9	29.4	265

phase shift between these incidence conditions is the main cause of depolarization.

The polarization dependence is defined as follows: first suppose that all particles have all their longest 1 axes fully aligned in one direction and all their shortest 3 axes aligned in an orthogonal direction. Let α_1, α_2, α_3 denote specific attenuations for waves polarized with the electric vector pointing along each of the axis directions. We now assume that, in the real storm, a fraction q_m of the suspended mass is carried by particles which are completely aligned as first stated, while the remaining fraction is in completely randomly aligned particles. The tabulated values are given by

$$\alpha_1' = q_m\alpha_1 + (1 - q_m)\frac{\alpha_1 + \alpha_2 + \alpha_3}{3}, \text{ etc.} \tag{6.2.2}$$

where the values of q_m are estimated in Table 6.3. As also discussed in section 6.1.5, it is believed that the aligned fraction will have their 3-axes vertical and their 1 axes horizontal and aligned in some azimuth direction by wind shear or electrostatic fields. Thus α_3' gives the attenuation for vertically polarized, horizontally propagating waves, while α_1' and α_2' give the attenuation for horizontally polarized, horizontally propagating waves, when the 1 axes are respectively broadside and parallel to the link.

Using the entries in the table, propagation effects can easily be calculated for any link geometry and any alignment state, by using the Rayleigh scattering approximation. (This takes the incident electric field to be resolved into components along the principal axes of the particles.) In the general case, the calculation involves finding the principal planes of the medium, which need not be vertical–horizontal, and finding the attenuation and phase shift for each of these planes.

6.2.3 Prediction of depolarization in a dust particle population

Effective path lengths and specific propagation factors from the previous sections may be used to calculate depolarization. McEwan *et al*. (1985) have made calculations at 14 GHz for circular polarization. In all practical situations the medium may be modelled by neglecting the effect of differential attenuation.

The improvement of linear polarization (XPD) over circular can be approximated (except for very large total differential phase shifts) as $-20 \log(\sin 2\delta)$ dB, where δ is the polarization tilt angle from vertical. It is, however, considered more likely that the particle azimuths are stochastic. McEwan *et al*. have calculated a stochastic improvement factor as the difference between the median linear XPD and the median circular XPD,

Table 6.8 Improvement in XPD for linear polarization compared with circular polarization

Path elevation angle	XPD improvement for the following polarization angles									
	0°	1°	2°	3°	5°	10°	20°	30°	40°	45°
0°	30	29	23	19.5	15	9	3.5	1	0	0
1°	30	29	23	20	15.5	9.5	4	1.5	0	0
3°	29	29	23	20	15	9.5	4	1.5	0	0
5°	24.5	26	23	20	15	9.5	4	1.5	0	0
10°	18.5	19	20	20	15.5	9.5	4	1.5	0.5	0
20°	12	12	12.5	13	14	10	4.5	2	0.5	0
30°	8.5	8.5	8.5	8.5	9	10.5	5.5	3	2	1
45°	5	5	5	5	5	5.5	7	6	4.5	3.5
60°	3.5	3.5	3.5	3.5	3.5	3.5	3.5	4	4	4
90°	3	3	3	3	3	3	3	3	3	3

To obtain linear XPD at given exceedance, add tabulated value (decibels) to circular XPD at that exceedance. Use linear interpolation for intermediate values. For polarization angle $\tau > 45°$, use $90° - \tau$.

again on the assumption of fully aligned particle azimuth which in the long term takes all values equiprobably. This improvement depends in a complex way on elevation and polarization angle. Illustrative values are shown in Table 6.8. The relations between XPD and elevation and visibility are shown in Figs 6.2 and 6.3.

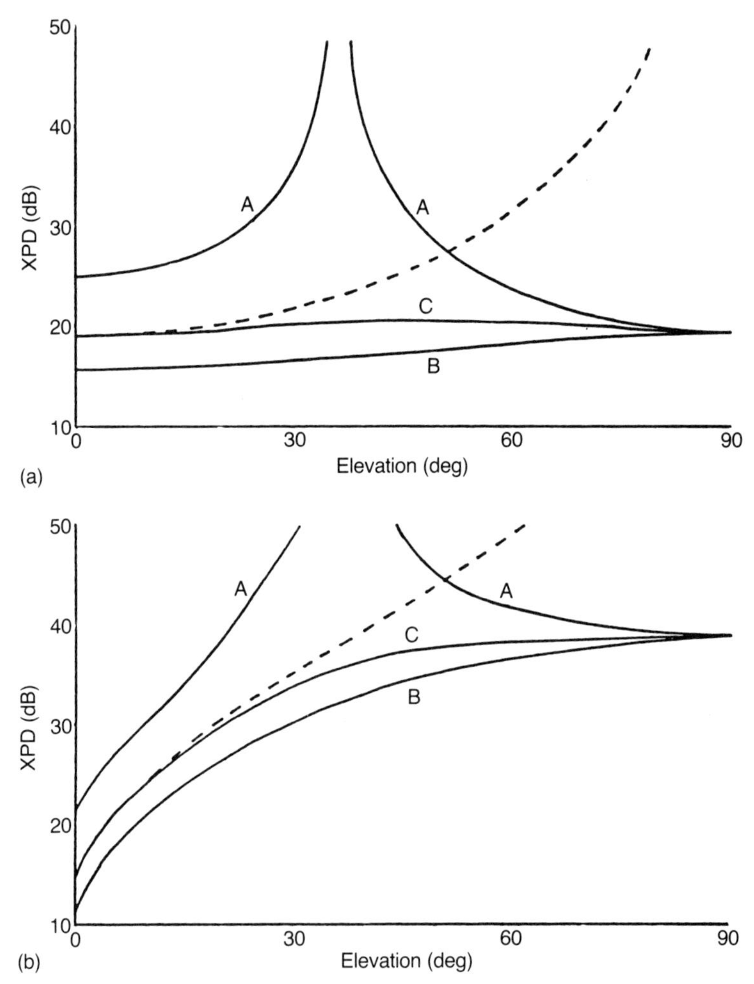

Fig. 6.2 Dependence of XPD on elevation at constant visibility (McEwan *et al.*, 1985): (a) path length held constant at 3° elevation value to show effects of scattering geometry alone; (b) dependence of path length on elevation included (14 GHz circular polarization; humidity, 0%; antenna height, 20 m; visibility, 10 m); ---, non-aligned particle azimuths; ——, values for fully aligned azimuths; curves A, long axis azimuth is link azimuth; curves B, long axes broadside to link; curves C, median values assuming equiprobable azimuths.

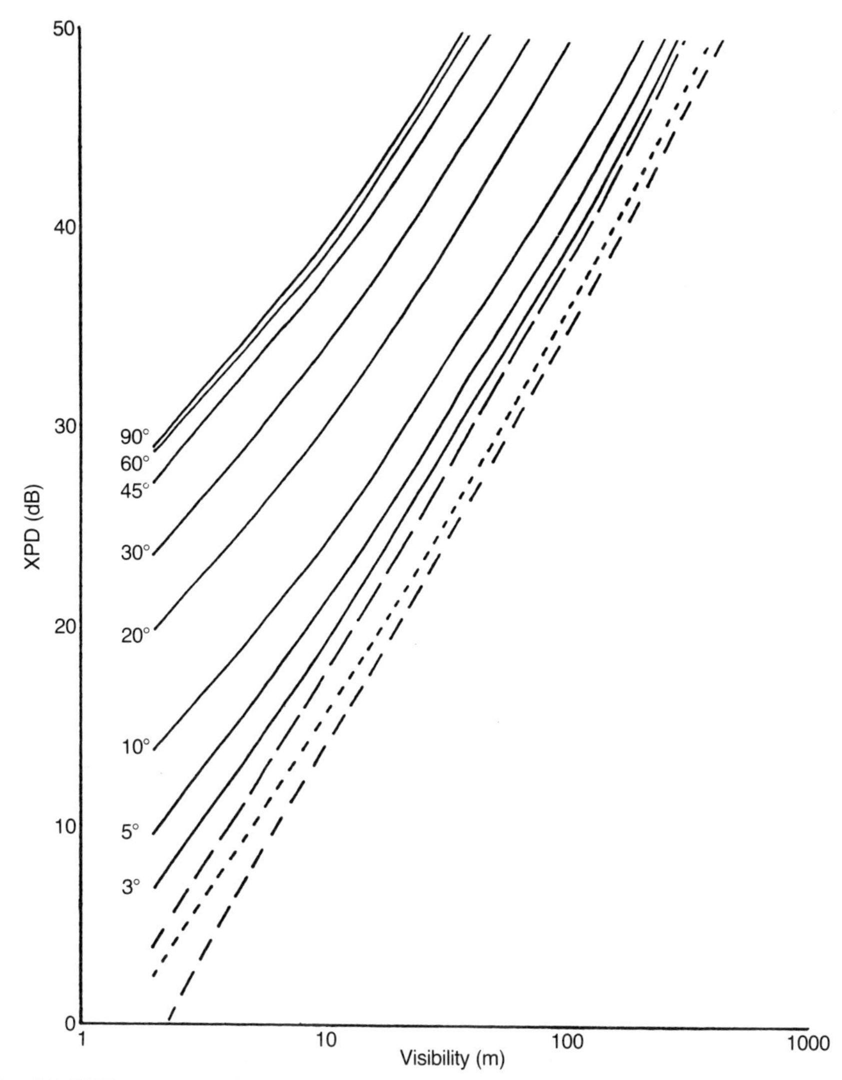

Fig. 6.3 XPD versus visibility (14 GHz circular polarization; fully aligned and equiprobable particle azimuths) (McEwan *et al.*, 1985): ——, 0% humidity for the path elevations shown and 20 m antenna height; — —, 3° elevation, 0% humidity and 2 m antenna height; - - - - -, 3° elevation, 60% humidity and 20 m antenna height; – – –, 3° elevation, 60% humidity and 2 m antenna height.

6.2.4 Consideration of scattering at frequencies > 40 GHz

Figure 6.4 illustrates calculations for a typical dust sample (namely 3/KH/82 in Table 6.3), assuming a 6% moisture regain and an exponential size distribution for particles up to 100 μm as suggested by McEwan *et al*. (1985).

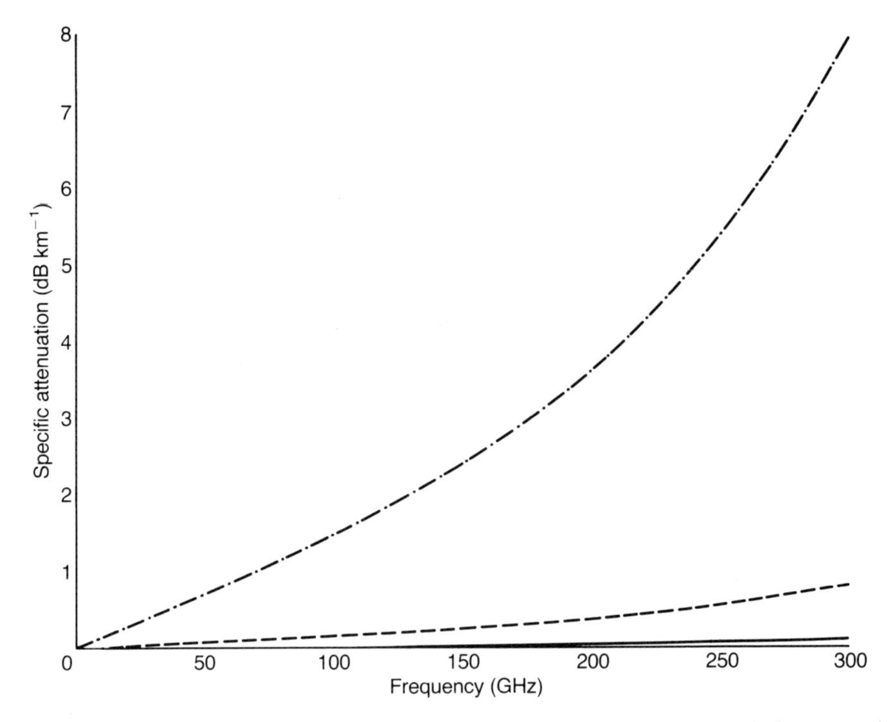

Fig. 6.4 Variation of specific attenuation with frequency for a typical dust sample throughout the millimetre wave range (sample 3/KH/82 in Table 6.3 is selected; permittivity remains at 3.292 − 0.417j for all frequencies): ——, high visibility, > 1000 m; – – –, medium visibility, < 100 m; – · – · –, low visibility, < 10 m.

The dust particles have been modelled as spheres using Mie scattering theory. Calculations have been made for a sparse sample (high visibility), a sample of intermediate concentration and for a dense sample. Generally these calculations gave reassuringly low values of specific attenuation, even for the dense sample at 300 GHz. It should also be noted that the nearly linear relationship between specific attenuation and frequency lends support to the extension of the Rayleigh scattering hypothesis up to frequencies as high as 300 GHz.

The Rayleigh scattering model should thus, in principle, be usable up to at least 300 GHz for the majority of dust particle populations. Nevertheless, uncertainty resides in the electric properties of dust particles in this range, especially in the loss term.

As regards depolarization, differential phase is expected to increase with frequency, and hence in the millimetre wave range we can expect increasing depolarization with frequency. (In contrast to rain, we do not expect a significant reduction in eccentricity with reduction in particle size.)

At some point, probably in the IR, we do of course expect scattering effects to become significant in contributing to extinction. There may also be dielectric resonance effects that need to be taken into account.

6.3 CONCLUSIONS

The models developed by McEwan *et al.* (1985) are adequate in the frequency range up to about 40 GHz, where it is seen that for all practical purposes attenuation from dust storms will be negligible. Depolarization, however, appears to have some significance in that it reaches practically important levels (XPD is ≈20 dB at 0.01% annual exceedance) in unfavourable cases, namely a combination of a poor location, low elevation, circular polarization and high frequency.

At higher frequencies (above 30 GHz) depolarization increases in significance and, as frequency increases into the higher millimetre wave and IR bands, attenuation from enhanced scattering will eventually become significant.

REFERENCES

Ansari, A.J. and Evans, B.G. (1982) Microwave propagation in sand and dust storms. *Proc. IEE F*, **129**(5), 315–22.

Bashir, S.O. (1984) The effects of sand and dust storms on microwave antennas and propagation. PhD Thesis, University of Bradford.

Looyenga, H. (1965) Dielectric constant of heterogeneous mixtures. *Physica*, **31**, 401–6.

McEwan, N.J., Bashir, S.O., Connolly, C. and Excell, D. (1985) The effect of sand and dust particles on 6/4 and 14/11 GHz signals on satellite to earth paths. *University of Bradford Rep. 379* (Final Report to INTELSAT under Contract INTEL-349).

Stratton, J.A. (1941) *Electromagnetic Theory*, McGraw-Hill, New York.

Van de Hulst, H.C. (1957) *Light Scattering by Small Particles*, Wiley, New York.

7

Attenuation by clouds and the melting layer

7.1 INTRODUCTION

The attenuation of clouds assumes a greater significance with increasing frequency as illustrated in Table 7.1. Hence in the millimetre wave range, especially for communications systems with low availability ($\leqslant 99.9\%$ of the year) or for remote sensing systems, the effects of clouds must be quantified for system design.

The melting layer at the base of stratiform clouds begins to make a non-negligible contribution to slant-path attenuation for frequencies above 10 GHz; however, the proportional contribution of the melting layer to attenuation is at its highest in the 10–20 GHz range. The two effects will be considered separately in this chapter.

7.2 PROBLEMS WITH THEORY AND PREDICTION

The theory of scattering from single particles forms the basis for the prediction of attenuation (and backscattering) by clouds. Although, with some exceptions (notably for ice–water mixtures and large irregular particles), the single particle scattering theory gives accurate predictions, application to clouds is not straightforward owing to the complex physical modelling of clouds and the lack of accurate data.

In view of the complexity of cloud modelling, several attempts have been made to relate cloud attenuation to easily available or measurable parameters in a semi-empirical fashion. Some of these will be summarized later. We will first review application of the basic scattering theory to cloud attenuation prediction and then attempt to summarize the current position on relevant aspects of cloud modelling, illustrating some of the uncertainties and giving some example calculations.

In view of the interest in cloud liquid-water characterization for global

Table 7.1 Attenuation by atmospheric gases, clouds and rain (Guissard, 1980)

f (GHz)	A (dB)		
	Gases; $\rho = 7.5\ \mathrm{g\,m^{-3}}$	Clouds; thickness, 1 km; $T = 0\ ^\circ\mathrm{C}$; $\rho = 1\ \mathrm{g\,m^{-3}}$	Rain; height, 1 km; $R = 12.5\ \mathrm{mm\,h^{-1}}$
5	0.0314	0.0230	0.0313
10.7	0.0409	0.106	0.249
15.4	0.066	0.217	0.528
23.8	0.449	0.507	1.114
31.4	0.179	0.859	1.574
90	0.793	4.74	3.17

climate modelling and the relative importance of cloud attenuation in millimetre wave link budget calculations, it is likely that our knowledge in this area will improve significantly over the next few years.

7.3 APPLICATION OF SINGLE-PARTICLE SCATTERING THEORY

7.3.1 Multiple scattering

The question must be raised of the importance (or otherwise) of multiple scattering in clouds in view of the higher particle density than in rain and the occurrence of ice particles.

Ishimaru (personal communication) and Guissard (1980) observe that if the scattering albedo

$$\Phi_0 = \frac{\sigma_s}{\sigma_e} \tag{7.3.1}$$

is $\leqslant 0.5$ then multiple scattering is probably negligible. For water clouds this is fulfilled for the whole millimetre wave frequency range. At 300 GHz the scattering albedo for a 50 μm radius water drop is ≈ 0.05. For ice spheres, Papatsoris (1993) has calculated the critical radii above which the scattering albedo becomes greater than 0.5 over the 60–300 GHz range (Table 7.2). Evidently, for frequencies greater than 100 GHz, the critical radii are quite small; on the other hand, the concentrations of particles observed in some ice clouds such as cirrus are also small (typically less than 1 $\mathrm{cm^{-3}}$) and, therefore, should not give rise to multiple scattering. A question must nevertheless remain on the importance of multiple scattering in ice clouds at frequencies above 100 GHz.

Table 7.2 Critical radii for ice particles above which the scattering albedo becomes greater than 0.5

Frequency (GHz)	60	100	150	300
Critical radius (μm) ($\omega_0 = 0.5$)	70	37	22	9.5

7.3.2 Single-particle scattering

Since it appears that multiple scattering can be neglected, we can adopt the 'coherent scattering' approximation for calculation of cloud attenuation (section 4.4), based on a simple summation of the single particle properties.

For most cloud particles, lying in the size range up to ≈ 50 μm, it is attractive to use the Rayleigh approximation for simplicity of calculation and ease of introducing polarization properties.

The condition for the Rayleigh approximation may be given as

$$\frac{2\pi a}{\lambda} \ll 1 \text{ and } \frac{2\pi a n}{\lambda} \ll 1 \tag{7.3.2}$$

where n is the complex index of refraction of the particle and a is the (equivalent) radius. This condition defines the maximum, i.e. critical, radius of a droplet for which the Rayleigh approximation holds. The critical radius of a droplet for a number of frequencies is shown in Table 7.3.

(a) Approaches to calculation for water clouds

Comparison of the critical radius with the maximum radii of water droplets observed in clouds (models are described in later sections of this chapter) leads to the conclusion that the Rayleigh approximation should hold up to approximately 100 GHz.

The Rayleigh and Mie scattering approaches have been compared for water clouds by Liebe, Manabe and Hufford (1989) for a size distribution from Falcone, Abreu and Shettle (1979) representing the case of a heavy cumulus cloud ($\rho = 1\ \text{g m}^{-3}$, $T = 20$ °C, modal radius of 6 μm but sizes $\geqslant 50$ μm

Table 7.3 Critical radius a_c of droplets (Guissard, 1980)

f (GHz)	λ (cm)	ε_r'	ε_r''	$\|n\|$	a_c (μm)
10	3	60	36	8.36	571
30	1	24	36	6.58	242
60	0.5	11	19	4.69	170
90	0.33	7.2	13.5	3.91	134

contributing < 1% to ρ). The results are shown in Table 7.4. These results suggest that the Rayleigh approximation can be used for frequencies up to 300 GHz for attenuation from water clouds.

(b) Approaches to calculations for ice clouds

Attenuation from absorption in pure ice populations is small owing to very low imaginary component of the refractive index for ice. Nevertheless, for the higher millimetre wave range (above 100 GHz) scattering becomes significant and the attenuation from ice particles is no longer negligible.

Rayleigh–Mie comparisons in the millimetre wave range for ice particles reveal good agreement for the real part of the complex scattering amplitude up to at least 100 GHz, but significant departure for the imaginary part at 100 GHz (A.R. Holt, personal communication). However, in view of the relative insignificance of ice particle absorption at 100 GHz this should not detract from using Rayleigh scattering for small particles up to about 100 GHz.

A major complication for ice particle attenuation calculation is of course the fact that ice particles in clouds are often highly eccentric (Table 13.7) and may be comparatively large. This means that neither Mie nor Rayleigh scattering calculations can be applied for accurate results for a range of particle types. Nevertheless, where absorption is negligible, the Rayleigh approximation can be applied for clouds with small particle populations (for example cirrus) especially for polarization properties (Chapter 10) and, when polarization properties are of secondary interest, a useful assessment of attenuation properties may be obtained from Mie theory over the whole frequency range.

(c) Path length dependence of cloud attenuation

For clear sky conditions the path attenuation is proportional to the distance through the atmosphere. For a uniform cloud it is reasonable to assume that

Table 7.4 Comparison of attenuation values obtained from the Mie scattering theory and Rayleigh approximation for droplets with a radius $a \leq 50$ μm (Liebe, Manabe and Hufford, 1989)

f (GHz)	α_m (Mie) (dB km^{-1} (g m^{-3})$^{-1}$)	α_r (Rayleigh) (dB km^{-1} (g m^{-3})$^{-1}$)	$\alpha_r/\alpha_m - 1$ (%)
300	15.0	15.0	0
400	20.5	20.3	−1
500	26	25	−4
800	41	38	−7
1000	50	43	−14

the attenuation is proportional to the distance through the cloud. Thus the total attenuation should be proportional to the cosecant of the elevation angle. Altshuler and Marr (1989) found that about 80% of the cloud attenuation data and over 90% of the mixed cloud attenuation data were essentially proportional to the distance through the atmosphere, and thus, for these conditions, the attenuation at any angle can be estimated from the zenith attenuation. Figure 7.1 shows the angle and slant-path distance dependence of attenuation for uniform cloud and clear sky conditions.

7.4 CLOUD CHARACTERISTICS AND TYPES

7.4.1 Introduction

For the prediction of the influence of clouds on millimetre wave propagation, information must be found on the following characteristics:

- size distribution of particles;
- particle phase (ice–water);
- temperature of droplets or particles;
- shape of particles;
- vertical and horizontal extent;
- spatial distributions within a cloud type of the first four of these characteristics;
- relationship of the first five of these characteristics to cloud type;
- occurrence of cloud types in a particular climate.

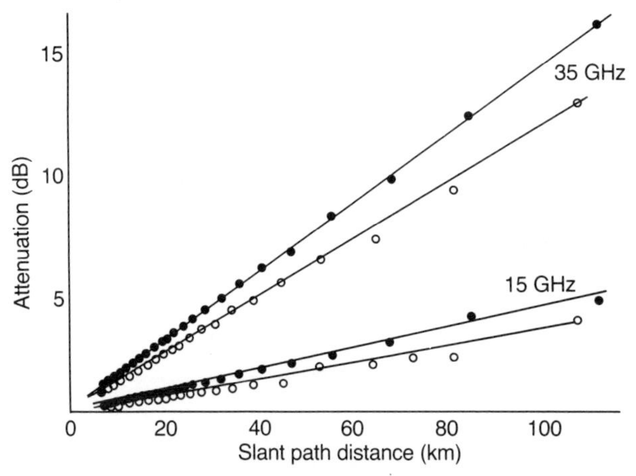

Fig. 7.1 Angle and slant path distance dependence of attenuation for uniform cloud (●) and clear sky (○) conditions, for the same surface humidity $\rho = 13.4\ \text{g m}^{-3}$ (Altshuler and Marr, 1989).

Also, in many approaches to prediction cloud liquid water content is often used as an input parameter.

The characteristics listed above have been exemplified in a number of well-known cloud models, especially those of Carrier, Cato and von Essen (1967), Silverman and Sprague (1970) and Falcone, Abreu and Shettle (1979) (the AFGL models). There are significant variations in the model parameters for a particular cloud type and the data need to be used with caution. Owing to the improving instrumentation with time, greater weight perhaps ought to be given to later results. A useful review of the AFGL models has been given recently by Shettle (1990), including recommendations on situations for which a particular model is most appropriate.

We will examine some of the data available and look at the implications in terms of attenuation for particular instances and examples, before discussing semi-empirical approaches to prediction (section 7.5).

7.4.2 Cloud particle size distributions

The distribution of sizes found in the data of Carrier, Cato and von Essen (1967) are shown in Fig. 7.2 and Table 7.5. All the models adopt a size

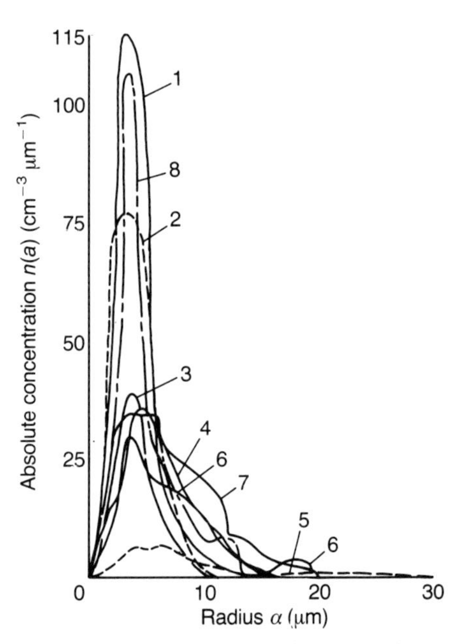

Fig. 7.2 Size distribution of droplets in clouds (Carrier, Cato and von Essen, 1967): curve 1, stratus I; curve 2, stratocumulus; curve 3, fair-weather cumulus; curve 4, stratus II; curve 5, cumulonimbus; curve 6, cumulus congestus; curve 7, nimbostratus; curve 8, altostratus.

Table 7.5 Model cloud drop size and concentration (Carrier, Cato and von Essen, 1967)

Cloud type	M_0 (cm^{-3})	a_{mo} (μm)	a_{mi} (μm)	a_{ma} (μm)	Δr (μm)
Stratus I	464	3.5	0	16.0	3.0
Altostratus	450	4.5	0	13.0	4.5
Stratocumulus	350	3.5	0	11.2	4.4
Nimbostratus	330	3.5	0	19.8	9.5
Fair-weather cumulus	300	3.5	0.5	10.0	3.0
Stratus II	260	4.5	0	20.0	5.7
Cumulus congestus	207	3.5	0	16.2	6.7
Cumulonimbus	72	5.0	0	30.0	7.0

M_0, total concentration; a_{mo}, mode radius, i.e. radius corresponding to the maximum number of droplets; a_{mi}, minimum radius; a_{ma}, maximum radius; Δr, width of the drop size distribution at half-value points.

distribution of the form $n(a) = da^{\alpha} \exp(-ba)$ which poses the question of the truncation of the upper limit on size. Flight instrumentation for drop-size spectral measurements always has an upper limit on size and will not usually detect the presence of superlarge drops in the tail of the distribution. Recently Ajvazyan (1991a, b) and Papatsoris (1993) have examined the effect of the occurrence of superlarge drops on extinction and radar backscatter. Papatsoris (1993) concludes that whilst superlarge drops in the tail of an exponential-type distribution will have negligible effect on attenuation, the radar reflection coefficient may be increased by two orders of magnitude and must be taken into account. These conclusions were based on Mie scattering theory.

7.4.3 Cloud particle temperatures and phases

The temperature of particles in clouds differs from the temperature of the surrounding air usually by only a few tenths of a degree. Only in dense cumulus clouds can the difference rise to more than a degree.

Whilst clouds at a temperature greater than 0 °C can generally be assumed to consist of water droplets, the cloud phase structure for temperatures below 0 °C shows considerable variation. Figures 7.3 and 7.4 show the behaviour of the freezing temperature of droplets in clouds. The presence of supercooled droplets in clouds cannot be attributed to the presence in the droplets of dissolved substances, because the concentration of these is much too low. For droplets to freeze, it is necessary that a solid-phase embryo (which can be created by impurities) forms in these droplets. According to observational

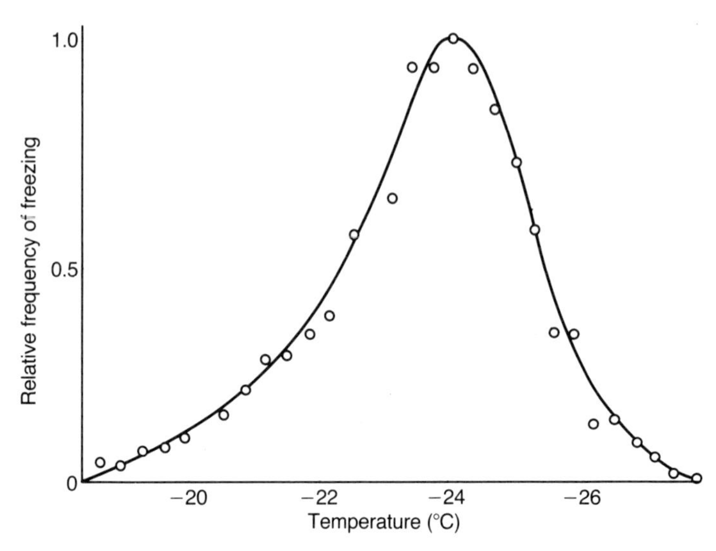

Fig. 7.3 The distribution of freezing temperatures of 1127 drops of 1 mm diameter (Mason, 1971).

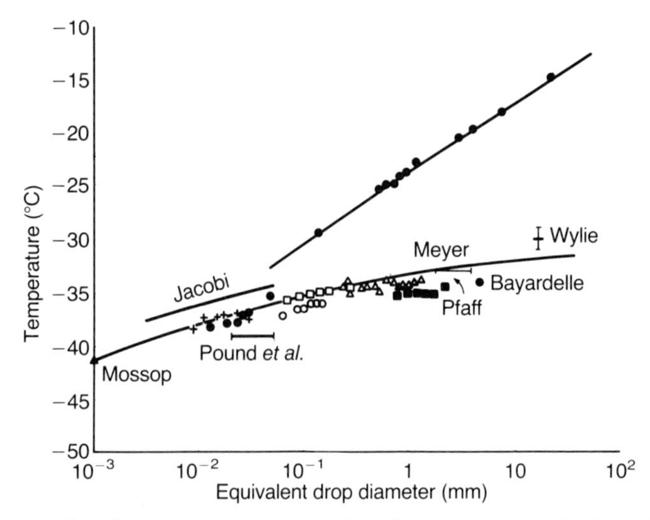

Fig. 7.4 The median freezing temperature, i.e. the temperature below which half of the drops froze, as a function of their equivalent drop diameter obtained by different experimenters (Mason, 1971).

data the greater part of a cloud consists of liquid droplets down to temperatures of −12 to −16 °C (Tverskoi, 1962).

Note also that the presence of liquid drops in clouds above the 0 °C isotherm is not related to the decrease of pressure with altitude, since the triple point for water lies at a pressure of 611.2 Pa and a temperature of 0.0098 °C. This means that the freezing temperature of water remains almost constant to very high altitudes.

Frequencies of occurrence of different phases in clouds over the former Soviet Union (as a function of temperature) are illustrated in Table 7.6 from Feigelson (1984).

7.4.4 Vertical extent of clouds

Data on the typical heights of cloud base and thickness are given in the review by Guissard (1980) and summarized in Table 7.7. Also, Falcone and Abreu (1979) give some characteristics of typical fog and cloud models which are representative of mid-latitude conditions and the liquid water content (Table 7.8).

7.4.5 Water content of clouds

The water content of cloud is defined as the mass of water in the condensed state per unit volume. It refers to the total water content and not only the water in droplet form. It is nevertheless sometimes referred to as the liquid-water content. It is often used as an input parameter to cloud models (section 7.5).

The variation in average water content with height in stratiform and cumuliform clouds is illustrated in Figs 7.5 and 7.6 (from Feigelson, 1984).

7.4.6 Frequency of occurrence of different cloud types

To analyse the impact of clouds on propagation links, information is needed about the frequency of occurrence of different cloud types. For this purpose present weather reports could be used as an input. The advantage of using meteorological parameters is that these are available for a large number of places on earth for long measuring periods (up to 100 years). However, in these reports, cloud cover is not expressed in tenths or eighths, but only the presence or absence of clouds is recorded. This may be relevant information for optical wavelengths, but it is of little use in the microwave range. Moreover, information about cloud type is often lacking.

It is thus difficult to translate the information in weather reports to information for propagation models. Recently Salonen *et al.* (1990) have attempted to apply the middle and low altitude cloud type classification of Warren *et al.* (1986) to synoptic weather reports to give monthly statistics of occurrence for low and middle cloud types.

Table 7.6 Frequencies of occurrence of different phases of clouds over the former Soviet Union (after Feigelson, 1984)

	Frequencies (%) for the following temperature ranges											
Phase of cloud	0 to −2 °C	−4 to −6 °C	−8 to −10 °C	−12 to −14 °C	−16 to −18 °C	−20 to −22 °C	−24 to −26 °C	−28 to −30 °C	−32 to −34 °C	−36 to −38 °C	−40 to −42 °C	−44 to −46 °C
Droplet	84	69	54	37	23	17	10	6	3	2	1	0
Mixed	14	26	35	42	40	35	31	25	17	11	7	6
Crystal	2	5	11	21	37	48	59	69	80	87	92	94

Table 7.7 Characteristics of clouds (Guissard, 1980)

			Microstructure[a]							
				Drops diameter (μm)			Water content[b] ($g\,m^{-3}$)			
Form[a]	Height of base[a] (km)	Thickness[a] (km)	Crystal (C) or Liquid (L)	Mean	Limits	Precipitation[a]	Mean	Maximum	σ_{AV}	Temperature at base[c] (°C)
High clouds										
Cirrus (Ci)	7–10	Hundreds of metres to several kilometres	C			Hardly ever reaches ground				
Cirrocumulus (Cc)	6–8	0.2–0.4	C			None				
Cirrostratus (Cs)	6–8	0.1 to several kilometres	C			Hardly ever reaches ground				

Middle clouds										
Altocumulus (Ac)	2–6	0.2–0.7	L (+C)	5–7	3–24	None	(0.2[c])			−7.8
Altostratus (As)	3–5	1–2	L + C			Precipitation present not to ground	0.15	0.70	0.014	
							(0.3[c])			−10.9
Low clouds										
Stratocumulus (Sc)	0.6–1.5	0.2–0.8	L	5–7	1–60	None	}0.26	0.88	0.06	
Stratus (St)	0.1–0.7	0.2–0.8	L	2–5	1–29	None	(0.23[c])			−8.6
Nimbostratus (Ns)	0.1–1	To several kilometres	C + L	7–8	2–72	Rain or snow	(0.33[c])			4.6
Clouds of vertical development										
Cumulus (Cu)	0.8–1.5 or higher	Hundreds of metres to several kilometres	L	6–11		None				8
				4[d]	0.5–14[d]		0.49	5.5	0.08	
Cuhum		1[c]					(0.15[c])			
Cumed		2[c]					(0.26[c])			
Cocong		4.5[c]					(1.04[c])			
Cumulonimbus (Cb)	0.4–1 or lower	To several kilometres (sometimes reach tropopause)	L C (in upper region)			Torrential rain	0.51	1.72		

[a]Tverskoi (1962).
[b]Fletcher (1962).
[c]Akvilonova and Kutuza (1979).
[d]Deirmendjian (1969).

Table 7.8 Typical characteristics of some fog and cloud models, valid for mid-latitude conditions: liquid water content, bottom and top (Falcone and Abreu, 1979) (these typical values are not always averages, but sometimes 'worst-case' typical values)

Cloud type	ρ ($g\,m^{-3}$)	Bottom (m)	Top (m)
Heavy fog 1	0.37	0	150
Heavy fog 2	0.19	0	150
Moderate fog 1	0.06	0	75
Moderate fog 2	0.02	0	75
Cumulus	1.00	660	2700
Altostratus	0.41	2400	2900
Stratocumulus	0.55	660	1320
Nimbostratus	0.61	160	1000
Stratus	0.42	160	660
Stratus	0.29	330	1000
Stratus–stratocumulus	0.15	660	2000
Stratocumulus	0.30	160	660
Nimbostratus	0.65	660	2700
Cumulus–cumulus congestus	0.57	660	3400

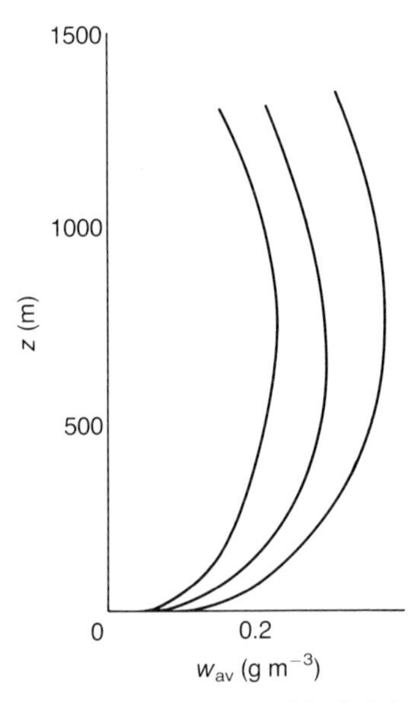

Fig. 7.5 Average variation of water content with height in stratiform clouds for different temperatures (after Feigelson, 1984): curve 1, −10 to −5 °C; curve 2, −5 to +5 °C; curve 3, +5 to +10 °C.

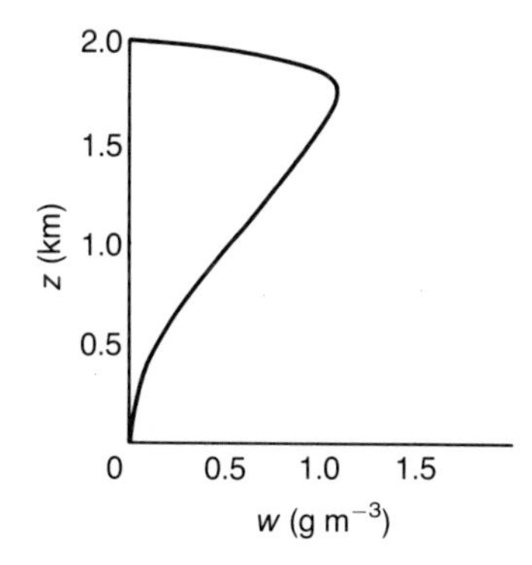

Fig. 7.6 Variation of water content with height in a typical cumuliform cloud (from Feigelson, 1984).

7.5 EXAMPLES OF ATTENUATION CALCULATED FROM VARIOUS CLOUD MODELS

It is of interest to consider the variation in attenuation that results from applying the cloud models published in the literature for specific cloud types. Such variation would represent only one component of uncertainty in any semi-empirical cloud attenuation prediction method.

Figures 7.7 and 7.8 show the application of Mie theory to the cloud models of Carrier, Cato and von Essen (1967) and Falcone and Abreu (1979) (AFGL). The differences are considerable.

7.6 MODELS FOR PREDICTING CLOUD ATTENUATION

7.6.1 Introduction

The following models show three different approaches to the prediction of cloud attenuation.

- Liebe, Manabe and Hufford try to give an exact and complete description of cloud attenuation; this results in a complex model, which requires several parameters which are not easy to measure.
- The models of Guissard, Goldstein and Staelin *et al.* are models based on curve fitting of experiments and scatter calculations; these result in simple semi-empirical models with restricted applicability.
- Altshuler and Marr relate cloud attenuation to the surface absolute humidity, which is easy to measure, but is not a parameter which can completely and exactly describe the cloud attenuation process.

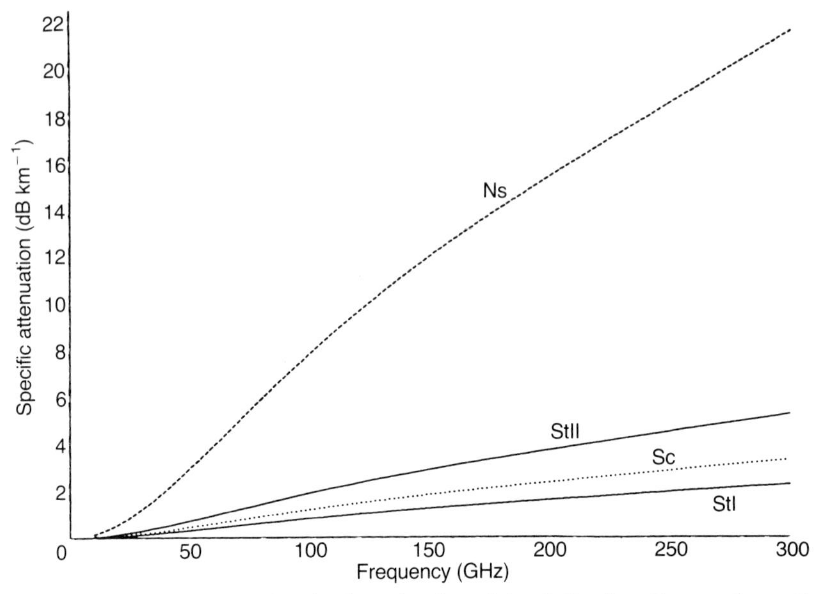

Fig. 7.7 Specific attenuation for the low cloud models of Carrier, Cato and von Essen (1967).

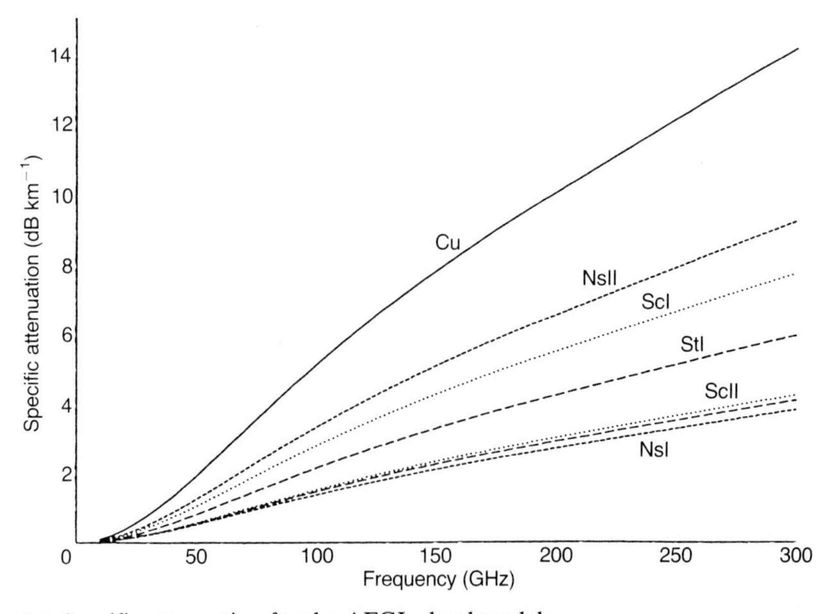

Fig. 7.8 Specific attenuation for the AFGL cloud models.

7.6.2 The model of Liebe, Manabe and Hufford

The basic interactions between radio waves and atmospherically suspended particles for the model of Liebe, Manabe and Hufford (1989) are

$$N = N_0 + N'(f) - \mathrm{j}N''(f) \tag{7.6.1}$$

$$\alpha\,(\mathrm{dB\,km^{-1}}) = 0.1820 f N''(f) \tag{7.6.2}$$

$$\beta\,(\mathrm{deg\,km^{-1}}) = 1.2008 f[N_0 + N'(f)] \tag{7.6.3}$$

$$\tau\,(\mathrm{ps\,km^{-1}}) = 3.3356[N_0 + N'(f)] \tag{7.6.4}$$

where N (ppm) is the complex refractivity, N_0 is the part independent of frequency, $N'(f)$ is the refractivity spectrum, $N''(f)$ is the loss spectrum, α is the power attenuation coefficient, β is the phase shift coefficient and τ is the propagation delay coefficient.

The refractivity model for suspended water drops (SWDs) and suspended ice crystals (SICs) is derived from the Rayleigh approximation of Mie's forward scattering function (Van de Hulst, 1981) which provides both amplitude and phase information independent of the particle size distribution:

$$N_{\mathrm{w,i}}\,(\mathrm{ppm}) = \rho \frac{3}{2\rho_{\mathrm{w,i}}} \frac{\varepsilon_{\mathrm{rw,i}} - 1}{\varepsilon_{\mathrm{rw,i}} + 2} \tag{7.6.5}$$

where ρ is the mass content per unit of air volume, $\rho_{\mathrm{w}} = 1.000\ \mathrm{kg\,m^{-3}}$ is the specific mass of water, $\rho_{\mathrm{i}} = 0.916\ \mathrm{kg\,m^{-3}}$ is the specific mass of ice and $\varepsilon_{\mathrm{rw,i}} = \varepsilon_{\mathrm{r}}'(f) - \mathrm{j}\varepsilon_{\mathrm{r}}''(f)$ is the complex relative permittivity of liquid water or ice. Expressions for the permittivity of liquid water are given in section 13.2.2. The permittivity of ice is derived from Hufford (1989):

$$\varepsilon_{\mathrm{ri}}' = 3.15 \tag{7.6.6a}$$

$$\varepsilon_{\mathrm{ri}}'' = b_{\mathrm{i}}/f + c_{\mathrm{i}}f \tag{7.6.6b}$$

$$b_{\mathrm{i}} = [50.4 + 62(t-1)]10^{-4}\,\mathrm{e}^{-22.1(t-1)} \tag{7.6.6c}$$

$$c_{\mathrm{i}} = (0.663/t - 0.131)10^{-4} + \left(\frac{7.36 \times 10^{-4} t}{t - 0.9927}\right)^2 \tag{7.6.6d}$$

with $t = 300/T$ where T is in kelvin and f is in gigahertz. Haze, fog and non-precipitating clouds are assumed to be composed of spherical water droplets with radii small enough ($a \leqslant 50\ \mu\mathrm{m}$) to keep them suspended in air by microturbulence.

Over a limited temperature range (0–20 °C) the following simple approximation is used to estimate SWD attenuation rates:

$$\alpha(\mathrm{dB\,km^{-1}}) = pbt^{c}. \tag{7.6.7}$$

The frequency-dependent coefficient $b(f)$ and exponent $c(f)$ were determined from fits of the above model predictions to polynomials in powers of frequency (gigahertz) as follows:

$$b(f) = x_0 + x_1 f + x_2 f^2 \tag{7.6.8a}$$

$$c(f) = y_0 + y_1 f + y_2 f^2 \tag{7.6.8b}$$

where x_n and y_n ($n = 0, 1, 2$) are given in Table 7.9.

The accuracy of attenuation predictions, using this model, is limited by uncertainties in the permittivity data for water. New measurements of ε_r are needed above 100 GHz at atmospheric temperatures.

7.6.3 The models of Guissard, Goldstein and Staelin *et al.*

Guissard (1980) has reviewed the models of Goldstein (1965) and Staelin *et al.* (1975, 1976) and proposed an extension. The semi-empirical relations of Goldstein for liquid water and ice crystal clouds in the range 3 GHz $\leq f \leq$ 90 GHz are

$$\alpha\,(\mathrm{dB\,km^{-1}}) = 0.438 s(T, \lambda)\frac{\rho}{\lambda^2} \tag{7.6.9}$$

for liquid-water clouds and

$$\alpha\,(\mathrm{dB\,km^{-1}}) = u\frac{\rho}{\lambda} \tag{7.6.10}$$

for ice crystal clouds, where ρ (g m^{-3}) is the water content, T is the temperature, λ (cm) is the wavelength, $s(T, \lambda)$ is a correction factor and u is a

Table 7.9 Coefficients to estimate SWD attenuation over the temperature range 0–20 °C

f range (GHz)	Coefficient	$n = 0$	$n = 1$	$n = 2$
0–100	x_n	0.00	2.18×10^{-3}	3.90×10^{-4}
100–1000	x_n	−2.24	7.02×10^{-2}	-2.05×10^{-5}
0–250	y_n	9.73	-8.92×10^{-2}	1.73×10^{-4}
250–1000	y_n	−1.12	-3.04×10^{-3}	3.60×10^{-7}

shape factor. The coefficient $s(T, \lambda)$ is unity at $T = 18$ °C and remains of the order of unity for other temperatures. It is given in the form of a table for a number of frequencies and temperatures in the range $0\ °C \leqslant T \leqslant 40\ °C$. The coefficient u depends on the shape of the ice crystals and is of the order of 5×10^{-3} at $T = 0$ °C and 5×10^{-4} at $T = -40$ °C. The water content ρ in liquid water clouds can exceed $1\ \mathrm{g\,m^{-3}}$; in ice clouds it rarely exceeds $0.5\ \mathrm{g\,m^{-3}}$.

Similar formulae with λ-dependent terms slightly different from equations (7.6.9) and (7.6.10) are proposed in Benoit (1968). Equation (7.6.9) is used by Thrane (1975), where the correction factor s is assumed to be independent of λ and is given for some temperatures in the range $8\ °C \leqslant T \leqslant 30\ °C$. Staelin *et al.* (1975, 1976) introduce another similar formula for the frequency range 22–59 GHz:

$$\alpha = 0.438 s(T) \frac{\rho}{\lambda^2} \tag{7.6.11}$$

where $s(T) = 10^{0.0122(291-T)}$ and is independent of frequency. T is given in kelvin. Their results differ from those of Goldstein.

The approximations given above are valid for limited frequency ranges, and it is not evident that they can be used outside these ranges. The correction factor s is not specified for negative temperatures, although clouds of liquid droplets contain water droplets with temperatures well below 0 °C.

Guissard (1980) derives an analytical expression for $s(\lambda, T)$ using the Rayleigh approximation and the Stogryn model for the dielectric constant of water:

$$s(\lambda, T) = \frac{\epsilon_r'' \lambda}{|\epsilon_r + 2|^2}. \tag{7.6.12}$$

Guissard notices that, for frequencies up to 90 GHz and a temperature range which also includes negative temperatures, this function cannot be replaced by a constant. The attenuations calculated by Guissard are compared with those calculated by Gunn and East (1954) in Table 7.10.

7.6.4 The model of Altshuler and Marr

Altshuler and Marr (1988, 1989) propose an attenuation model based on the measurement of the surface absolute humidity. They based their model on measurements carried out at 15 and 35 GHz. From Figs 7.9–7.11 it can be concluded that there is a correlation between attenuation and surface absolute humidity.

Zenith attenuation was fitted to sets of data which were considered

Table 7.10 Comparison of the attenuation computed from equation (7.6.12) derived by Guissard (1980) with attenuations obtained by Gunn and East (1954)

λ (cm)		Attenuation ($dB\ km^{-1}$) for the following temperatures			
		−8 °C	0 °C	10 °C	20 °C
3.2	Guissard	0.103	0.081	0.061	0.047
	Gunn and East	0.112	0.086	0.063	0.048
1.24	Guissard	0.648	0.524	0.397	0.306
	Gunn and East	0.684	0.532	0.406	0.311
0.9	Guissard	1.165	0.962	0.738	0.575
	Gunn and East	1.25	0.990	0.681	0.647

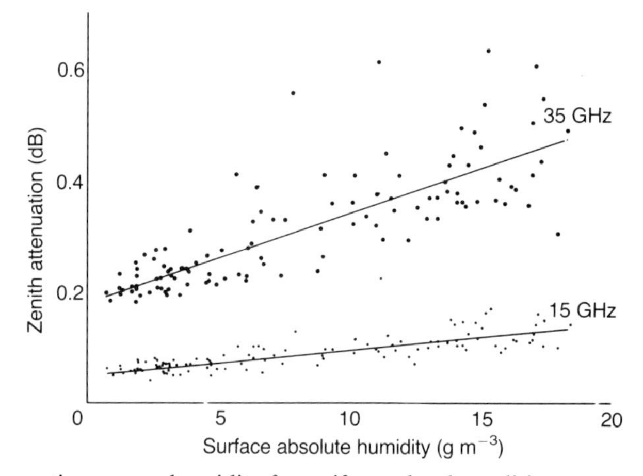

Fig. 7.9 Attenuation versus humidity for uniform cloud conditions.

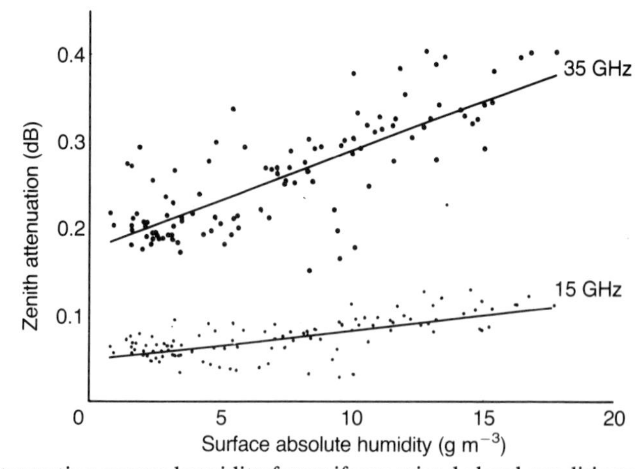

Fig. 7.10 Attenuation versus humidity for uniform mixed cloud conditions.

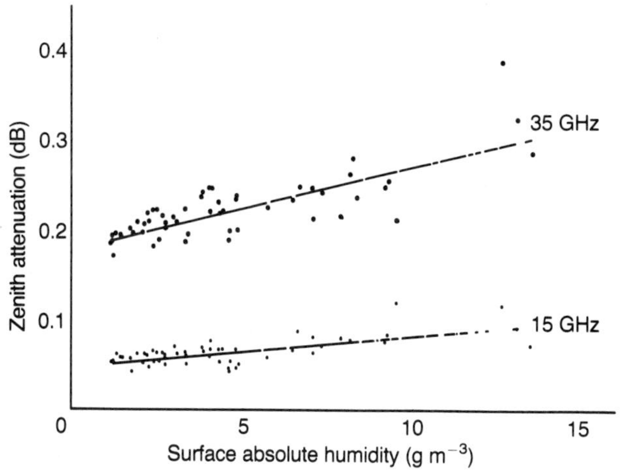

Fig. 7.11 Attenuation versus humidity for clear sky conditions.

representative of uniform cloud cover, mixed cloud conditions and clear sky condition. The regression lines are of the form

$$A = A_\mathrm{d} + A_\mathrm{w}\rho \tag{7.6.13}$$

where A (dB) is the attenuation, A_d is a dry term corresponding to the oxygen attenuation and A_w (dB g^{-1} m^{-3}) is a wet term corresponding to the water vapour and liquid water attenuation.

In order to arrive at a model, they assume that

- Rayleigh approximation is valid for the drops considered,
- the refractive index of water is a function of frequency and temperature as given by Rozenberg (1974),
- cloud attenuation has the same frequency dependence as fog since cloud composition below the 0 °C isotherm is similar to fog, and
- the temperature of the liquid water content of the cloud is 10 °C.

The normalized attenuation in decibels per gram per cubic metre was then calculated at 13 wavelengths in the window regions of the millimetre wave spectrum from 3 mm to 2 cm. The following regression line for attenuation as a function of wavelength was derived:

$$A = -1.20 + 0.0371\lambda + \frac{19.96}{\lambda^{1.15}} \tag{7.6.14}$$

where λ is in millimetres. This wavelength dependence was then fitted to the results of the measurements for zenith attenuation for complete cloud cover at 35 GHz, yielding

$$A_z(\lambda, \rho)\ (\text{dB}) = \left(-0.0242 + 0.00075\lambda + \frac{0.403}{\lambda^{1.15}}\right)(11.3 + \rho). \quad (7.6.15)$$

This expression plotted in Fig. 7.12 and is believed to be valid in the window regions for frequencies from 15 to 100 GHz with a better fit for the higher frequencies. The complete algorithm for cloud attenuation, as a function of elevation, can be written

$$A_{\text{cl}}(\epsilon, \lambda, \rho) = \left(-0.0242 + 0.00075\lambda + \frac{0.403}{\lambda^{1.15}}\right)(11.3 + \rho)D(\epsilon) \quad (7.6.16\text{a})$$

$$D(\epsilon) = \begin{cases} [(a_e + h_e)^2 - a_e^2 \cos^2 \epsilon]^{1/2} - a_e \sin \epsilon & \epsilon \leqslant 8^\circ \quad (7.6.16\text{b}) \\ \operatorname{cosec} \epsilon \text{ (flat earth approximation)} & \epsilon > 8^\circ \quad (7.6.16\text{c}) \end{cases}$$

where ϵ is the elevation angle, $a_e = 8497$ km is the effective earth radius and h_e is the effective height of the attenuating atmosphere.

The effective height is correlated with the surface absolute humidity ρ and can be approximated by (Altshuler and Marr, 1988)

$$h_e = 6.35 - 0.302\rho. \quad (7.6.17)$$

From their measurements, Altshuler and Marr concluded that the mixed cloud attenuations are approximately 85% of the attenuation for full cloud cover at both 15 and 35 GHz. The expression for mixed cloud attenuation was

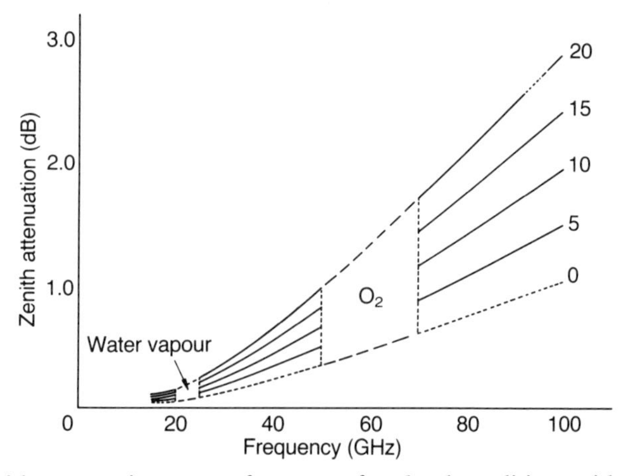

Fig. 7.12 Zenith attenuation versus frequency for cloud conditions with various values of ρ.

thus written as:

$$A(\epsilon, \lambda, \rho)_{\text{mixed cloud}} = 0.85A(\epsilon, \lambda, \rho)_{\text{cl}}. \qquad (7.6.18)$$

It should be noted that the Altshuler and Marr model is only valid in the window regions, it is based on measurements at only two frequencies (15 and 35 GHz) and it can only be used for climates similar to that occurring in the Boston area of the USA. Applicability to other regions of the world would depend on their climatic conditions. The correlation coefficients for the model are nevertheless good; of the order of 0.7–0.8.

7.7 ATTENUATION IN THE MELTING LAYER

7.7.1 Introduction

The assumptions made in the physical models of the melting layer are described in section 13.5. Although some of the models are based on fairly crude assumptions, they nevertheless do lead to a better prediction of attenuation on a slant path than if the presence of the melting layer is ignored.

The most realistic model in physical terms is probably that of Klaassen (1988, 1989) (section 13.5.5) which has more recently been applied directly to attenuation prediction (Klaassen, 1990).

Prediction of attenuation in the melting layer on a slant path requires meteorological input parameters which are relevant and available for prediction. The simplest approach to modelling for prediction purposes is to include an extra attenuation for the zone, derived from on-average thickness (as observed perhaps by radars), and an average specific attenuation as predicted from theoretical models.

A better approach is to add a rain rate related contribution for the melting zone, which is possible from the model of Klaassen (1990), the model of Kharadly and Choi (1988) or the model of Dissanayake and McEwan (1978) as extended by Cherry, Goddard and Hall (1981).

The question of when the melting band contribution should be added to slant path attenuation is not simple to answer, especially in middle latitude maritime climates. The separation of rain types into widespread and showery, whilst useful for modelling of the convective-type structures, can ignore the fact that very frequently convective showers can be embedded in a stratiform structure (Fig. 7.13 (Leitao, 1984)). On a slant path this can mean that the radio path will pass through a small convective shower and simultaneously intercept a melting region in a more stable air mass a short distance away. Even in quite clearly convective storms (thunderstorms) stable melting regions are frequently seen in the tail of the structure, giving rise to the same effect on a slant path.

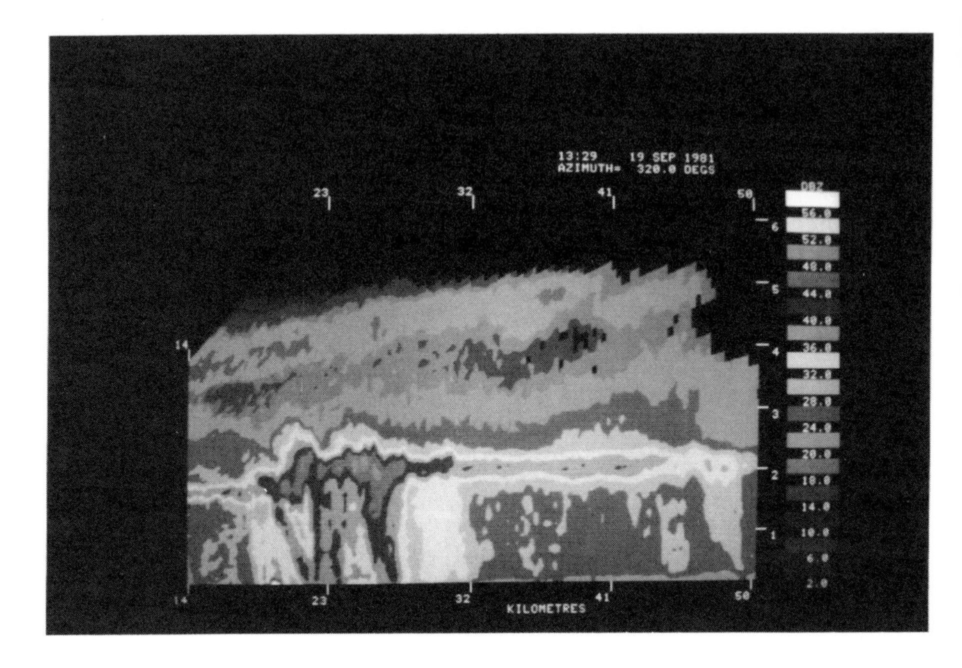

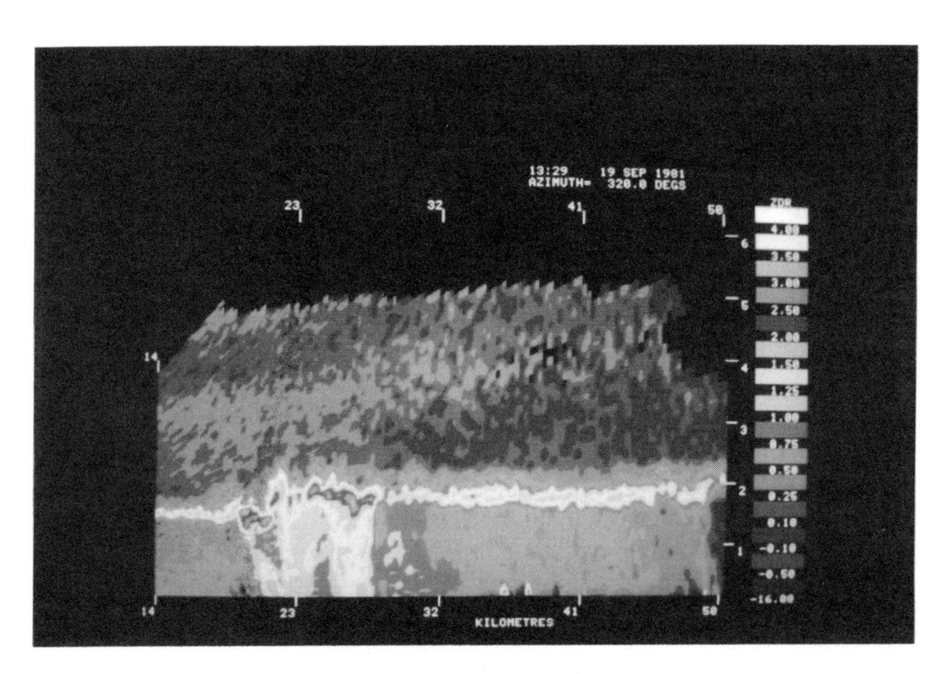

Fig. 7.13 Widespread rain event showing embedded convection (Leitao, 1984): radar scan on RAL dual-polarization radar showing (a) range height–reflectivity and (b) range height–differential reflectivity.

The simplest approach in middle latitudes and especially in maritime climates may thus be to add attenuation contribution for the melting zone, irrespective of the type of rainfall.

7.7.2 Quantification of attenuation in the melting zone

(a) Dissanayake and McEwan model

This model has been successfully applied by a number of authors to give improvements in attenuation prediction on slant paths. Details of the model are given in section 13.5.2. Dissanayake and McEwan (1978) proposed the following relationship between average specific attenuation α and average reflectivity Z through the melting layer:

$$\alpha = b Z^c \tag{7.7.1}$$

where b and c are frequency-dependent parameters. Cherry, Goddard and Hall (1981) and Leitao (1984) modified this relationship to make use of dual-polarization radar data to help predict the bright band attenuation. They used the observed differential reflectivity to characterize M_0 of the drop-size distribution beneath the bright band and rewrote equation (7.7.1) as follows:

$$\alpha = (M_0/8.10^4)^{1-c} b Z^c. \tag{7.7.2}$$

Equation (7.7.2) holds only if the drop-size distribution assumed is exponential.

By making use of dual-polarization radar data in this way with the Dissanayake and McEwan (1978) model, Cherry, Goddard and Hall (1981) and Leitao (1984) were able to report clear improvements in attenuation prediction on a slant path. Nevertheless Leitao (1984) reports a tendency for underestimation (at 11 GHz) when the melting region contribution is noticeable ($>20\%$), suggesting that the modelling of melting particles could be improved.

A development of the Dissanayake and McEwan model to include the polarization properties of the melting zone (using a coated spheroid representation) was given by Dissanayake, Chandra and Watson (1983).

(b) The melting zone model of Jain

Jain (1984) with an improved melting zone model (including coalescence) gives predictions of specific attenuation at various heights throughout the zone and also gives average values of attenuation for a path intercepting the zone at a shallow angle. Table 7.11 gives average values of propagation

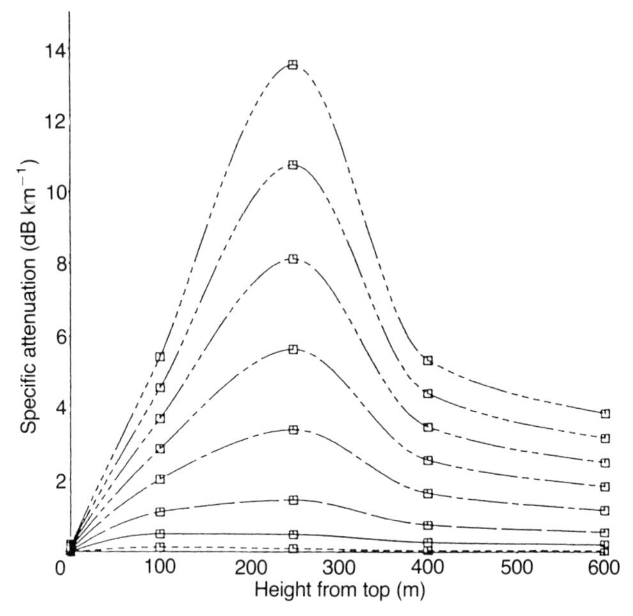

Fig. 7.14 Specific attenuation at 36.25 GHz for horizontal polarization versus height from the top of the melting zone for various rain rates (Jain and Watson, 1985): - - - - -, 0.25 mm h^{-1}; ——, 1.0 mm h^{-1}; — —, 2.5 mm h^{-1}; — -, 5.0 mm h^{-1}; — - —, 7.5 mm h^{-1}; — - - —, 10.0 mm h^{-1}, — - - - —, 12.5 mm h^{-1}; — - - - - —, 15.0 mm h^{-1}.

Table 7.11 Average specific attenuation for horizontal polarization predicted for a path at shallow incidence (R = 10 mm h^{-1} and three-fold coalescence intensity (Jain, 1984)

f (GHz)	α (dB km^{-1})
6.0	0.323
11.575	1.039
15.0	1.531
20.00	2.179
28.56	3.287
30.00	3.421
36.25	4.250

factors over the 6–36 GHz range for a 10 mm h^{-1} rain rate. (It should be noted that Jain's model uses oblate spheroidal particle representations in the bright band from Dissanayake, Chandra and Watson (1983), hence permitting polarization properties to be included.) A brief description of the Jain model and its consequences for microwave terrestrial links is given by Jain and Watson (1985).

(c) The melting zone model of Klaassen

Klaassen (1988, 1989, 1990) in a series of papers has developed a model based on the assumption that a melting particle is a heterogeneous mixture of ice, liquid water and air, with the melted water surrounding the individual ice crystals. Maxwell-Garnett dielectric mixing formulae have been used in the extension given by Bohren and Batten (1982) for elliptical inclusions. This probably gives a more realistic physical representation of a melting particle than the assumption that melted water remains outside the snowflake.

Klaassen's approach has been applied only to spherical particles and thus does not give a prediction of the polarization properties of the medium. Nevertheless in his most recent paper Klaassen (1990) reports a statistical analysis which permits the deduction of an attenuation–rain rate relationship for the melting zone, which is particularly useful for on-average predictions.

7.7.3 On-average predictions

Methods for giving on-average predictions of the attenuation of the melting zone are the subject of current research. Nevertheless we can give some suggestions as to approaches.

In the frequency range 10–14 GHz Leitao and Watson (1986) have shown that the melting zone can simply be modelled as rain, with little error. This means that, by using the height of the 0 °C isotherm (with suitable seasonable and climatic corrections) as the effective rain height, the contribution of the melting zone is then included. For higher frequencies this approach overestimates attenuation in the melting zone.

However, it should be noted that the Leitao and Watson approach for 10–14 GHz includes an empirical compromise which leads to the overestimate if the approach is used for higher frequencies. In particular Leitao and Watson found that the specific attenuation in rain cannot be taken as uniform up to the melting zone, but decreases with height (Chapter 8), resulting in an overestimate of attenuation. This overestimate of attenuation by rain at 10–14 GHz is less than the added attenuation of the melting zone and, by a fortunate coincidence, if the melting zone is modelled as a region of rain, then for 10–14 GHz the excess attenuation of the melting zone (over and above that given by modelling the region as rain) equals or cancels the reduced attenuation from the non-uniform vertical profile of specific attenuation.

For frequencies above 20 GHz, the proportional contribution of the melting zone is reduced, such that at 30 GHz the excess zenith attenuation of the melting zone is less than the underestimate introduced by the non-uniform rain profile. It is thus evident that the concept of 'excess zenith attenuation' for the melting zone should be used with caution.

Taking all these factors into account, an approach to on-average prediction of the attenuation in the melting zone could nevertheless be based on the attenuation rain rate relationship given by Klaassen, modified to give an absolute value of attenuation (not an excess attenuation over that for the region modelled as rain). Attenuation in the rain region would then have to be modelled up to a height related to the 0 °C isotherm less the thickness of the melting zone (also available from Klaassen's work). The model for the rain region beneath should take into account the on-average decrease in rain intensity with height. Such an approach is necessary to develop a physically based model with realistic frequency dependence, in view of the markedly different attenuation–frequency relationships for melting particles and for rain.

7.8 CONCLUSIONS

Prediction of attenuation from clouds and the melting layer in the millimetre wave range is set about by uncertainty. In particular current knowledge of cloud particle size distributions is sufficiently ill defined so as to lead to significant uncertainty in attenuation for specific cloud types.

The melting zone, whilst not making a major contribution to attenuation in the millimetre wave bands in comparison with that observed in the rain beneath, is exceedingly difficult to model and can at present be represented only by simplistic on-average semi-empirical corrections to rain attenuation.

Little is known about the occurrence of the melting layer without rain falling on the ground, although several experiments with radars have confirmed this phenomenon. The existence of such a layer without rain can play an important role in attenuation statistics for communication systems in the high millimetre wave region. Moreover, at this moment insufficient information is available on the frequency of occurrence of the melting layer, especially in combination with different cloud and weather types.

REFERENCES

Ajvazyan, H.M. (1991a) Extreme values of extinction and radar reflection coefficients for mm and sub-mm waves in clouds. *Int. J. Infrared Millimetre Waves*, **12**(2), 157–75.

Ajvazyan, H.M. (1991b) Extinction and radar reflection coefficients of mm and sub-mm waves as a function of cloud drop temperature. *Int. J. Infrared Millimetre Waves*, **12**(2), 177–90.

Akvilonova, A.B. and Kutuza, B.G. (1979) Microwave radiation of clouds. *Radio Eng. Electron. Phys. (Engl. Transl.)*, **23**(9), 12–24 (*Radiotekh. Elektron.*, **23**(9) (1978), 1792–1806).

Altshuler, E.R. and Marr, R.A. (1988) A comparison of experimental and theoretical values of atmospheric absorption at the longer millimeter wavelengths. *IEEE Trans. Antennas Propag.*, **36**(10), 1471–80.

Altshuler, E.R. and Marr, R.A. (1989) Cloud attenuation at millimeter wavelengths. *IEEE Trans. Antennas Propag.*, **37**(11), 1473–9.

Benoit, A. (1968) Signal attenuation due to neutral oxygen and water vapour, rain and clouds. *Microwave J.*, **11**(11), 73–80.

Bohren, C.F. and Batten, L.J. (1982) Radar backscatter by spongy ice spheres. *J. Atmos. Sci.*, **39**, 2623–9.

Carrier, L.W., Cato, G.A. and von Essen, K.J. (1967) The backscattering and extinction of visible and infrared radiation by selected major cloud models. *Appl. Opt.*, **6**(7), 1209–16.

Cherry, S.M., Goddard, J.W.F. and Hall, M.P.M. (1981) Use of dual-polarisation radar data for evaluation of attenuation on a satellite-to-earth path. *Ann. Telecommun.*, **36**(1–2), 33–9.

Deirmendjian (1969) *Electromagnetic Scattering on Spherical Polydispersions*, Elsevier, New York.

Dissanayake, A.W. and McEwan, N.J. (1978) Radar and attenuation properties of rain and bright band, Int. Conf. on Antennas and Propagation, November 1978. *IEE Conf. Publ.*, **169**, 125–9.

Dissanayake, A.W., Chandra, M. and Watson, P.A. (1983) Prediction of differential reflectivity due to various types of ice particles and ice–water mixtures. *IEE Conf. Publ. ICAP*, **219**(Part 2), 56–9.

Falcone, V.J. and Abreu, L.W. (1979) Atmospheric attenuation of millimeter and submillimeter waves. EASCON '79 Record. *IEEE Publ. 79CH 1476-1 AES*, New York.

Falcone, V.J., Abreu, L.W. and Shettle, E.P. (1979) Atmospheric attenuation of millimeter and submillimeter waves: models and computer code. *AFGL Tech. Rep. TR-79-0253* (*AFGL Res. Pap. 679*), US Air Force Geophysics Laboratory, Hanscom Air Force Base, MA.

Feigelson, E.M. (1984) *Radiation in a Cloudy Atmosphere*, Atmospheric Sciences Library, Reidel, Dordrecht.

Fletcher (1962) *The Physics of Rain Clouds*, Cambridge University Press, Cambridge.

Goldstein, H. (1965) Attenuation by condensed water, in *Attenuation of Short Radio Waves* (ed. D.E. Kerr), Dover Publications, New York.

Guissard, A. (1980) Study of the influence of the atmosphere on the performance of an imaging microwave radiometer. *ESTEC Contract 4124/79/NL/DG (SC) Rep.*, Université Catholique de Louvain, European Space Agency.

Gunn, L.K.S. and East, T.W.R. (1954) The microwave properties of precipitation particles. *Q. J. R. Meteorol. Soc.*, **80**, 522–45.

Hufford, G.A. (1989) A model for the permittivity of ice from 0 to 1000 GHz. URSI National Radio Science Meet., Boulder, CO, p. 14.

Jain, Y.M. (1984) Microwave scattering from melting zone particles and oscillating raindrops. PhD Thesis, University of Bradford, Postgraduate School of Electrical and Electronic Engineering.

Jain, Y.M. and Watson, P.A. (1985) Attenuation in melting snow on microwave and millimetrewave terrestrial radio links. *Electron. Lett.*, **21**(2), 68–9.

Kharadly, M.M.Z. and Choi, A.S.V. (1988) A simplified approach to the evaluation of EMW propagation characteristics in rain and melting snow. *IEEE Trans. Antennas Propag.*, **36**(2), 282–96.

Klaassen, W. (1988) Radar observations and simulation of the melting layer of precipitation. *J. Atmos. Sci.*, **45**(24), 3741–53.

Klaassen, W. (1989) From snowflake to raindrop, Doppler radar observations and simulations of precipitation. PhD Thesis, University of Utrecht.

Klaassen, W. (1990) Attenuation and reflection of radio waves by a melting layer of precipitation. *Proc. IEE, Part H*, **137**(1), 39–44.

Lcitao, M.J.M. (1984) Propagation factors affecting the design of satellite communication systems. PhD Thesis, University of Bradford, Postgraduate School of Electrical and Electronic Engineering.

Leitao, M.J.M. and Watson, P.A. (1986) Method for prediction of attenuation on earth–space links based on radar measurement of the physical structure of rainfall. *Proc. IEE F*, **133**(4), 429–40.

Liebe, H.J., Manabe, T. and Hufford G.A. (1989) Millimeter-wave attenuation and delay rates due to fog/cloud conditions. *IEEE Trans. Antennas Propag.*, **37**(12), 1617–23.

Mason, D.J. (1971) *The Physics of Clouds*, Clarendon, Oxford.

Papatsoris, A.D. (1993) Improvements to the modelling of radiowave propagation at millimetre wavelengths. PhD Thesis, University of Bradford.

Rozenberg, V.I. (1974) Scattering and attenuation of electromagnetic radiation by atmospheric particles. *NASA TT F-771*.

Salonen, E., Karhu, S., Jokela, P., Zhang, W., Uppala, S., Aulamo, H. and Sarkkula, S. (1990) Study of propagation phenomena for low availabilities. *Final Rep. ESA/ESTEC*, Contract 8025/88/NL/PR, 316 pp.

Shettle, E.P. (1990) Models of aerosols, clouds and precipitation for atmospheric propagation studies. AGARD Conf. on Atmospheric Propagation in the UV, Visible, IR and MMW Regions and Related Systems Aspects, CP-454, pp. 15.1–15.12.

Silverman, B.A. and Sprague, E.D. (1970) Airborne measurements of in-cloud visibility. National Conf. on Weather Modification, American Meteorological Society, Santa Barbara, CA, April 6–9.

Staelin, D.H., Cassel, A.L., Künzi, K.F., Pettyjoh, R.L., Poon, R.K.L., Rosenkrantz, P.W. and Waters, J.W. (1975) Microwave atmospheric temperature sounding effects of clouds on Nimbus-5 satellite data. *J. Atmos. Sci.*, **32**(10), 1970–6.

Staelin, D.H., Künzi, K.F., Pettyjoh, R.L., Poon, R.K.L., Wilcox, R.W. and Waters, J.W. (1976) Remote-sensing of atmospheric water-vapour and liquid water with the Nimbus-5 microwave spectrometer. *J. Appl. Meteorol.*, **15**(11), 1204–14.

Thrane, L. (1975) Final study report of PAMIRASAT, Vol. 1. Executive Summary (Appendix A, Modelling and simulation of oceanographic radiometry). *Rep. ESS/SS 690* (*ESA CR(P) 839*), British Aircraft Corporation and Technical University, Denmark.

Tverskoi, P.N. (1962) *Physics of the Atmosphere – A Course in Meteorology* (ed. E.S. Selezneva), Gimiz, Leningrad, Israel Program for Scientific Translations, Jerusalem, 1965.

Van de Hulst, H.C. (1981) *Light Scattering by Small Particles*. Dover Publications, New York.

Warren, S.G., Hahn, C.J., London, J., Chervin, R.M. and Jenne, R.L. (1986) Global distribution of total cloud cover and cloud type overland. *NCAR Tech. Note 273*, STR Bolder Co., 199 pp.

8

Attenuation due to rain

8.1 APPROACHES TO PREDICTION

All available methods attempt to relate cumulative statistics of attenuation on a slant path to cumulative statistics of ground rainfall intensity. The coherent scattering model gives quite straightforwardly a relationship between specific attenuation α and rainfall rate R as follows:

$$\alpha\,(\mathrm{dB\,km^{-1}}) = aR^b \tag{8.1.1}$$

where a and b are frequency-dependent constants, b being close to unity. The assumptions on which this key formula is based, and for which a and b are derived, are as follows:

- coherent scattering in a slab of sparsely populated particles;
- an assumed drop size distribution (usually Laws–Parsons);
- an assumed drop shape–size relation (usually Pruppacher–Pitter);
- a model for drop orientation (usually all symmetry axes vertical);
- homogeneous liquid-water drops;
- uniform drop temperature;
- average fall velocity distribution (usually Gunn–Kinzer);
- dielectric properties given by the Ray formula.

Perhaps the most questionable assumption for millimetre wave systems is that of a Laws and Parsons drop size distribution. Measurements at 60 GHz (Humpleman and Watson, 1978) showed that for high availability ($\geqslant 99.99\%$) short terrestrial links, the Laws and Parsons distribution still gives remarkably good average predictions. However, this experiment did not give data relevant to lower availabilities. A more accurate and more comprehensive set of measurements is under way to address this problem (Gibbins, 1992), with initial results interpreted in terms of a gamma distribution.

The increased importance of the smaller drops (diameter < 1 mm) also means that assumptions on fall velocities (Gunn–Kinzer) will probably not apply. All of the other modelling assumptions are believed to be good up to at least 200 GHz for practical radio links. Above 200 GHz, questions on the importance of incoherent scattering are yet to be resolved (section 4.3).

Significantly greater difficulties in prediction are encountered in representing the macroscopic structure of rainstorms, expressing the changes of particle type and number density in various regions of a storm. Problems are also encountered in representing the scattering from ice–water mixtures and irregular ice particles, when these are large compared with the wavelength.

A more detailed critique of approaches to attenuation prediction for millimetre wave systems is given after a description of established techniques for the range 10–30 GHz.

8.2 REPRESENTATION OF RAINSTORM STRUCTURE: SPATIOTEMPORAL AVERAGING

Since radio link predictions are in general made on a long-term average basis, with an interest in relating annual cumulative statistics of ground rainfall intensity measured at a point to the annual cumulative statistics of slant path attenuation, the following assumptions are made by most prediction methods.

1. Rain is vertically homogeneous up to some height, known as the effective rain height h_r.
2. The contribution from ice and melting particles above the effective rain height is either neglected or included in a simple way by adjusting the effective rain height.
3. The average variation of attenuation along a slant path may be expressed by the attenuation along a horizontal projection, multiplied by a simple geometrical correction ($\sec \epsilon$ for $\epsilon \geqslant 5°$).
4. An average relationship may be extracted for the horizontal variations in rainfall intensity, which may be used to make point rainfall to path attenuation transformations.

Leitao and Watson (1986) have tested assumptions 1, 2 and 3 in middle latitudes and found that, although these are reasonable approximations, a systematic tendency for specific attenuation to increase towards the ground was observed, giving an overestimate of attenuation on a slant path of 5–10%. This effect was explained in terms of variation in particle fall velocity owing to increasing pressure towards the ground (as observed by Foote and Du Toit, 1969). If a true physical modelling approach to attenuation prediction is to be taken (as discussed later), then this effect should be taken into account.

The major uncertainties in prediction arise through climatic variability in the effective rain height, and the difficulty and climatic variability associated with assumption 4. The simplest approaches to assumption 4 use an empirical distance dependence of total attenuation A based on a 'reduction factor', i.e.

$$A\ (\mathrm{dB}) = aR^b sL, \tag{8.2.1}$$

where L is the path length through the rain (to the effective height of the rain) and s is a reduction factor, representing the departure from horizontal homogeneity in the rain structure. In such straightforward methods, the reduction factor is often determined empirically in various climates. In more sophisticated approaches model storms are used and, in the most accurate methods, radar data on storm structure are used.

We will describe some of the better-known approaches to prediction under two headings:

- empirical reduction factor methods (section 8.5);
- rain structure based methods (section 8.6).

8.3 RAINSTORM TYPES AND STRUCTURES

The identification of storm types and associated structures permits prediction in different climates, with different fractional occurrences of particular storm types, provided that prediction techniques can be evolved for each storm type.

In a global survey of atmospheric precipitation systems, Houze (1981) concludes that the traditional concepts of stratiform and convective precipitation still seem to be valid, although the two types of structures often occur in complex combinations. Also, it appears that stratiform structures are more prevalent, particularly in tropical regions, than had previously been supposed. However, the review by Houze gives rather limited attention to tropical rainstorm types.

In middle latitudes, the dominant drop growth and precipitation mechanism is believed to involve condensation on ice particles in supercooled water populations (known as the Wegener–Bergeron mechanism). However, even in middle latitudes, radar evidence has been published for the growth of drops by coalescence (Mason, 1971).

In tropical latitudes, Mason states that there may be a good reason to suppose that drop growth by coalescence is the predominant mechanism. Leopold and Halstead (1948) state categorically that 'the majority of tropical convective showers come from clouds which do not reach sub-freezing temperatures'. Also, Virgo (1950) reported in the Bahamas appreciable amounts of rainfall from cloud layers that are a mixture of cumulus and stratocumulus and have little vertical development. He also reports having rain in summer from cumulonimbus clouds that do not reach freezing level.

The occurrence of light rain has been reported from very thin stratiform clouds only a few hundred feet thick in Puerto Rico (Schaefer, 1951), while showers were observed to fall from clouds lying entirely below the 0 °C level at 4300 m (Alpert, 1955).

Whilst we can recognize with Houze that the traditional concepts of stratiform and convective precipitation still appear to be valid, we are left with a very significant problem from the radio-science viewpoint, of quantitative knowledge on the height of liquid water particles and the spatial distribution of ice particles, especially in tropical regions. A semantic problem does appear in this context. Reference to 'stratiform and convective' rain appears to bring with it the underlying assumption that the ice phase is present (as in the review by Houze), although it is known that in tropical climates stratiform and convective structures can exist without an ice phase. For this reason, some authors refer to a third category of rain, 'warm' rain, although, strictly speaking, this simply refers to the coalescence mechanism and not to the overall structure.

In the more rigorous physical structure approaches to rain attenuation prediction (section 8.6.4) an attempt is made to separate the rain types. This is particularly important for millimetre wavelengths.

8.4 EFFECTIVE HEIGHT OF RAIN

8.4.1 Definition

The effective rain height h_r is taken to mean the height of the liquid-water region of a rainstorm. As such it may be equal to (as in many tropical storms) or less than (as in middle latitudes) the vertical storm structure (including ice populations). It, or a variant of it, is used in all approaches to prediction on a slant path.

8.4.2 Relationship to 0 °C isotherm

In middle and high latitudes, a close relationship between the effective height of rain and the altitude h_f of the 0 °C isotherm is observed. In stratiform rain, a melting zone of thickness varying with rain intensity is seen below the 0 °C isotherm. An additional factor is included in rain height models to allow for this (section 8.4.4).

Watson *et al*. (1987) show that for the European region the yearly mean freezing height is given by

$$h_{fy}(\varphi)\ (\mathrm{m}) = 5800 - 72.6\varphi \qquad (8.4.1)$$

where φ is the station latitude in degrees. Since, in many climates, rainfall (1) is seasonal and (2) involves characteristic temperature changes (for example associated with convection), two corrections must normally be made to the yearly mean freezing height to relate it to the effective rain height, namely (1) a seasonal correction and (2) a climatic correction. Watson *et al*. have shown that, in Europe, the mean monthly heights can be predicted from

$$h_{\mathrm{fm}}(\varphi) = h_{\mathrm{fy}}(\varphi) + \Delta h_{\mathrm{fm}} \tag{8.4.2}$$

where Δh_{fm} is given in Table 8.1. If a seasonal rainfall pattern can be related to particular months, then equation (8.4.2) can be used for the seasonal correction. Climatic corrections can vary from +300 m in European regions with strong convective rain to −500 m in maritime regions affected by frontal rains. Watson *et al*. have given an effective rain height contour map including both of these corrections.

8.4.3 CCIR approach

The CCIR (1990, report 564-3) approach includes an empirical effective rain height model as follows:

$$h_{\mathrm{r}}(\mathrm{m}) = \begin{cases} 3000 + 28\varphi & 0° < \varphi < 36° \\ 4000 - 75\,(\varphi - 36°) & \varphi \geqslant 36°. \end{cases} \tag{8.4.3}$$

In the European region, this is not as accurate as the approach described earlier. For $\varphi < 36°$, the approach is believed to be much too simplistic.

Table 8.1 Variation Δh_{m} of difference between yearly average 0° isotherm and monthly average (typical RMS error in these values, ≈150 m)

Month	Δh_{m}
January	−990
February	−895
March	−750
April	−575
May	+85
June	+770
July	+1150
August	+1165
September	+835
October	+340
November	−375
December	−750

8.4.4 Relationship between effective rain height and 0 °C isotherm height

(a) Physical modelling approach

The 0 °C isotherm represents the most readily available modelling input parameter, which will inevitably be used as a starting point in all prediction models. However, this represents the height at which melting starts in stratiform rain, such that lying beneath there is a region of melting with specific attenuation properties quite different to that of rain. Furthermore, the frequency dependence of specific attenuation in the two regions is quite different. A true physical modelling approach would thus separate as far as possible the two regions.

The literature does not yet include such an approach, although what would be required is the rain rate dependence of the thickness of the melting zone, from which a simple implementation of the physical modelling approach is possible.

(b) Empirical adjustments for variations in rain intensity profile

From investigations of attenuation on a slant path at 11 GHz, using a beacon and data from a dual-polarized radar, Leitao and Watson (1986) found the following.

1. The assumption of a uniform vertical attenuation profile leads to an overestimate of attenuation when it is predicted from the horizontal path projection for a slant path up to the true physical rain height (excluding the melting region). That overestimate of attenuation of course increases with frequency.
2. If the melting zone is modelled as a region of rainfall of equivalent thickness, then, at 11 GHz, the attenuation in the region is underestimated.
3. At 11 GHz the underestimate of the attenuation in the melting zone region (2) almost exactly cancels the overestimate in the rain region 1 and attenuation for 11 GHz can thus be accurately predicted using the 0 °C isotherm as the effective rain height.
4. At 30 GHz Leitao and Watson predicted that the overestimate in the rain region would be so great that it would almost equal the attenuation observed in the melting zone at that frequency. Hence the effective rain height should then be the 0 °C isotherm height minus the melting zone thickness with no extra allowance for the melting zone.

Evidently such compromises must be applied with caution and, for a true physical model, the effects of the regions should be separated.

(c) Concept of zenith excess attenuation

Klaassen (1990) has introduced the concept of zenith excess attenuation, representing the excess attenuation observed in the melting zone over and above that when the region is represented in rain, i.e.

$$A_c = (\gamma_{av} - \gamma_R)\,\Delta h \tag{8.4.4}$$

where γ_R is the specific attenuation of rain with the drop size distribution that results from melting, γ_{av} is the average specific attenuation for the melting zone and Δh is the melting zone thickness. This concept also has to be applied with caution, since the frequency dependences of γ_{av} and γ_R are quite different. Since the true height of the rain region beneath ought to be estimated (in order to correct for vertical intensity profile variations), it may be better to take the true physical modelling approach and use only $\gamma_{av}\,\Delta h$ for the melting zone.

(d) Consequences for horizontal paths

While knowledge of the effective height of rain has greatest consequence for slant paths, it should not be entirely ignored on horizontal paths. If, as is often the case, microwave or millimetre links are operated on mountainous sites which lie above the snow line for much of the winter, but not in the summer, then the melting process will contribute excess attenuation for significant periods of the year. This possibility has been considered by Jain and Watson (1985) and Kharadly and Choi (1988).

8.5 EMPIRICAL REDUCTION FACTOR METHODS FOR SPATIAL INHOMOGENEITY

8.5.1 Introduction

It is emphasized that the empirical reduction factor methods have inherent weakness when it is necessary to make predictions in new circumstances, i.e. other climatic conditions or frequency bands to those in which the methods were evolved and tested. They must therefore be applied with caution.

Most of these techniques have been developed in the microwave frequency range. Their relevance to the frequency range above 30 GHz must therefore be questioned. A large number of techniques have been described in the literature. These include Lin (1975, 1977), Crane (1980) and Bothias (1989).

A distinction is also drawn between techniques which use measured spatial rain characteristics as a basis for modelling and those which derive empirically modelled rain cells from comparisons between measured attenuation and single-site rain intensity. The latter are considered to be closely related to the empirical reduction factor techniques (e.g. Stuzman and Dishman, 1982, 1984; and Misme and Waldteufel, 1980).

Only the CCIR (1990) approach, which represents an accumulation of several techniques will be described here.

8.5.2 The CCIR (1990) method

This method has been gradually developed from a large number of studies and represents an average that can be applied with reasonable accuracy for climates and availabilities dominated by showery or convective rain.

The following formula is recommended (CCIR, 1990, report 564-3; Karasawa, 1987) for the attenuation exceeded for 0.01% of the average year:

$$A_{0.01}\,(\mathrm{dB}) = \alpha_{0.01} L s_{0.01} \tag{8.5.1}$$

where the reduction factor $s_{0.01}$ is given by

$$s_{0.01} = \frac{1}{1 + (L/L_0)\cos\varepsilon} \tag{8.5.2}$$

and

$$L_0 = 35\,\mathrm{e}^{-0.015}\,R_{0.01}. \tag{8.5.2}$$

The specific attenuation α is found from

$$\alpha_{0.01}\,(\mathrm{dB\,km^{-1}}) = aR_{0.01}^{b} \tag{8.5.3}$$

where a and b are found at a given frequency from regression relationships for the theoretical scattering coefficients for Laws and Parsons rainfall. $R_{0.01}$ is the rain intensity (integrated over 1 min) exceeded for 0.01% of the average year. The attenuation for other exceedances may be found from

$$\frac{A}{A_{0.01}} = 0.12\;P^{-(0.546+0.043\log P)} \tag{8.5.4}$$

where P is the probability of exceedance, in the range from 0.001% to 1%.

An advantage of this method over the others is that it requires a minimum of information on the rain rate statistics, making it easy to apply.

8.6 RAIN STRUCTURE BASED METHODS

8.6.1 Introduction

Only those techniques which use independently derived data on the horizontal structure of rain will be described in detail. Such data may have been derived from a network of rain gauges or radars. The rain structure may be described in terms of 'rain cells' as an intermediate modelling concept.

Modelled storm approaches which adjust model cell parameters to give a best fit between attenuation and point rainfall intensity are not included here. Such approaches are very limited in their ability to predict attenuation outside of the climatic zone in which they were developed. Other approaches, especially the Excell method (Capsoni *et al.*, 1989) and the Leitao and Watson (1986) method are based entirely on radar data without subsequent empirical correction.

It is to be noted that the concept of a rain cell is in itself questionable, since rainstorms are rarely structurally symmetric, do not necessarily contain a single cell of peak intensity and do not have well-defined boundaries of rain intensity. In some methods (e.g. Leitao–Watson) it is not necessary to extract a rain cell as an intermediate parameter. It may be argued that by adopting a model rain cell (e.g. with circular symmetry and defined cutoff), and by postulating a single cell or single cell plus debris model, the model so produced has become qualitative and strictly non-physical.

The importance of separating rainstorm types for prediction in the millimetre wave region has already been emphasized. This is especially true for low availability ($\leqslant 99.9\%$) systems where widespread rather than convective or showery rain will dominate. Care should therefore be taken in applying any of the rain cell based techniques for availabilities less than 99.9% to ensure that widespread rain is also considered.

8.6.2 Rain structure from ground rain intensity measurements

(a) Harden, Norbury and White

Harden, Norbury and White (1974) give the following statistical profiles for rain cells as observed in the UK in showery rain:

$$R(r) = (0.46\,\mathrm{e}^{-3.37r} + 0.54\,\mathrm{e}^{-0.52r})R_\mathrm{m} \quad \text{(mean profile)} \tag{8.6.1a}$$

$$R(r) = (0.17\,\mathrm{e}^{-2.66r} + 0.83\,\mathrm{e}^{-0.38r})R_\mathrm{m} \quad \text{(mean + } \sigma'' \text{ profile)} \tag{8.6.1b}$$

$$R(r) = \mathrm{e}^{-2.46r}R_\mathrm{m} \quad \text{(mean} - \sigma'' \text{ profile)} \tag{8.6.1c}$$

where $R_\mathrm{m}(r)$ is the peak rain intensity and r (km) is the distance from the cell centre. The authors suggest application to prediction by using 700 mbar wind speed data in association with these profiles.

(b) Morita and Higuti

Morita and Higuti (1971) estimate the spatial correlation of 1 min rainfall from a network of rain gauges and also compare the spatial correlation with that estimated from the autocorrelation and velocity of motion of a shower.

They conclude that the spatial correlation of rain in Japan can be estimated by e^{-ax}, where a is in the range 0.08–0.11 (average 0.093).

Rain intensity distributions are seen to be of the gamma type, and a method is given for estimating the attenuation distribution function using the spatial correlation function. The method requires a lengthy mathematical description, which is given in full by Morita and Higuti. A simplified treatment, in which a reduction factor is extracted for various percentages, is also presented.

8.6.3 Rain cells derived from radar data

(a) Introduction

In deriving an intermediate concept (the 'rain cell') which is inevitably characterized by circular symmetry, these approaches necessarily approximate the physical structure of rain. Nevertheless, techniques so derived, which are independently taken from radar data, are an improvement over the simple reduction factor or empirically modelled rain cell techniques. Two models have become well known in the literature, the Crane (1982) two-component model and the Excell model (Capsoni *et al.*, 1987). Both approximate the rain structure to cells plus debris.

(b) The Crane two-component model

The two-component model (Crane, 1982) is based on the concept of components contributing to attenuation from (1) convective cells and (2) widespread debris. The two components are handled independently, i.e.

$$P(R) = P_c(R) + P_d(R) \tag{8.6.2}$$

where $P(R)$ is the total cumulative probability of rain rate in rain cells and $P_c(R)$ and $P_d(R)$ are the component cumulative probabilities in convective cells and debris respectively.

The path integrated rain rate produced by the convective cell (called 'volume' cell by Crane) is approximated by

$$\int R \, dz = \frac{R_m l_c}{s} \tag{8.6.3}$$

where R_m is the peak rain rate in the cell, l_c (km) is the average dimension of a volume cell and s is a correction factor intended to represent the contribution of the debris close to a cell but outside the region enclosing the −3 dB reflectivity value. The rain rate within a volume cell required to produce the specified path integrated rain rate is modelled by

$$R = \frac{R_m l_c}{l_e} \tag{8.6.4}$$

where l_e is the minimum cell dimension (or minimum path length, L). The correction factor, s, is given by

$$s = \frac{1 + 0.7(L - l_c)}{1 + L - l_c} \quad \text{for } L > l_c \tag{8.6.5}$$

and

$$s = 1 \quad \text{for } L \leqslant l_c. \tag{8.6.6}$$

Under the assumption that only one volume cell can affect the path at any time, the exceedance probability of path integrated rain rate is

$$P_c\left(\int R\,\mathrm{d}z\right) = \left(1 + \frac{L}{l_c}\right)P_c(R) = \left(1 + \frac{L}{l_c}\right)\mathrm{e}^{-R/R_c} \tag{8.6.7}$$

with R_c the average cell rain rate. Since the size and shape of the cells were uncertain, they were approximated by a circular cell with $l_c = \sqrt{5}$ km. A similar expression was used for the debris region, i.e.

$$P_d\left(\int R\,\mathrm{d}z\right) = \left(1 + \frac{L}{l_d}\right)P_d(R) \tag{8.6.8}$$

with l_d the average dimension of debris.

$P_d(R)$ results in a normal distribution function of ln R, with R obtained from equation (8.6.3) where l_e again is $\min(L, l_d)$. The scale length parameters l_c and l_d are assumed to apply universally to all climatic regions.

Crane gives in his paper a straightforward step-by-step approach to apply his model and tabulates values for the parameters of $P_c(R)$ and $P_d(R)$ for various climate zones in Europe and the USA.

(c) The Excell model

The exponential cell model (Excell model) (Capsoni *et al*. 1987; Paraboni personal communication) uses a population of cells with exponential profile and rotational symmetry, derived from the best fit to radar data, and where $R(l)$ is the rainfall rate at a distance l from the centre given by

$$R(l) = R_m \exp\left(-\frac{l}{l_0}\right) \tag{8.6.9}$$

where l_0 is the cell radius, i.e. the distance at which the intensity falls by a factor 1/e.

The cumulative probability for attenuation is given by

$$P(A) = \int_{\ln R_e}^{\infty} \exp\left(\frac{-l_{0,\min}}{\bar{l}_0}\right)\left[\frac{1}{2}\ln^2\left(\frac{R_m}{R_e}\right) + \frac{h/\sin\epsilon}{4\pi\bar{l}_0}\ln\left(\frac{R_m}{R_e}\right)\right][-P^3(R_m)]\,\mathrm{d}(\ln R_m) \tag{8.6.10}$$

where $l_{0,\min}$ is the minimum cell radius that can cause an attenuation A, $\bar{l}_0$ is the conditional average cell radius, $P^{(3)}(R_m)$ is the third-order derivative of the point rain rate intensity cumulative distribution $P(R)$ with respect to $\ln(R)$, evaluated for $R = R_m$ and changed in sign, R_e is the equivalent rain rate, given by

$$R_e^b = \frac{A - aR_d^bL}{aL} \quad (\text{for } A > aR_d^bL), \tag{8.6.11}$$

where a and b are the parameters in equation (8.1.1) and R_d is the rain rate in debris around the rain cell, assumed to be 4 mm h^{-1}, and L is the total path length ($L = h/\sin\epsilon$ for a slant path).

8.6.4 Rain structure taken directly from radar data

(a) Introduction

Here the approach is to take rain intensity structure directly from radar data and to produce point-to-path transformations without introducing an intermediate modelling parameter (i.e. the rain cell). Only dual-polarization radars can do this with sufficient accuracy and then the frequency range over which useful conclusions can be drawn is limited (perhaps up to 50 GHz) owing to the inability of the radar to give sufficient weight to the smaller drops (Jones and Watson, 1992).

(b) Leitao and Watson

Leitao and Watson (1986) take conversion factors of point rainfall to path attenuation directly from dual-polarization radar measurements, with no hypotheses on the size or shape of rain cells. Only dual-polarization rain data give sufficiently quantitative estimates of rain rate and attenuation; single-polarization radars only give qualitative estimates. No empirical 'best fit' adjustments are made retrospectively.

The cumulative distribution of attenuation $A(P)$ is calculated from the cumulative distribution of rain intensity at a point $R_p(P)$ with the following equation:

$$A(P)\,(\mathrm{dB}) = \alpha(P)\frac{h_r - h}{\sin\epsilon} \tag{8.6.12}$$

where the effective height h_r of rain can be taken from the CCIR method or, more accurately, from the method described by Watson *et al.* (1987), given in section 8.4.2. Also,

$$\alpha(P) = \begin{cases} a\left[\dfrac{R_p(P)}{20}\right]^{b_1} & \text{for } R_p(P) < 20\ \mathrm{mm\,h^{-1}} \\ a\left[\dfrac{R_p(P)}{20}\right]^{b_2} & \text{for } R_p(P) \geqslant 20\ \mathrm{mm\,h^{-1}}. \end{cases} \tag{8.6.13}$$

The factors a, b_1 and b_2 are given by

$$a_{h,v} = \frac{C_{1h,v}}{1 + 0.01L\cos\epsilon} \tag{8.6.14}$$

$$b_{1h,v} = C_{2h,v}\log(1 + L\cos\epsilon) + C_{3h,v} \tag{8.6.15}$$

$$b_{2h,v} = C_{4h,v}\log(1 + L\cos\epsilon) + C_{5h,v} \tag{8.6.16}$$

in which $L\cos\epsilon$ is the horizontal projection of the slant path and the constants $C_{1\text{-}5}$ are dependent on frequency and weather type (climate). Values for widespread and showery rain in the UK are given by Leitao and Watson, and interpolation formulae for intermediate frequencies are given in the original reference.

Also, to obtain a, b_1 and b_2 for the required elevation and polarization, the following relationships are used:

$$a = |a_h + a_v + (a_h - a_v)\cos^2\epsilon\cos 2\delta|/2 \tag{8.6.17}$$

$$b_1 = |a_h b_{1h} + a_v b_{1v} + (a_h b_{1h} - a_v b_{1v})\cos^2\epsilon\cos 2\delta|/2a \tag{8.6.18}$$

$$b_2 = |a_h b_{2h} + a_v b_{2v} + (a_h - a_v C_{2v})\cos^2\epsilon\cos 2\delta|/2a \tag{8.6.19}$$

where δ is the polarization angle with respect to the horizontal.

In applying the techniques to predictions in the 11–14 GHz range, Watson *et al.* (1987) used only the convective rain model. For application to predictions at frequencies above 20 GHz and especially for availabilities $\leqslant 99.9\%$, both convective and showery rain should be considered.

(c) Extension to millimetre-wave prediction

Recently Salonen *et al*. (1990), Karhu *et al*. (1993), and Watson, Glover and Hu (1993) have taken the Leitao–Watson model and proposed techniques for applying it to millimetre wave prediction. Salonen *et al*. (1990) have taken just the widespread rain model, but Karhu *et al*. (1993) and Watson, Glover and Hu (1993) have proposed using a combination of widespread and showery. The method is yet to be fully developed and tested against 20–30 GHz propagation data.

8.7 CLIMATIC ZONES

At least three factors are significantly climate dependent, so that they require climatic correction. These are

- rainfall intensity,
- effective rain height, and
- rain horizontal structure.

Currently, climatic zones are defined for rainfall intensity (CCIR, 1990). In addition, it may be useful to define effective rain height and a rain horizontal structure factor for each climatic zone or alternatively separate climatic zones for each of these factors. Prediction may then be achieved by overlaying all three aspects.

The climatic zones of rainfall intensity given in the CCIR have been gradually evolved over some 20 years and are the subject of continual review, especially in the tropical regions where data are still sparse.

8.8 RELEVANCE OF MODELS AND METHODS TO PREDICTION FOR 30–300 GHz

All of the prediction techniques developed to date (with the exception of those mentioned in section 8.6.4(c)) concern the situation where convective rain dominates the yearly attenuation statistics. The more accurate and sophisticated approaches, e.g. the Excell model or the radar method of Leitao and Watson, depend on radar observations of storm structure, with the inherent sixth-power dependence of backscattering on drop size mitigating against accurate use above 30 GHz. The dual-polarization radar approach (on which the Leitao–Watson method is based) does better in this regard; nevertheless it is well recognized that the eccentricities and differential reflectivity of the smaller drops are less than expected, so that this technique fails for drop sizes of ≈1 mm diameter or less (Jones and Watson, 1993).

Nevertheless, as has been suggested in section 8.6.4(c), the application of both widespread and showery rain models combined via some index of convection should allow the techniques to be applied up to at least 50 GHz.

In the frequency range at and above 30 GHz, especially for low availability link design, it becomes important to consider appropriate methods for combining the components of rain attenuation, including clouds, the melting zone and rain. Also a component of background water vapour must be added, which during rainfall may be regarded as saturated.

For the higher frequencies improved drop-size modelling is also essential (Gibbins, 1992). Unfortunately as the smaller drops become more important with increasing frequency it becomes more difficult to relate attenuation to ground rainfall intensity. With the emphasis on the smaller particles and clouds, interest should perhaps switch to 'top side prediction': based on remote sensing of clouds tops from space, the building up of a global database and appropriate models for ground rainfall prediction (e.g. Atlas, 1989; Nakajima and King, 1990).

As for the height of rain, evidently as frequency increases into the high millimetre wave range, attenuation from ice particles will become significant and the effective rain height should not be modelled only up to the 0 °C isotherm.

REFERENCES

Alpert, L. (1955) Notes on warm cloud rainfall. *Bull. Am. Meteorol. Soc.*, **36**, 64.

Atlas, D. (1989) The estimation of convective rainfall by storm area integrals and height. Proc. URSI Symp., France, pp. 4.1.1–4.1.8.

Bothias, L. (1989) Prediction of the attenuation due to rain. CCIR IWP 5/2 Document 89/5.

Capsoni, C., Fedi, F., Magistroni, C., Paraboni, A. and Pawlina, A. (1987) Data and theory for a new model of the horizontal structure of raincells for propagation applications. *Radio Sci.*, **22**(3), 395–404.

CCIR (1990) Propagation in non-ionized media, in *Recommendations and Reports of the CCIR*, Vol. V, Düsseldorf.

Crane, R.K. (1980) Prediction of attenuation by rain. *IEEE Trans. Commun.*, **29**, 1717–33.

Crane, R.K. (1982) A two component rain model for the prediction of attenuation statistics. *Radio Sci.*, **17**, 1371–87.

Foote, G.B. and Du Toit, P.S. (1969) Terminal velocities of raindrops aloft. *J. Appl. Meteorol.*, **8**, 249–53.

Gibbins, C.J. (1992) Studies of millimetre-wave propagation and related meteorology over a 500 m path. Proc. URSI Commission F Int. Symp., Ravenscar, pp. 10.6.1–10.6.5.

Harden, B.N., Norbury, J.R. and White, W.J.K. (1974) Model of intense convective rain cells for estimating attenuation on terrestrial millimetre radio links. *Electron. Lett.*, **10**(23), 483–4.

Houze, R.A. (1981) Structures of atmospheric precipitation systems. *Radio Sci.*, **16**(5), 671–90.

Humpleman, R.J. and Watson, P.A. (1978) Investigation of attenuation by rainfall at 60 GHz. *Proc. IEE*, **125**, 85–91.

Jain, Y.M. and Watson, P.A. (1985) Attenuation in melting snow on microwave and millimetre-wave terrestrial radio links. *Electron. Lett.*, **21**(2), 68–9.

Jones, S.M.R. and Watson, P.A. (1992) A study of attenuation and countermeasures in millimetre-wave point-to-multipoint networks. Proc. URSI Int. Symp., Ravenscar, pp. 10.2.1–10.2.8.

Jones, S.M.R. and Watson, P.A. (1993) Attenuation and countermeasures in millimetre-wave point-to-multipoint networks. *Radio Sci.*, **28**(6), 1057–69.

Karasawa, Y. (1987) An improved prediction method for rain attenuation on earth–space paths. CCIR IWP 5/2 Document TG2-87/3.

Karhu, S., Salonen, E., Hgvmen, R., Uppala, S. and Poiares Baptista, J.P.V. (1993) Prediction of rain attenuation at low availabilities using models and data of widespread and convective rains. *IEE Conf Publ.*, **370**(1), 56–9.

Kharadly, M.M.Z. and Choi, A.S.V. (1988) A simplified approach to the evaluation of EMW propagation characteristics in rain and melting snow. *IEEE Trans. Antennas Propag.*, **36**(2), 282–96.

Klaassen, W. (1990) Attenuation and reflection of radio waves by a melting layer of precipitation. *Proc. IEE H*, **137**(1) 39–44.

Leitao, M.J. and Watson, P.A. (1986) Method for prediction of attenuation on earth–space links based on radar measurement of the physical structure of rainfall. *Proc. IEE F*, **133**(4), 429–40.

Leopold, L.B. and Halstead, M.H. (1948) First trials of the Schaefer–Langmuir dry-ice cloud seeding technique in Hawaii. *Bull. Am. Meteorol. Soc.*, **29**, 525.

Lin, S.H. (1975) A method for calculating rain attenuation distributions on microwave paths. *Bell Syst. Tech. J.*, **54**(60), 1051–86.

Lin, S.H. (1977) Nationwide long term rain rate statistics and empirical calculations of 11 GHz microwave rain attenuation. *Bell Syst. Tech. J.*, **56**, 1581–604.

Mason, D.J. (1971) *The Physics of Clouds*, Clarendon, Oxford.

Misme, P. and Waldteufel, P. (1980) A model for attenuation by precipitation on a microwave earth–space link. *Radio Sci.*, **15**(3), 655–65.

Morita, K. and Higuti, I. (1971) Statistical studies in electromagnetic wave attenuation due to rain. *Rev. Electr. Commun. Lab. Jpn*, **19**(7–8), 798–842.

Nakajima, T. and King, M.D. (1990) Determination of the optical thickness and effective particle radius of clouds from reflected solar radiation measurements. Part 1: theory. *J. Atmos. Sci.*, **47**(15), 1878–93.

Salonen, E., Karhu, S., Jokela, P., Zhang, W., Uppala, S., Aulamo, H. and Sarkkula, S. (1990) Study of propagation phenomena for low availabilities. *ESA Contract Rep. 8025/88/NL/PR*.

Schaefer, V.J. (1951) Report on cloud studies in Puerto Rico. *Project Cirrus Rep. 11*.

Stutzman, W.L. and Dishman, W.K. (1982) A simple model for the estimation of rain induced attenuation along earth–space paths at millimetre wavelengths. *Radio Sci.*, **17**, 1465–76.

Stutzman, W.L. and Dishman, W.K. (1984) Corrections to a simple model for the estimation of rain induced attenuation along earth–space paths at millimetre wavelengths. *Radio Sci.*, **19**, 946.

Virgo, S.E. (1950) Tropical rainfall from cloud which did not extend to the freezing level. *Meteorol. Mag.*, **79**, 237.

Watson, P.A., Glover, I.A. and Hu, Y.F. (1993) Models of hydrometeors at ground level and aloft for application to centimetre and millimetre wave propagation. Proc. Olympus Utilisation Conf., Seville, *ESA Conf. Publ.*, WPP-60, pp. 647–53.

Watson, P.A., Leitao, M.J., Sathiaseelan, V., Gunes, M., Poiares Baptista, J.P.V., Potter, B.A., Sengupta, N., Turney, O. and Brussaard, G. (1987) Prediction of attenuation on satellite–earth links in the European region. *Proc. IEE F*, **134**(6), 583–96.

9

Depolarization by rain

9.1 INTRODUCTION

By using orthogonal polarizations, two independent information channels using the same frequency band can be transmitted over a single link. This technique is used in satellite communication systems effectively to increase the available spectrum. While the orthogonally polarized channels are completely isolated in theory, some degree of interference between them is inevitable, owing to less than theoretical performance of spacecraft and earth station antennas and depolarizing effects on the propagation path. The main sources of this depolarization at millimetre wave frequencies are hydrometeor absorption and scattering in the troposphere. This chapter reviews the depolarization caused by rain.

Most prediction methods now in use are semi-empirical. They are modelled on general theories for rain with randomly canted raindrops and with parameters chosen to give approximate agreement with experimental data. A two-parameter model based on a Gaussian distribution of raindrop canting angles is commonly assumed.

9.2 THEORY OF DEPOLARIZATION

9.2.1 Introduction

Depolarization occurs because of the lack of spherical symmetry of the drops (the top and bottom are flattened) along with their tendency to have a preferred orientation. The effects of a medium filled with rain on a wave propagating through it are dependent on the orientation of the electric field vector with respect to the preferred drop orientation.

Depolarization occurs as a result of the differential attenuation and differential phase shift of the components of the electromagnetic wave along two symmetry axes of the medium. No depolarization occurs if the polarization state of the electromagnetic wave is such that the polarization vector can be projected on only one symmetry axis.

Above 15 GHz, the depolarizing effect of rain is mainly produced by differential attenuation; below 10 GHz, it is mainly caused by differential phase shift. Nevertheless, in both cases, rain depolarization and attenuation are relatively well correlated.

9.2.2 Polarization state of a wave

The most general case of polarization is the elliptical polarization. In the following, the definitions of NASA (1989) are used, in general. The electric field vector $\boldsymbol{E}(t)$ is composed of two sinusoidal components, which have different amplitudes $|E_x|$ and $|E_y|$ and a phase difference $\phi = \arg(E_y/E_x)$:

$$\boldsymbol{E}(t) = \mathrm{Re}(\boldsymbol{E}\,\mathrm{e}^{\mathrm{j}\omega t}) = \mathrm{Re}[(\boldsymbol{u}_x E_x + \boldsymbol{u}_y E_y)\,\mathrm{e}^{\mathrm{j}\omega t}]$$

$$= \boldsymbol{u}_x |E_x| \cos(\omega t) + \boldsymbol{u}_y |E_y| \cos(\omega t + \phi) \qquad (9.2.1)$$

where $\boldsymbol{u}_x$ and $\boldsymbol{u}_y$ are unit vectors in the x and y directions, ω is the angular frequency, t is the time and the phase is taken relative to the phase of E_x.

The polarization ellipse is fully described by the angle between the ellipse major axis and the x axis and the ratio of the major and minor axes of the ellipse (Fig. 9.1). This ratio is the magnitude of an important parameter, known as the axial ratio, and is the ratio of the maximum to the minimum magnitude of the electric field vector. The axial ratio's sign is assigned to be positive if the vector rotation has a left-hand sense and negative for rotation with a right-hand sense (Fig. 9.2). Linearly polarized waves have an infinite axial ratio; circularly polarized waves have an axial ratio of ± 1, corresponding

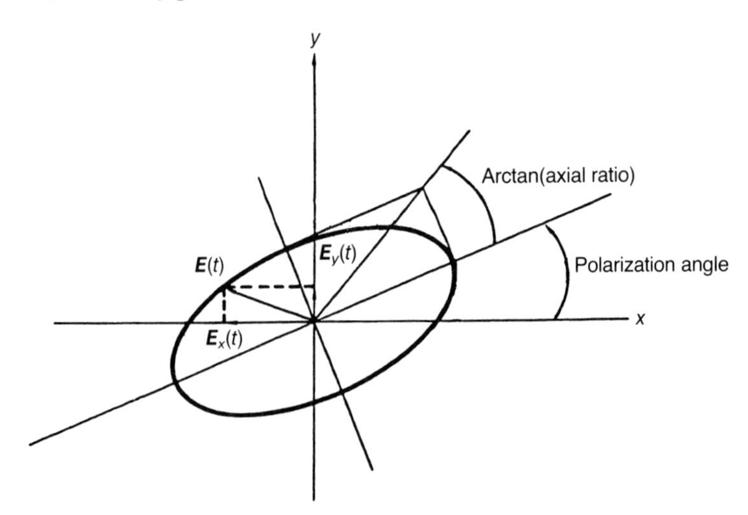

Fig. 9.1 Polarization ellipse (NASA, 1989).

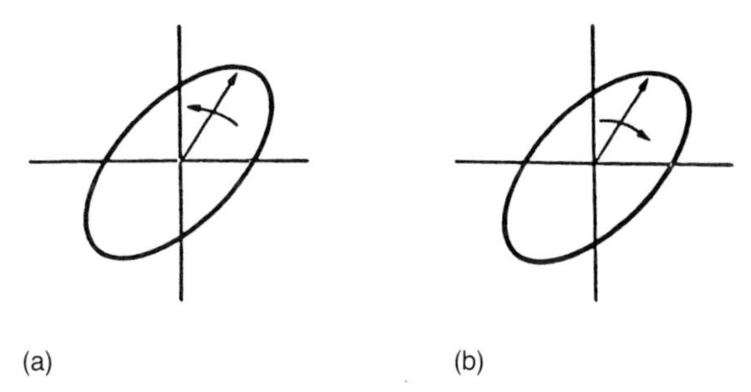

Fig. 9.2 Definition of sign of axial ratio (propagation out of paper, i.e. wave approaching) (NASA, 1989): (a) right-hand sense, axial ratio < 0; (b) left-hand sense, axial ratio > 0.

to either left-hand circular polarization (LHCP) or right-hand circular polarization (RHCP).

There are also other methods to specify the polarization state (Stutzman, 1977).

9.2.3 Cross-polarization discrimination and isolation

Let E_{ij} be the magnitude of the electric field at the receiver that is transmitted in polarization state i and received in an orthogonal polarization state j $(i, j = x, y)$. E_{xx} and E_{yy} denote the copolar waves and E_{xy} and E_{yx} refer to the cross-polar waves (Fig. 9.3).

Cross-polarization discrimination XPD is the ratio (in decibels) of the power in the copolarized wave to the power in the cross-polarized wave that was transmitted in the same state:

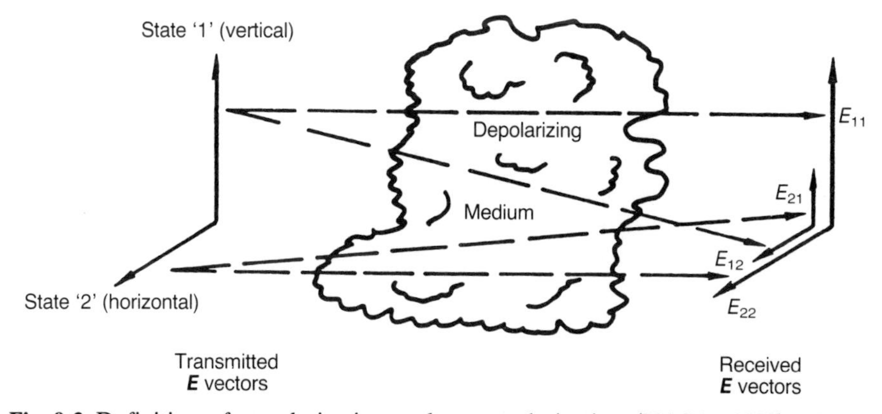

Fig. 9.3 Definition of copolarization and cross-polarization (NASA, 1989): E_{11} and E_{22} represent copolarized waves and E_{12} and E_{21} represent cross-polarized waves.

$$\text{XPD} = 20\log\left|\frac{E_{xx}}{E_{xy}}\right|. \tag{9.2.2}$$

Cross-polarization isolation XPI is the ratio (in decibels) of the power in the copolarized wave to the power of the cross-polarized wave that is received in the same polarization state:

$$\text{XPI} = 20\log\left|\frac{E_{xx}}{E_{yx}}\right| \tag{9.2.3}$$

XPI is more meaningful to system engineers, since it directly gives the carrier-to-interference ratio in a received channel. However, XPD is the parameter that is most easily measured in experiments, and hence the quantity XPD is often used in the literature.

It has been shown (Watson and Arbabi, 1973) that XPI and XPD are equal if the hydrometeors that are responsible for the depolarization have certain symmetry properties. The geometric models that have been used for raindrops and ice crystals have the necessary symmetry, so XPI = XPD in theory. In practice, it has been found that there is no significant difference between XPI and XPD.

9.2.4 Geometry

Raindrops are assumed to be oblate spheroids. As the size of the raindrops increases, their shape departs from spherical and is similar to that of oblate spheroids, with an increasingly pronounced flat base, in which a concave depression develops for very large drop sizes (Pruppacher and Pitter, 1971) (section 13.2).

Raindrops may also be inclined from the horizontal (Saunders, 1971). The phenomenon of raindrop canting may be explained in terms of vertical wind gradients (Brussaard, 1976) (section 13.2).

On the basis of measurements below 40 GHz, it appears that depolarization due to rain is primarily a coherent phenomenon (Oguchi, 1981; Olsen, 1981). In the generally accepted model of coherent propagation, depolarization occurs as a result of differential attenuation and differential phase shift between two eigendirections of the medium. The angle of these directions with respect to the horizontal and the vertical (both mutually perpendicular and perpendicular to the propagation path) is often termed the effective canting angle (CCIR, 1990, report 722-3). If the effective canting angle (given by the path-averaged raindrop canting angle) is 0°, no depolarization occurs for horizontal and vertical polarizations on a horizontal path. Since the effective canting angle is near zero, the cross-polarization isolation is better on horizontal paths using vertical and horizontal polarizations than on such paths using tilted linear polarizations or circular polarization.

The relative contribution of differential attenuation and phase shift is different at different frequencies. Differential phase shift appears to be the dominant factor in rain-induced depolarization at frequencies below 10 GHz (for which significant depolarization is possible even for low values of attenuation), and differential attenuation becomes increasingly important at higher frequencies.

Figure 9.4 shows the geometry for a dual linearly polarized (LP) wave incident on an oblate spheroidal raindrop. The raindrop is at an arbitrary orientation with respect to the direction of propagation of the wave. The orientation is specified by the angle of incidence ξ between the propagation vector and the symmetry axis of the raindrop. The plane containing the propagation vector and the raindrop's symmetry axis will be referred to as the plane of incidence.

$E_x \boldsymbol{u}_x$ and $E_y \boldsymbol{u}_y$ are electric field vectors of two orthogonal LP waves. They are in a plane normal to the propagation vector, and each one can be resolved into two components: a component in the plane of incidence and a component normal to it. Parallel to these components, two symmetry axes are defined, labelled 1 and 2 in Fig. 9.4. The projection of the raindrop into the plane containing the electric field vectors is an ellipse and axes 1 and 2 are its minor and major axes. Figure 9.5 shows this ellipse and how the electric fields are resolved into their 1 and 2 components. The total electric field magnitudes in the 1 and 2 directions are given by

$$\begin{bmatrix} E_1 \\ E_2 \end{bmatrix} = \begin{bmatrix} \cos\theta & -\sin\theta \\ \sin\theta & \cos\theta \end{bmatrix} \begin{bmatrix} E_x \\ E_y \end{bmatrix} = \boldsymbol{R} \begin{bmatrix} E_x \\ E_y \end{bmatrix} \tag{9.2.4}$$

where the canting angle θ is the angle between the axes x and 1.

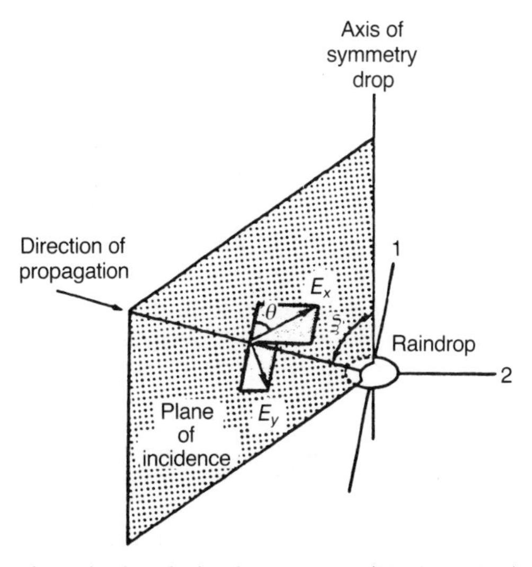

Fig. 9.4 Geometry for rain depolarization analysis (NASA, 1989).

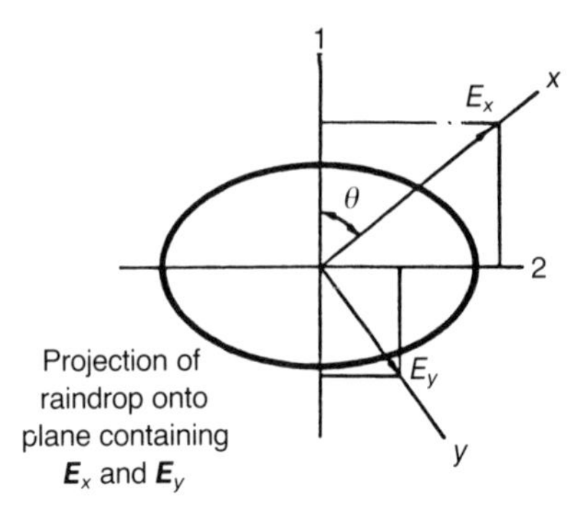

Fig. 9.5 Resolution of electric fields into components 1 and 2 (NASA, 1989).

Consider a region of space containing many identical raindrops with the same orientation distributed throughout it. According to scattering theory, the effect of many scatterers along the propagation path of a wave is to multiply the electric field vector by a transmission coefficient T (section 4.6.3) of the form

$$T = \mathrm{e}^{-(\alpha \ln(10^{0.05}) - \mathrm{j}\beta\pi/180)L} \tag{9.2.5}$$

where α is the specific attenuation of power flux density of the wave (the factor $\ln(10^{0.05})$ allows α to be expressed in decibels (power) per kilometre), β is the specific phase lag of the wave (the factor $\pi/180$ allows β to be expressed in degrees per kilometre), where the phase lag is in addition to the normal free-space phase retardation of the fields, and L (km) is the path length through the scattering region. A region filled with oblate spheroidal raindrops may be characterized by two transmission coefficients: T_1 applied to the 1 component of the electric field and T_2 applied to the 2 component.

Denoting the fields of the wave incident on the scattering region by a subscript i and the fields of the wave propagating from the region by s (for 'scattered'), it is possible to write

$$\begin{bmatrix} E_{1\mathrm{s}} \\ E_{2\mathrm{s}} \end{bmatrix} = \begin{bmatrix} T_1 & 0 \\ 0 & T_2 \end{bmatrix} \begin{bmatrix} E_{1\mathrm{i}} \\ E_{2\mathrm{i}} \end{bmatrix} = \boldsymbol{T} \begin{bmatrix} E_{1\mathrm{i}} \\ E_{2\mathrm{i}} \end{bmatrix}. \tag{9.2.6}$$

Now the coordinate rotation $\boldsymbol{R}$, defined in equation (9.2.4), can be applied to obtain an expression for the effect of the scattering medium on the field vectors in the directions x and y (Fig. 9.6):

$$\begin{bmatrix} E_{x\mathrm{s}} \\ E_{y\mathrm{s}} \end{bmatrix} = \boldsymbol{R}^{-1}\boldsymbol{T}\boldsymbol{R} \begin{bmatrix} E_{x\mathrm{i}} \\ E_{y\mathrm{i}} \end{bmatrix} = \boldsymbol{T}' \begin{bmatrix} E_{x\mathrm{i}} \\ E_{y\mathrm{i}} \end{bmatrix}. \tag{9.2.7}$$

Chu (1974) gives expressions for the components of the rotated transmission matrix $\boldsymbol{T}'$ in terms of α and β. The expressions for XPD are

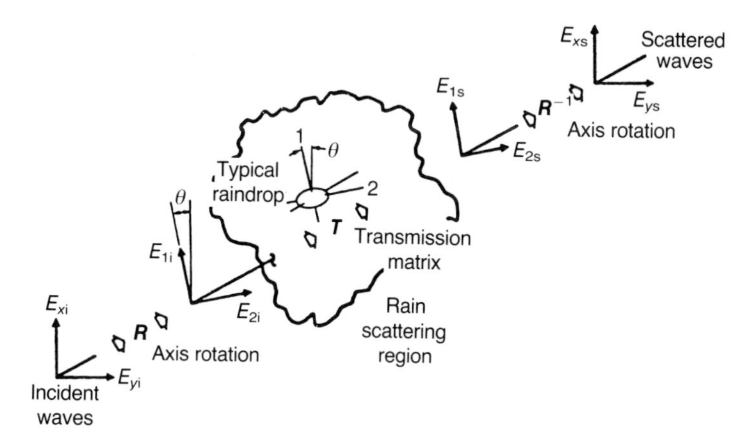

Fig. 9.6 Components of overall transformation matrix $\boldsymbol{T}'$ describing rain depolarization.

$$\mathrm{XPD}_x = 20 \log \left| \frac{E_{xs}}{E_{ys}} \right|_{(E_{yi}=0)} = 20 \log \left| \frac{T'_{xx}}{T'_{yx}} \right| \tag{9.2.8}$$

and

$$\mathrm{XPD}_y = 20 \log \left| \frac{E_{ys}}{E_{xs}} \right|_{(E_{xi}=0)} = 20 \log \left| \frac{T'_{yy}}{T'_{yx}} \right|. \tag{9.2.9}$$

For the case of circular polarization, Chu shows that

$$\mathrm{XPD}_c = 20 \log \left| \frac{T'_{xx}}{T'_{yx}} \right|_{(\theta=45°)} \tag{9.2.10}$$

with T'_{xx} and T'_{yx} the components of $\boldsymbol{T}'$. This expression is independent of the sense of rotation of the copolarized wave.

Any prediction model must account for the distribution of the sizes and shapes of raindrops and the distribution of the angles θ and ξ that are present in the rain along the path.

9.2.5 Derivation of path parameters from depolarization measurements

For an arbitrary configuration of path and polarization, the complete characterization of coherent transmission through the atmosphere requires the measurement of the four complex elements of the transmission matrix:

$$\begin{bmatrix} E_{1r} \\ E_{2r} \end{bmatrix} = \begin{bmatrix} T_{11} & T_{12} \\ T_{21} & T_{22} \end{bmatrix} \begin{bmatrix} E_{1t} \\ E_{2t} \end{bmatrix} \tag{9.2.11}$$

where the subscripts t and r denote transmitted and received waves respectively.

Normalizing the transmission matrix with respect to T_{11} (total attenuation and phase shift), three complex numbers remain to characterize the properties of the path:

$$\frac{1}{T_{11}}\begin{bmatrix} E_{1r} \\ E_{2r} \end{bmatrix} = \begin{bmatrix} 1 & t_{12} \\ t_{21} & t_{22} \end{bmatrix}. \tag{9.2.11}$$

Hence, six scalar parameters need to be measured, which can be associated directly with relative cross-polar level and phase and differential copolar attenuation and phase.

In general, the matrix cannot be converted into a diagonal matrix by a rotation operation such as in equation (9.2.7). It was shown in the European Olympus propagation experiment project OPEX (*Handbook for Data Analysis*) that this situation may occur if the medium contains two distinctly different particle populations along the path, such as rain and ice (clouds) or two different raincells. However, realistic calculations using a two-medium model have shown that the asymmetry is too small to be measurable. Hence, in the derivation of path parameters, the particle distribution along (part of) the path must be assumed to be homogeneous. For linear polarization this 'longitudinal homogeneity' implies that $t_{21} = t_{12}$. The 'effective canting angle' θ of the medium may then be obtained from

$$\tan(2\theta) = \frac{2t_{12}}{1 - t_{22}}. \tag{9.2.12}$$

θ is complex, in general.

Alternatively, the transmission matrix may be converted into that for circular polarization and an equivalent formulation found (Paraboni, Mauri and Martellucci, 1993).

In practice a population of only raindrops or only ice particles will result in a canting angle that is close to a real value. The occurrence of a complex effective canting angle may therefore be indicative of the existence of both ice and rain and be used to separate the effects of ice and rain in the derivation of cross-polarization statistics.

9.2.6 Modelling approach to the prediction of cross-polarization

Two approaches to modelling exist: the theoretical and semi-empirical approaches. Both methods result in the following XPD–copolar attenuation (CPA) relations:

$$\text{XPD (dB)} = U - V \log A \tag{9.2.13a}$$

and

$$U\ (\text{dB}) = S + C \log f \tag{9.2.13b}$$

where XPD and A are the values of cross-polarization discrimination and attenuation not exceeded for the same percentage of time. The relation (9.2.13b) is used for frequency scaling purposes (f in gigahertz). The constant U is typically found to be in the 30–50 dB range and V is usually around 20 dB. The exact evaluation of the coefficients U and V requires regression fitting of the theoretical or measured curves. A theoretical verification of this expression is given by Nowland, Olsen and Shkarofsky (1977) and by NASA (1989).

The empirical approach is straightforward; using a beacon signal, XPD and A are measured and regression fitting is used to determine the constants in equations (9.2.13).

In the theoretical approach, first the scattering effect of a single raindrop is determined as a function of a certain parameter (mostly the size). Then the distribution of that parameter over the population of raindrops as a function of rain rate is used in calculating the two orthogonal transmission coefficients. The transmission coefficients (or more exactly the specific attenuations and phase lags α and β) have been calculated in this manner as a function of rain rate by several authors:

- Chu (1974) and Watson and Arabi (1973) for oblate spheroidal raindrops, Laws and Parsons distribution and eccentricities directly related to sizes (largest drop most deformed);
- Oguchi (1977) and Morrison, Cross and Chu (1973) for the more realistic Pruppacher–Pitter drop shapes (Fig. 9.7).

Mostly, the results are presented in the form of differential attenuation or phase, meaning the attenuation or phase of one channel relative to the other. The differences in attenuation and phase shift for the two polarizations along the symmetry axes of the propagation medium are defined as the differential attenuation ΔA and the differential phase shift $\Delta\phi$. For depolarization analysis the medium can be characterized by ΔA, $\Delta\phi$ and the canting angle θ.

It can be seen from Fig. 9.7 that the worst case for differential attenuation and differential phase corresponds to $\xi = 90°$. This agrees with intuition, since the projection of the ellipsoidal drop into the plane containing the field vectors has the greatest eccentricity for that case.

The differential attenuation and phase for $\xi = 90°$ are mostly used to derive the differential attenuation and phase for other angles, for example (Chu, 1974):

$$\Delta A = \sin^2 \xi\, (\Delta A)_{\xi=90°}, \tag{9.2.14}$$

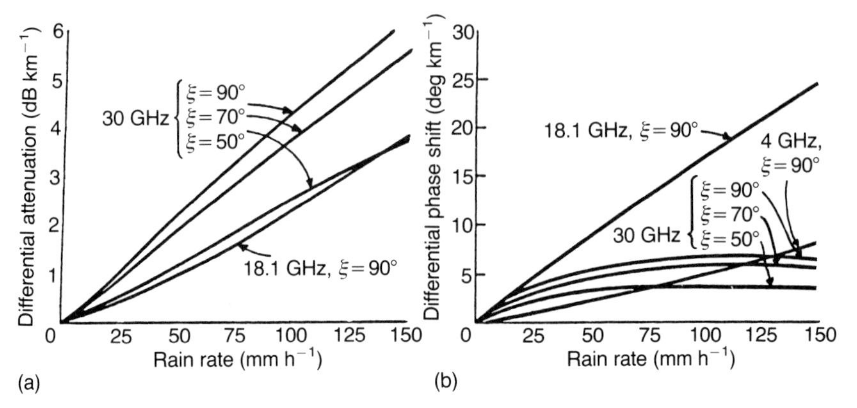

Fig. 9.7 (a) Differential attenuation ΔA and (b) differential phase $\Delta\phi$ for rain at different frequencies and incidence angles ξ (Morrison, Cross and Chu, 1973): $\theta = 25°$; drop temperature, 20 °C.

$$\Delta\phi = \sin^2 \xi \, (\Delta\phi)_{\xi=90°}. \tag{9.2.15}$$

The ξ component of drop orientation is usually considered to be equal to 90° for line-of-sight (horizontal) paths and the complement of the elevation angle for satellite (oblique) paths. In the literature, no allowance is made for the effect of a distribution of incidence angles. Oguchi (1977) calculated the effect of Gaussian distributed canting angles and angles of incidence on the differential attenuation and differential phase shift. Experimental evidence of the spreading of the drop axes orientation in slant paths is reported by Matricciani *et al.* (1981). They found an average spreading of some 25°.

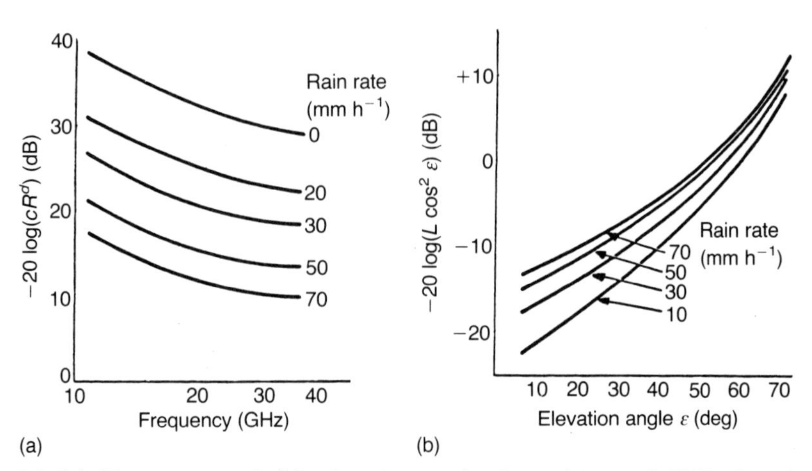

Fig. 9.8 (a) Frequency and (b) elevation angle dependence of XPD for circular polarization and various rain rates (NASA, 1989).

Knowledge of the dependence of XPD on rain rate, elevation angle and frequency is quite valuable, because it allows us to extend the use of measurements, which are often time consuming and costly. Unfortunately, the present limited body of experimental evidence does not overwhelmingly support the theoretical scaling relations, so their accuracy may be limited. Figure 9.8 shows the dependence of XPD for circular polarization on the parameters mentioned.

9.3 DEPOLARIZATION MECHANISMS AND BEHAVIOUR

9.3.1 Introduction

Models of depolarization behaviour, based on theory, have been developed by Ajayi, Owolabi and Adimula (1987) and by Stutzman and Runyon (1984). The model of Ajayi, Owolabi and Adimula was primarily developed for a tropical environment with a different raindrop size distribution than the one used for the European climate, namely the Ajayi and Olsen (AO) distribution (Ajayi and Olsen, 1985). Although this model is not intended for the European climate, the calculations made using this model give very good insight into the mechanisms and behaviour of depolarization. Therefore this model will be used in this section.

The model uses the forward scattering amplitude calculations of other authors (Table 9.1). Therefore, the calculations for spheroidal drops are made for a frequency range of 1–300 GHz, but the calculations for Pruppacher–Pitter drops are made for frequencies of 1–33 GHz.

Table 9.1 Parameters and numerical techniques used for the computation of the forward scattering amplitude for non-spherical drops

Numerical technique	Refractive index of water	T (°C)	Shape	Maximum f (GHz)
Point matching	Debye's relation	20	Spheroidal	50
Point matching	Debye's relation	20	Pruppacher–Pitter	34.8
Unimoment method	Ray's relation	10	Pruppacher–Pitter	33
Point matching	Ray's relation	20	Spheroidal	400

From Ajayi and Olsen (1985) and references cited therein.

9.3.2 Variation of XPD with rain rate

It can be seen from Fig. 9.9 that XPD varies linearly with $\log R$ at all frequencies and elevation angles for horizontal polarization. XPD for vertical polarization does not show this kind of relationship. A linear relationship is still observed between XPD_v and $\log R$ at 4 and 11 GHz; at higher frequencies and high rain rates, the linear relationship no longer holds.

9.3.3 Variation of XPD with frequency

Precise calculations for the Laws and Parsons (LP) drop size distribution are also given by Ajayi and Olsen. These calculations show that XPD for frequencies up to 100 GHz based on that distribution are the same as for the AO drop size distribution, for the same rain rate (Fig. 9.10).

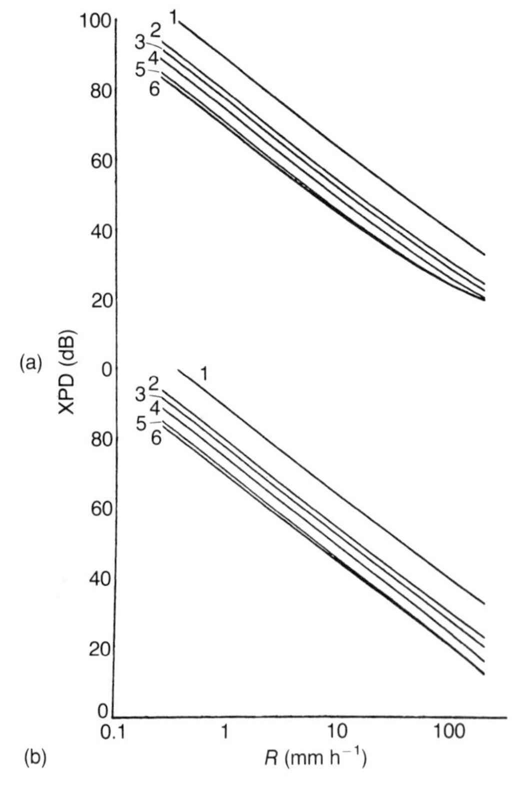

Fig. 9.9 Variation of XPD with rainfall rate R, computed for an elevation angle of 30° and Pruppacher–Pitter drops (Ajayi, Owolabi and Adimula, 1987): (a) vertical cross-polarization; (b) horizontal cross-polarization; curves 1, 4 GHz; curves 2, 11 GHz; curves 3, 14 GHz; curves 4, 19 GHz; curves 5, 28 GHz; curves 6, 33 GHz.

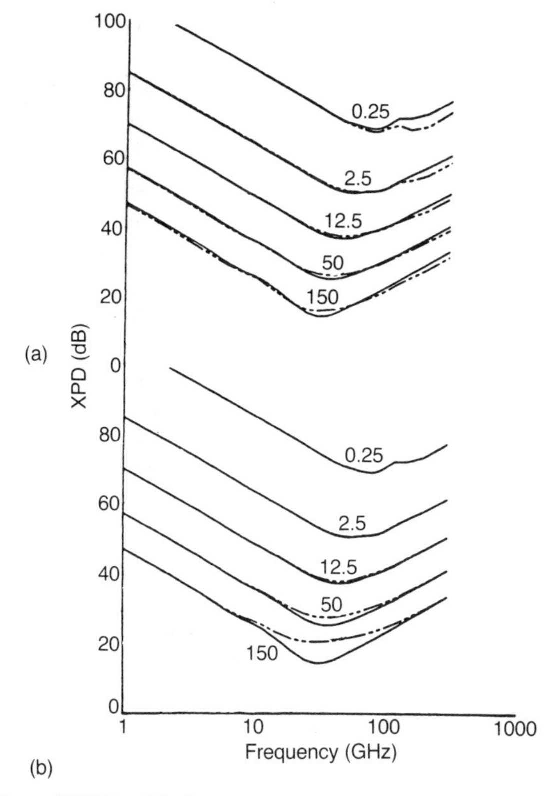

Fig. 9.10 Variation of XPD with frequency at fixed rainfall rates R (mm h^{-1}) (Ajayi, Owolabi and Adimula, 1987). (a) Comparison of models for cross-polarization: ——, AO drop size model; — -, LP drop size model. (b) Horizontal (——) and vertical (— -) cross-polarization for the AO drop size model.

9.3.4 Variation of XPD with attenuation

The variations of XPD with copolar attenuation A show a remarkably good logarithmic behaviour, as can be seen in Fig. 9.11. Therefore, the variation of XPD with A can be represented as equation (9.2.13a)

$$\text{XPD (dB)} = U - V \log A.$$

For spheroidal drops and frequencies of 1–400 GHz, it can be shown that

- V depends mainly on frequency and is almost independent of the canting angle of the drops, and
- U depends on both frequency and canting angle.

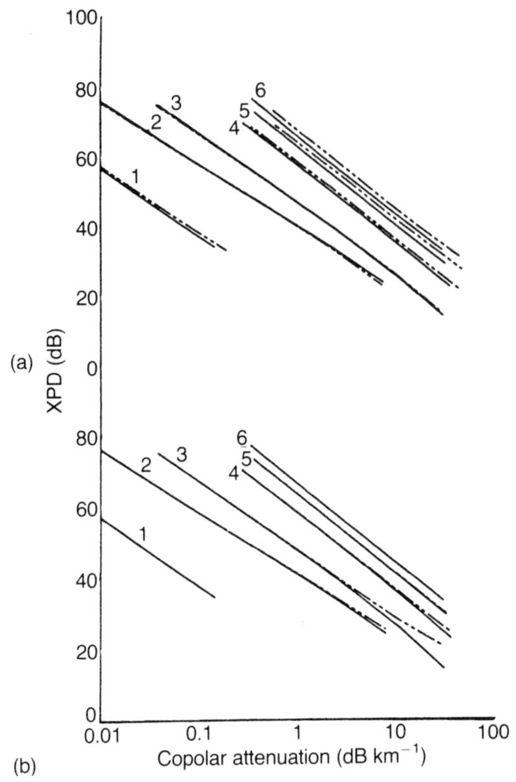

Fig. 9.11 Variation of XPD with A at various fixed frequencies: (a) AO (——) and LP (— -) models for horizontal cross-polarization; (b) horizontal (——) and vertical (— -) cross-polarization in the AO model; curves 1, 4 GHz; curves 2, 12 GHz; curves 3, 30 GHz; curves 4, 100 GHz; curves 5, 200 GHz; curves 6, 300 GHz.

Over a frequency range of 6–300 GHz, U and V, obtained from regression analysis using AO drop size distribution, are

$$U = 3.5 + 21.1 \log f + 17.3 \log \operatorname{cosec} \theta \tag{9.3.1}$$

and

$$V = 14.6 + 3.9 \log f \tag{9.3.2}$$

for $5° \leqslant \theta \leqslant 40°$ where f (GHz) is the frequency and θ is the canting angle of the raindrops. Similar results were obtained for U and V using the LP drop size distribution. This shows that U and V are almost independent of the assumed raindrop size distribution. A similar regression analysis, carried out for Pruppacher and Pitter drops, gave the expressions for U and V over the frequency range from 6 GHz to 33 GHz as

$$U = -6.8 + 30 \log f + 17.6 \log \operatorname{cosec} \theta \tag{9.3.3}$$

and

$$V = 14.5 + 6.5 \log f \tag{9.3.4}$$

for $5° \leqslant \theta \leqslant 40°$.

9.3.5 Variation of XPD with drop shape

The effect of drop shape on the variation of XPD with A is shown in Fig. 9.12. The Pruppacher–Pitter drop shape gives a higher XPD than the spheroidal drop shape especially for $A < 1$ dB, which corresponds to low rainfall rates. This can be explained as follows. At low rainfall rates, the rain contains relatively more small drops, which are less deformed in the Pruppacher and Pitter drop shape model than in the spheroidal drop shape model. The difference in XPD increases with frequency; this can be explained by the greater influence of smaller drops for higher frequencies.

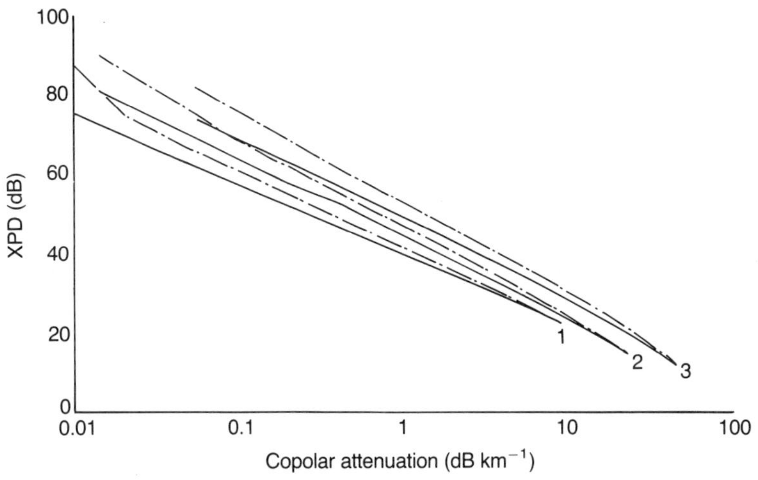

Fig. 9.12 Variation of XPD (horizontal) with A at three frequencies showing the effect of raindrop shape: ——, spheroidal drops; — · —, Pruppacher–Pitter drops; curves 1, 11 GHz; curves 2, 19.3 GHz; curves 3, 34.8 GHz.

9.3.6 Variation of XPD with wind velocity

From the expressions in section 9.3.4, it has already been shown that the XPD depends on the canting angle of the raindrops. This is also illustrated in Fig. 9.13, in which the regression curve of XPD versus A is plotted for six different events, observed on a short terrestrial line-of-sight path at 30 GHz. The corresponding mean wind velocity measured for each event is also included in the figure. From this figure it can be concluded that for a given attenuation the depolarization tends to increase with increasing wind velocity. This agrees with theory, which predicts an increase of the expected canting angle with growing wind velocity (Brussaard, 1976) (section 13.2).

9.3.7 Conclusions

XPD exhibits a linear variation with $\log R$ for all elevation angles at 4 and 10 GHz. This linear relationship does not hold for vertical polarization at higher frequencies and high rain rates, especially for low elevation angles.

At a constant rain rate for frequencies up to 100 GHz, XPD for LP and AO drop size distributions are almost equal. Therefore, the model of Ajayi, Owolabi and Adimula can be regarded as also valid for the European climate.

The Pruppacher–Pitter drops have been shown to give higher XPD than the spheroidal drops, especially for low values of A. Therefore, the spheroidal drops can be used for worst-case predictions, but the Pruppacher–Pitter drops are more realistic.

An empirical relationship has been developed between the XPD and A, the frequency and the canting angle of the raindrops. The relationship for spheroidal raindrops is applicable for frequencies of 6–300 GHz and canting

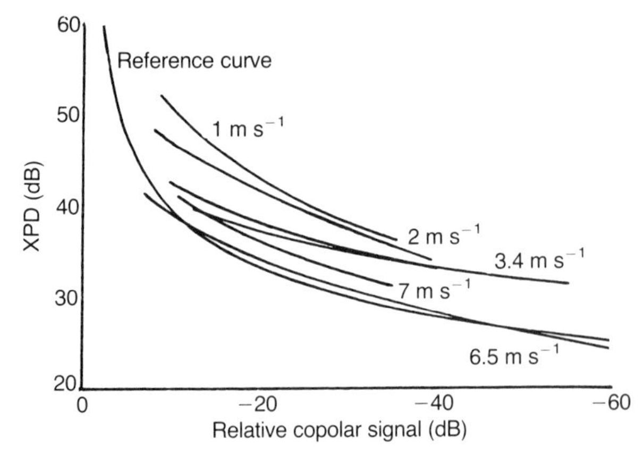

Fig. 9.13 Variation of XPD with A for six events with different mean wind velocities (Herben, 1984).

angles of 5–40°, while the relation for Pruppacher and Pitter drops is only applicable to frequencies from 6 to 33 GHz.

9.4 COMPARISON OF DIFFERENT MODELS

9.4.1 Introduction

Recent measurements at the ground station of the Eindhoven University of Technology (EUT) have been used to validate some of the most commonly used models (Hogers, Herben and Brussaard, 1991). The measurements were done using the single-polarized 12.5 and 30 GHz beacon signals of the Olympus satellite. The signals were received using a single 5.5 m Cassegrain antenna. The system-induced depolarization of the 12.5 and 30 GHz beacon signals was cancelled by vector cancellation.

9.4.2 XPD–CPA relations

Virtually every well-known theoretical or empirical XPD–CPA relation for rain depolarization is written in the form of equations (9.2.13). Table 9.2 shows values for U and V, recommended for five well-known XPD–CPA models (Van de Kamp, 1989), and the experimentally obtained values for U and V.

The mean 'short-term' values were calculated by averaging the values obtained by curve fits on single events. The large standard deviations of these short-term values indicate that U and V can change significantly from event to event. The 'long-term' values for U and V agree fairly well with the theoretical values of the Chu (1982) and SIM (Stutzman and Runyon, 1984) models. The absolute values for U and V predicted by the CCIR and the Nowland, Olsen and Shkarofsky (1977) (NOS) model are too high for the events observed at Eindhoven. Thus these models predict higher XPD values for low CPA levels and show a faster decay. In Table 9.2 'DHW' refers to the model by Dissanayake, Haworth and Watson (1980).

The values for S and C, which describe the observed frequency dependence of U, were calculated from the 12.5 and 30 GHz values for U, as given in the table. It can be seen that they differ from theoretical values. The reliability of the S and C values can be increased using measurements at other frequencies.

9.4.3 Differential attenuation and phase shift

From tables by Chu (1974) it may be concluded that a rain medium can be assumed to be causing purely differential phase shift ($\Delta A = 1$) for frequencies

Table 9.2 The values for U, V, S and C for different rain depolarization models and experimentally obtained values at EUT

	V (dB)			U (dB)			S (dB)	C (dB)
	12.5 GHz	20 GHz	30 GHz	12.5 GHz	20 GHz	30 GHz	12.5–30 GHz	12.5–30 GHz
CCIR model	−20	−21	−23	43.99	50.11	55.4	11.08	30
DHW model	−20	−20	−20	40.92	45.2	48.9	17.88	21
Chu model	−20	−20	−20	39.8	43.89	47.42	17.87	20
SIM model	−19	−19	−19	−39.45	42.98	46.03	20.47	17.3
NOS model	−20.7	−22.6	−22.6	42.27	50.93	55.5	$0.7(V-20)$ $+ 15.28$	26
Mean 'short-term' value for EUT	−14.52	NA	−19.44	37.98	NA	48.38	7.98	27.35
Standard deviation of 'short-term' value	5.58	NA	9.01	3.28	NA	9.93	–	–
'Long-term' value for EUT	−18.12	NA	−19.00	38.20	NA	46.97	12.90	23.07

NA, not analysed.

below 15 GHz and for frequencies above 15 GHz causing purely differential attenuation ($\Delta\phi = 0°$).

Using the theory of Fukuchi, Awaka and Oguchi (1985) the behaviour of ΔA and $\Delta\phi$ can be predicted. This can be compared with the measured differential phase and attenuation (Figs 9.14(a)–(d)). It can be seen from these figures that the experimental values follow the theoretical curves fairly well.

Using for 12.5 GHz the assumption that the medium only introduces differential phase shift, the measured values were transformed to differential phase only. These values can also be compared with the predicted behaviour (Fig. 9.14(e)). From this figure, it can be seen that the assumption $\Delta A = 1$

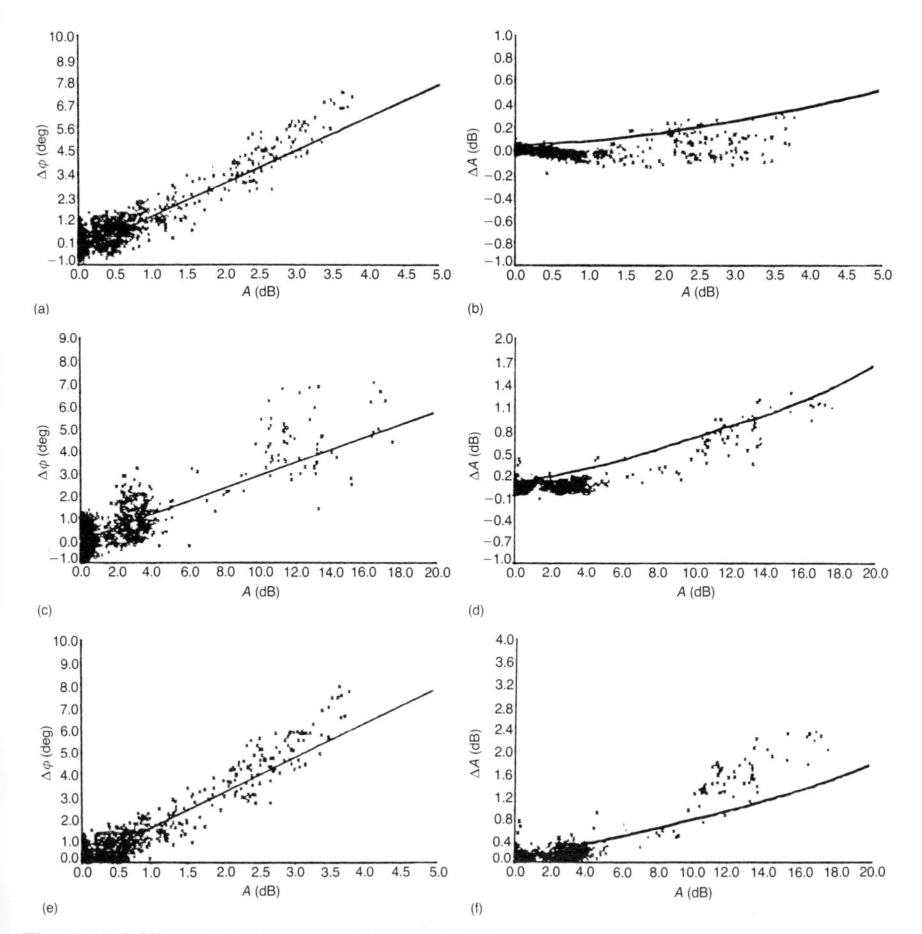

Fig. 9.14 Differential phase shift $\Delta\phi$ and differential attenuation ΔA versus copolar attenuation (Hogers, Herben and Brussaard, 1991): (a), (b) 12.5 GHz; (c), (d) 30 GHz; (e) 12.5 GHz with $\Delta A = 1$; (f) 30 GHz with $\Delta\phi = 0°$; ——, predictions of Fukuchi, Awaka and Oguchi (1985); ×, values based on measurements.

holds for 12.5 GHz. Using for 30 GHz the assumption that the medium only causes differential attenuation, the results are shown in Fig. 9.14(f). The assumption $\Delta\phi = 0°$ is much worse.

9.4.4 Frequency scaling of copolar attenuation

The frequency dependence of the copolar attenuation was determined from scatter plots of 30 GHz copolar attenuation versus 12.5 GHz copolar attenuation. The frequency scaling relation for copolar attenuation of Boithias and Battesti, which was adopted by the CCIR (1990, reports 721-3 and 338-6) was used. Figure 9.15 demonstrates the very good agreement between theory and measurements.

9.4.5 Conclusions

Curve fits on scatter plots of corrected XPD versus CPA agree with theoretical rain XPD–CPA curves as given by the SIM and the Chu model. The CCIR rain XPD–CPA model predicts too high absolute U and V values for the events observed at Eindhoven.

Calculated ΔA and $\Delta\phi$ values show a dependence on the copolar attenuation and beacon frequency that agrees fairly well with the theoretical dependence given by Fukuchi, Awaka and Oguchi (1985).

If no phase information is available, the 12.5 GHz differential phase shift can be estimated fairly well if the assumption is made that the depolarization is caused purely by a differential phase shift. An estimation of the 30 GHz

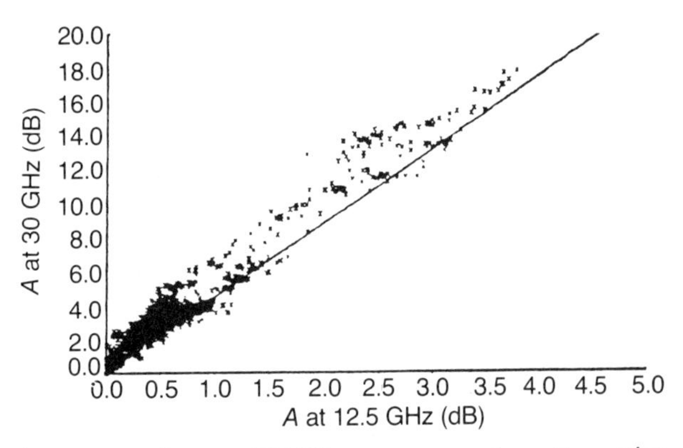

Fig. 9.15 Copolar attenuation at 30 GHz versus copolar attenuation at 12.5 GHz (Hogers, Herben and Brussaard, 1991): ——, theoretical model of CCIR; ×, values based on measurements.

differential attenuation based on the assumption that the depolarization is caused by a differential attenuation is inaccurate.

9.5 CONCLUSIONS

Although relationships of the form of equations (9.2.13) are being used to predict the total cumulative distributions of XPD from cumulative distributions of A, more generally a joint distribution of XPD and A may be desired in future design applications. This appears to be particularly important for earth–space links with small attenuation margins, since XPD varies over a wider range for small values of A. However, a model for XPD which is based not on A, but on physical parameters, has not yet been developed. Furthermore, a joint distribution of XPD and A has the advantage that the XPD predictions are not influenced by uncertainties in the attenuation measurements and prediction models.

Ajayi, Owolabi and Adimula (1987) and Stutzman and Runyon (1984) have made models based on theory. The evaluation of these models is difficult because of the lack of experimental data, especially for higher frequencies. Ajayi, Owolabi and Adimula did not evaluate their model with experimental data. Stutzman and Runyon did evaluate their model with experimental data and with other models, but their model is only valid for frequencies between 10 and 30 GHz.

Stutzman and Runyon succeeded in developing a better model based on theory than the one the CCIR uses, based on existing theory and available data. This is another reason to develop a separate model for A, if better prediction models are necessary.

Note that the models apply only to depolarization induced by rain. For links in which antennas do not have good cross-polarization isolations, the models are only valid in the XPD range where the antenna residuals do not contribute significantly to the cross-polar level. In all other cases, prediction techniques, based on more general theory for propagation between imperfect antennas, must be used.

Some widely used XPD–CPA models are reviewed in section 9.4.2, where it is shown that the SIM and Chu models gives good results. In section 9.4.4, the frequency scaling relation of the CCIR is also checked for 12.5 GHz and 30 GHz: the model gives good results for these frequencies.

REFERENCES

Ajayi, G.O. and Olsen, R.L. (1985) Modeling of a tropical raindrop size distribution for microwave and millimeter wave applications. *Radio Sci.*, **20**(2), 193–202.

Ajayi, G.O., Owolabi, I.E. and Adimula, I.A. (1987) Rain induced depolarization from 1 GHz to 300 GHz in a tropical environment. *Int. J. Infrared Millimeter Waves*, **8**(2), 177–97.

Brussaard, G. (1976) A meteorological model for rain-induced cross-polarization. *IEEE Trans. Antennas Propag.*, **24**(1), 5–11.

CCIR (1990) *Recommendations and Reports of the CCIR*, Vol. V, Propagation in Non-ionized Media.

Chu, T.S. (1974) Rain-induced cross polarization at centimeter and millimeter wavelengths. *Bell Syst. Tech. J.*, **58**(8), 1557–79.

Chu, T.S. (1982) A semi-empirical formula for microwave depolarization versus rain attenuation on earth–space paths. *IEEE Trans. Communication*, **30**, 2550–4.

Dissanayake, A.W., Haworth, D.P. and Watson, P.A. (1980) Analytical model for cross-polarisation on earth–space paths for the frequency range 9–30 GHz. *Ann. Telecommun.*, **35**(11–12), 398–404.

Fukuchi, H., Awaka, J. and Oguchi, T. (1985) A theoretical formula for the prediction of crosspolar signal phase. *IEEE Trans. Antennas Propag.*, **33**(9), 1557–79.

Herben, M.H.A.J. (1984) The influence of tropospheric irregularities on the dynamic behaviour of microwave radio systems. PhD Thesis, Eindhoven University of Technology, p. 110 (ISBN 90-9000584-6).

Hogers, R., Herben, M. and Brussaard, G. (1991) Depolarisation analysis of the 12.5 and 30 GHz Olympus beacon signals. Proc. 1st Opex Workshop, April 23–24, 1991, ESTEC, Noordwijk, pp. 2.4.1–2.4.12.

Matricciani, E., Paraboni, A., Possenti, G. and Tirro, S. (1981) Determination of rain anisotropy and effective spreading in the orientation of ellipsoidal raindrops during intense rainfall. *IEEE Trans. Antennas Propag.*, **29**(4), 679–82.

Morrison, J.A., Cross, M.J. and Chu, T.S. (1973) Rain-induced differential attenuation and differential phase shift at microwave frequencies. *Bell Syst. Tech. J.*, **52**(4), 599–604.

NASA (1989) In *Propagation Effects Handbook for Satellite Systems Design*, 4th edn (ed. L.J. Ippolito), Reference Publication 1082(04).

Nowland, W.L., Olsen, R.L. and Shkarofsky, I.P. (1977) Theoretical relationship between rain depolarisation and attenuation. *Electron. Lett.*, **13**(22), 676–8.

Oguchi, T. (1977) Scattering properties of Pruppacher-and-Pitter form raindrops and cross polarization due to rain: calculations at 11, 13, 19.3 and 34.8 GHz. *Radio Sci.*, **12**, 41–51.

Oguchi, T. (1981) Scattering from hydrometeors – a survey. *Radio Sci.*, **16**(5), 691–730.

Olsen, R.L. (1981) Cross-polarization during precipitation on terrestrial links: review. *Radio Sci.*, **16**(5), 761–79.

Paraboni, A., Mauri, M. and Martellucci, A. (1993) The physical basis of depolarization. Proc. OLYMPUS Utilization Conf., Sevilla, April 20–22, pp. 573–81.

Pruppacher, H.R. and Pitter, R.L. (1971) A semi-empirical determination of the shape of cloud and raindrops. *J. Atmos. Sci.*, **28**, 86–94.

Saunders, M.J. (1971) Cross-polarization at 18 and 30 GHz due to rain. *IEEE Trans. Antennas Propag.*, **19**(2), 273–7.

Stutzman, W.L. (1977) Mathematical formulations and definitions for dual polarized reception of a wave passing through a depolarizing medium (a polarization primer). Virginia Polytechnic Institute and State University Report, prepared under NASA

Contract NAS5-22577, 1977.

Stutzman, W.L. and Runyon, D.L. (1984) The relationship of rain-induced cross-polarization discrimination to attenuation for 10 to 30 GHz earth–space radio links. *IEEE Trans. Antennas Propag.*, **32**(7), 705–10.

Van de Kamp, M.M.J.L. (1989) Software set-up for data processing of depolarisation due to rain and ice crystals in the Olympus project. *EUT Rep. 89-E-218*, April 1989 (ISBN 90-6144-218-4).

Watson, P.A. and Arbabi, M. (1973) Rainfall crosspolarisation at microwave frequencies. *Proc. IEE*, **120**(4), 413–8.

10

Depolarization by ice particles and the melting layer

10.1 DESCRIPTION OF MECHANISMS

Depolarization without significant attenuation can arise from small dielectric particles of low loss with pronounced eccentricity and orientation, such as ice crystals. The depolarization produced by small ice spheroids in the frequency range up to at least 30 GHz is mainly due to differential phase shift. Above 30 GHz, however, the situation is more complex. Calculations of attenuation for ice clouds (Papatsoris, 1993) show that, for very small particles, attenuation is negligible for much of the millimetre wave range. Hence, for such particles, differential phase will continue to dominate (and indeed to become more significant in absolute terms) as frequency increases. For larger ice particles above about 100 GHz, almost the same levels of attenuation and cross-polarization should be observed as for rain particles. The contribution of differential attenuation to ice particle depolarization, although increasing with frequency, may thus not be very significant in statistical terms, especially when low availability or diversity systems are considered.

The mechanisms and models developed to date are for the frequency range 4–30 GHz and will need considerable revision for higher frequencies.

Of the ice particles identified in Chapter 13, it is the needle-like and plate-like particles which have high enough eccentricity to cause significant depolarization with low attenuation, when present in considerable quantity ($\approx 10^4$ particles m^{-3}). Ice pellets and hail are only slightly eccentric, have relatively little orientation and are generally not present in sufficient numbers, or in sufficiently large volumes, to cause significant depolarization.

Ice depolarization will in general increase in significance with frequency in the millimetre wave region, in contrast to rain depolarization. (The smaller rain particles are nearly spherical, whereas ice needles and plates exhibit high eccentricity). However, we would expect that in the higher millimetre wave bands, such ice depolarization would be accompanied by more significant attenuation than in the lower bands.

The alignment mechanisms for ice particles include gravitational–aerodynamic effects and electrostatic effects. The former usually give rise to preferred orientations with major axes in the horizontal plane, whereas the latter can cause alignment both within and outside of the horizontal plane. Snowflakes, which are aggregates of ice crystals, show some alignment, but are nevertheless very irregular in shape. There is no direct evidence yet for significant depolarization in dry snow at frequencies $\leqslant$ 30 GHz.

The melting layer, consisting of melting snowflakes which are of irregular shape and lossy, exhibits both differential attenuation and differential phase. Because of the very thin extent of the depolarizing region of the melting layer (this region appears to be considerably thinner than the full extent of the layer), depolarization from melting particles is not significant to systems operating below 30 GHz.

10.2 EXPERIMENTAL EVIDENCE AND DEDUCTIONS FROM MODELS

Ice particle depolarization was first observed on satellite–earth paths in 1975–1976 during the European phase of ATS-6 ('Application Technology Satellite') (McEwan *et al.*, 1977; Shutie, Allnutt and MacKenzie, 1977). Models for its explanation quickly followed (Haworth, McEwan and Watson, 1977a; Haworth, Watson and McEwan, 1977) based on the presence of ice needles and plates. Several pieces of important experimental evidence fit together to give indirect substantiation of an explanation in terms of high altitude ice particles. This evidence and the deductions that follow are listed in Table 10.1.

It is interesting to note that all evidence for ice depolarization to depend mainly on the presence of high altitude particles is of an indirect nature. To date, no observations have been reported of high differential reflectivity Z_{dr} or high orientation from dual-polarization radars simultaneously with high depolarization on a satellite link traversing the same volume of scatterers as the radar. (Z_{dr} is the ratio in decibels of the reflectivity Z between horizontal and vertical polarization.) Measurements of Z_{dr} on the Chilbolton radar during the OTS ('Orbital Test Satellite') programme were not supported with adequate beacon depolarization measurements. At other sites and during ATS-6, only single-polarization radars were used.

The most substantive results reported to date appear to be those of Alves (1981), using a single-polarization radar alongside an OTS beam. Unfortunately, the radar used was not sensitive enough for the majority of events (it was unable to see the responsible populations of small ice particles), although in some events surprisingly high ice reflectivities were seen at distant ranges, and a correlation with XPD was observed.

Table 10.1 Experimental evidence and deductions on ice particle depolarization

Observation	Deduction
Severe depolarization in absence of attenuation and in absence of significant low altitude radar reflections (McEwan *et al.*, 1977)	High-altitude ice particles causing depolarization
Severe depolarization significantly displaced in time before or after period of heavy rain attenuation (McEwan *et al.*, 1977; Arnold *et al.*, 1980)	High-altitude ice particles causing depolarization
No significant depolarization associated with widespread rain events reported	Densities of required particle types not usually present in widespread rain
Very rapid and substantial changes in cross-polar level observed, correlating with lightning strokes (McEwan *et al.*, 1977; Cox and Arnold, 1979)	Rapid cross-polar level changes of substantial magnitude imply rapid change of orientation over a large volume. Electrostatic field can thus play an important role in orientation mechanism
Correlation observed between electric field at ground and depolarization (Haworth, McEwan and Watson, 1977b)	Evidence for electrostatic alignment
Cross-polar phasors during the course of lightning strokes frequently move along the quadrature axis (McEwan *et al.*, 1981)	Ice crystals can be pulled by electrostatic fields out of the horizontal plane, without an appreciable change of the apparent orientation angles in the polarization plane
High Z_{dr} seen in rain populations but low Z_{dr} usually seen in major ice populations of thunderstorms (Hall, Goddard and Cherry, 1982)	Major regions of thunderstorm with ice particle populations directly above intense rain often do not show high orientation or depolarization
High particle orientation seen on dual circularly polarized radars in thunderstorms, especially for high altitude particles (Hendry and McCormick, 1976; McCormick and Hendry, 1977)	High-altitude particles might cause ice depolarization
Widespread optical glint from the top side of thunderstorms changing rapidly in lightning	Presence of highly oriented ice particles at top of storms, influenced by electrostatic fields
High Z_{dr} observed in thin region below peak Z position in melting zone (Hall, Goddard and Cherry, 1982)	No significant depolarization seen in dry snow region, thin layer of high differential propagation properties, possibly giving rise to small depolarization contribution

10.3 ORIENTATION MECHANISMS

The orientation mechanisms for ice needles are considered to be almost identical to those for dust particles (section 6.1.5). Turbulence tends to destroy the alignment of particles, while an 'inertial torque', produced by air flow round a particle, tends to align it so that its shortest axis is vertical. Electrostatic fields tend to align a particle so that its longest axis is along the field. For small highly asymmetric particles an asymmetry torque may dominate, tending to make the long axes vertical.

The alignment theory leads to the conclusion that the largest particles will be aligned with their shortest axes vertical. This applies to needles and plates. For needles, the question also remains of whether the azimuths of the longest axes, lying in the horizontal plane, would be random or would be systematically aligned by wind shear or electrostatic forces.

Analogously to dust particles, the conclusions on ice particle alignment from the analysis of McEwan *et al.* (1985) are as follows:

- all particles with largest dimension $> 50\ \mu m$ are usually horizontally aligned;
- typically 15% of particles between 30 and 50 μm are aligned vertically and the rest horizontally;
- smaller particles are randomly aligned;
- in the presence of strong vertical electrostatic fields, a high percentage of particles can be aligned vertically;
- alignment of the azimuth directions of needle-like particles within the horizontal plane can occur as a result of wind shear and/or electrostatic effects.

10.4 PROPAGATION IN POPULATIONS OF ICE NEEDLES AND ICE PLATES

10.4.1 Single-particle scattering

For a description of shapes of ice crystals the reader is referred to section 13.3.3.

At frequencies up to 20 or perhaps 30 GHz, propagation through populations of ice needles and plates can be adequately modelled using Rayleigh scattering. Comparison with more accurate methods is possible for ice spheroids (for example with the collocation technique or the Fredholm integral technique), and it is seen that the imaginary part of the forward scattering coefficient differs by less than 1.5% from that calculated by Rayleigh scattering at frequencies ≈ 20 GHz and for particles < 1 mm (e.g. Dissanayake, 1978; Evans and Holt, 1978) (more general comparisons are given in Chapter 4). A Rayleigh scattering approximation for propagation through ice needles and plates is thus still useful for evaluation of depolarization with satellite beacon experiments (such as the ESA Olympus experiment), especially for measurements with the 20 GHz switched polarization beacon.

In the higher millimetre wave range (30–300 GHz), significant problems arise in calculating scattering coefficients for highly eccentric particles. Particles of modest eccentricity and smooth shape (e.g. similar in shape to raindrops) can be handled up to at least 100 GHz using the collocation or Fredholm integral techniques.

10.4.2 Populations of ice needles and plates lying in the horizontal plane

Taking needles and plates approximated by prolate and oblate spheroids of unity eccentricity, the Rayleigh scattering approximately gives intrinsic propagation constants (Haworth, Watson and McEwan, 1977) as

$$k_{\|,\perp} = k_0 + \frac{3\pi q}{\lambda} a_{\|,\perp} \tag{10.4.1}$$

where $k_{\|}$ and $k_{\perp}$ are the wavenumbers of the field polarized parallel and perpendicular to the symmetry axis of the particles, q is the fractional volume of ice, λ is the wavelength, k_0 is the free-space wavenumber and $a_{\|,\perp}$ are called the (frequency-independent) anisotropy terms, as given in Table 10.2. It should be noted that the propagation constants given in equation (10.4.1) are independent of the assumed size distribution. The propagation constants may be used directly in a coherent transmission matrix, in order to calculate depolarization per unit path length for ice needles and plates (Haworth, Watson and McEwan, 1977), and

$$\frac{\mathrm{d}}{\mathrm{d}l}\,\mathrm{XPD} = -20\log|T_{12}| \tag{10.4.2}$$

(for a pure phase shifting medium) where T_{12} is the diagonal term (equal to T_{21}) in the transmission matrix.

Table 10.2 Anisotropy coefficients for Rayleigh scattering (equation (10.4.1))

a	Prolates	Oblates
$\|$	0.72280 – j0.00285	0.22813 – j0.00090
$\perp$	0.34680 – j0.00137	0.72280 – j0.00285

Refractive index n of ice, 1.78 + j0.0024.

For ice plates with no orientation dependence in the azimuth plane,

$$T_{12} = \frac{-(k_{\parallel} - k_{\perp})}{2} \cos^2 \epsilon \sin 2\delta \tag{10.4.3}$$

where ϵ is the elevation angle and δ the polarization angle.

For ice needles, if we assume a symmetrical distribution about a mean orientation in the azimuth plane of $\bar{\psi}$, as in Fig. 10.1, then

$$T_{12} = -\frac{k_{\parallel} - k_{\perp}}{2}\left[\frac{\sin 2\delta}{2} \cos^2 \epsilon(\overline{\cos 2\psi} - 1) - \overline{\cos 2\psi} \sin 2\delta \right.$$
$$\left. - \overline{\sin 2\psi} \cos 2\delta \sin \epsilon \right]. \tag{10.4.4}$$

(Haworth, Watson and McEwan (1977) express $\overline{\cos 2\psi}$ and $\overline{\sin 2\psi}$ in terms of $\bar{\psi}$ and a 'degree of alignment'.)

10.4.3 Other ice particle populations

Comparing the depolarization from highly eccentric ice spheroids (needles and plates) with that of ice spheroids having the same eccentricities as raindrops, it is noted that, for the former, the particle concentrations necessary for the same XPD are approximately an order of magnitude less. Also, for a given particle concentration, plates are about 17 dB more depolarizing than needles, owing to their larger volume and eccentricity.

Hence, an explanation of the observed XPDs on ATS-6 and OTS by the presence of raindrop-shaped particles (of relatively low eccentricity) would imply densities aloft of (2–8) $\times 10^5$ particles m^{-3}, which are not likely.

10.5 PREDICTION OF ICE DEPOLARIZATION FROM METEOROLOGICAL FACTORS

The simple model of propagation in populations of ice needles and plates gives a good explanation of ice depolarization at frequencies up to at least 30 GHz, where attenuation is small and the Rayleigh scattering approximation holds for most particles of interest. The following conclusion can be drawn using this model.

1. The cross-polar field strength is proportional to the total volume of ice, and is almost independent of particle size.
2. Highly complex alignment mechanisms result in considerable variation in alignment and depolarization.

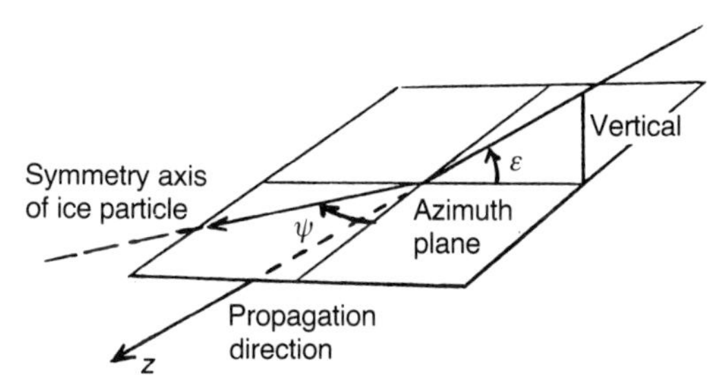

Fig. 10.1 Orientation of ice needles in the azimuthal plane for the model of Haworth, Watson and McEwan (1977).

3. Depolarization with low attenuation is most adequately explained by the occurrence of populations of needle- or plate-shaped crystals, frequently in the higher regions of storms.

While conclusions 1 and 3 might give some indication of a possible meteorologically based prediction method, this will be modified by the uncertainty mentioned in conclusion 2. Nevertheless, we might expect some broad climatological dependencies to be observed, for example, relating to different ice crystal growth in maritime and continental climates.

McEwan (personal communication) examined OTS depolarization data for evidence of such climatological links. Although the database examined was small, it appeared that there might be some evidence of less ice depolarization in continental with respect to maritime climates.

It may be the case that the frequency of occurrence of ice depolarization can be related to the frequency of occurrence of particular rainstorm types in particular climates. We will look to Olympus and other satellite beacon experiments to see whether any systematic ice depolarization prediction technique can be evolved.

As frequency increases into the high millimetre wave bands, we might expect that ice depolarization will occur with higher attenuation and that the semi-empirical laws of XPD versus A, currently used for rain depolarization, will need modifying.

10.6 DEPOLARIZATION IN THE MELTING ZONE

The melting zone exhibits a thin region of high Z_{dr} near the base. The region, when traversed on a slant path with elevation $> 10°$, is so thin that its contribution to depolarization is probably negligible. Attempts to measure this contribution have to date not been successful. Indeed Alves (1981), with a single-polarization radar to identify the presence of the melting zone and beacon depolarization measurements using OTS, verified on each occasion that depolarization was negligible. A similar observation was drawn from ATS-6 at 20 GHz but with much less data (Watson *et al.*, 1979).

As for models, the approaches of Dissanayake *et al.* and Jain (section 13.5) give quantitative predictions in reasonable agreement with observations. However, Klaassen (1988) has pointed out that a melting snowflake is not always properly represented as an ice core surrounded by a water shell.

REFERENCES

Alves, A.P. (1981) Slant path microwave propagation studies using the orbital test satellite. PhD Thesis, University of Bradford.

Arnold, H.W., Cox, D.C., Hoffman, H.H. and Leck, R.P. (1980) Characteristics of rain and ice depolarisation for a 19 and 28 GHz propagation path from a Comstar satellite. *IEEE Trans. Antennas Propag.*, **28**(1), 22–8.

Cox, D.C. and Arnold, H.W. (1979) Observations of rapid changes in the orientation and degree of alignment of ice particles along an earth–space radio propagation path. *J. Geophys. Res.*, **84**(C8), 5003–10.

Dissanayake, A.W. (1978) Cross-polarisation on satellite–earth radio links. PhD Thesis, University of Bradford.

Evans, B.G. and Holt, A.R. (1978) Forward and backscattering from liquid and solid hydrometeors of arbitrary orientation. IEE Int. Conf. on Antennas and Propagation Proc., pp. 116–18.

Hall, M.P.M., Goddard, J.W.F. and Cherry, S.M. (1982) Identification of hydrometeors and other targets by dual polarisation radar. URSI Open Symp., Bournemouth.

Haworth, D.P., McEwan, N.J. and Watson, P.A. (1977a) Cross-polarisation for linearly and circularly polarised waves propagating through a population of ice particles on satellite–earth links. *Electron. Lett.*, **13**(23), 703–4.

Haworth, D.P., McEwan, N.J. and Watson, P.A. (1977b) Relationship between atmospheric electricity and microwave radio propagation. *Nature (London)*, **266**(5604), 703–4.

Haworth, D.P., Watson, P.A. and McEwan, N.J. (1977) Model for the effect of electric fields on satellite–earth microwave radio propagation. *Electron. Lett.*, **13**(19), 562–4.

Hendry, A. and McCormick, G.C. (1976) Radar observations of the alignment of precipitation particles by electrostatic fields in thunderstorms. *J. Geophys. Res.*, **81**, 5353–7.

Klaassen, W. (1988) Radar observations and simulation of the melting layer of precipitation. *J. Atmos. Sci.*, **45**(24), 3741–53.

McCormick, G.C. and Hendry, A. (1977) Depolarisation by solid hydrometeors. *Electron. Lett.*, **13**, 83–4.

McEwan, N.J., Watson, P.A., Dissanayake, A.W., Haworth, D.P. and Vakili, V.T. (1977) Cross-polarisation from high altitude hydrometeors on a 20 GHz satellite radio path. *Electron. Lett.*, **13**(1), 13–14.

McEwan, N.J., Alves, A.P., Poon, H.W. and Dissnayake, A.W. (1981) OTS propagation measurements during thunderstorms. *Ann. Telecommun.*, **36**, 102–10.

McEwan, N.J., Bashir, S.O., Connelly, C. and Excell, D. (1985) The effect of sand and dust particles on 6/4 and 14/11 GHz signals on satellite to earth paths. *University of Bradford Rep. 379* (Final report to INTELSAT under Contract INTELSAT-349).

Papatsoris, A.D. (1993) Improvements to the modelling of radiowave propagation at millimetre wavelengths. PhD Thesis, University of Bradford.

Shutie, P.F., Allnutt, J.E. and MacKenzie, E.C. (1977) Satellite–earth signal depolarisation at 30 GHz in the absence of significant fading. *Electron. Lett.*, **13**(1), 1–2.

Watson, P.A., McEwan, N.J., Dissanayake, A.W. and Haworth, D.P. (1979) Attenuation and cross-polarisation measurements at 20 GHz using the ATS-6 satellite with simultaneous radar observations. *IEEE Trans. Antennas Propag.*, **27**, 11–17.

11

Scattering and interference

11.1 GENERAL

For definitions and notation used in this chapter reference is made to Chapter 5.

When a coherent field is incident on a rain volume, i.e. a 'tenuous polydispersion' of partially absorbing, partially scattering particles, the result is an average field which has a lower flux density. This average field is, by definition, the coherent field. The fluctuating part, with zero average field strength and induced voltage, but with non-zero average power, is the incoherent part. When a coherent plane wave is (normally) incident on a homogeneous slab containing a monodisperse distribution of particles, the attenuation of the wave is determined by the forward scattering theorem:

$$\sigma_{\mathrm{t}} = \frac{4\pi}{k_0}\,\mathrm{Im}\,f(\boldsymbol{u}_{\mathrm{i}}, \boldsymbol{u}_{\mathrm{i}}). \tag{11.1.1}$$

This theorem may be expanded to a polydispersion by integrating over all particle sizes. It may be used for the general case of a radio link, on the condition that the rain medium extends over several Fresnel zones.

The incoherent field is generally analysed using transport theory. For the analysis of (incoherent) bistatic scattering, forward scattering and backscattering, the bistatic cross-section σ_{bi} (equation (5.2.2)) is defined as the ratio of scattered power and incident power flux density for the hypothetical case that scattering in all other directions is the same as in the direction considered. The backscattering (radar) cross-section and the forward scattering cross-section are special cases of the bistatic cross-section.

11.2 INCOHERENT SCATTERING BY RAIN

11.2.1 Forward scattering

An important assumption underlying the transport theory for incoherent scattering is the assumption that all individual contributions from the scattering particles are uncorrelated and therefore add up to a field with zero mean value. For forward scattering, there is a conceptual problem. The assumption that the individual contributions are uncorrelated is certainly justified for multiple scattering but, for first-order scattering, this does not seem self-evident.

A clear mathematical treatment is given by Ishimaru *et al.* (1982), in terms of the scattering matrix of the Stokes parameters. This treatment, combined with the treatment of radiative transfer of thermal energy by Ishimaru and Cheung, referred to in Chapter 5, should form the basis for a consistent treatment of single and multiple scattering through rain.

When a plane wave is incident on a plane parallel slab of rain (not necessarily normal incidence), Ishimaru distinguishes the coherent part of the field, i.e. the 'reduced incident specific intensity', and the field which results from first-order scattering of this reduced intensity. It is the latter, first-order scattered, field which constitutes the source for the transfer equation giving the total incoherent field. The source intensity has a non-zero value in the forward direction.

The reduced intensity is determined by the extinction cross-section, which is directly related to the imaginary part of the scattering function in the forward direction (equation (11.1.1)). The first-order forward scattered intensity is determined by the differential cross-section. This is related to the product of the scattering function in the forward direction and its complex conjugate:

$$\sigma_d(0) = 4\pi|\boldsymbol{f}(\boldsymbol{u}_i, \boldsymbol{u}_i)|^2 = 4\pi\boldsymbol{f}(\boldsymbol{u}_i, \boldsymbol{u}_i)\boldsymbol{f}^*(\boldsymbol{u}_i, \boldsymbol{u}_i). \qquad (11.2.1)$$

However, as the first-order forward scattered field has a causal relation to the (reduced) incident field, the condition of statistical independence, and hence zero-average sum of individual contributions, appears not to be fulfilled.

This enigma reappears in many publications. Capsoni and Paraboni (1978) argue that the amplitude fluctuations caused by a slab of precipitation in the near-field region are due to the variation in number of the particles contained in a columnar region determined by the effective aperture of the receiving antenna. At the same time, the value for the first-order forward scattered, incoherent intensity is directly derived from equation (11.2.1) and the reduced incident field, integrating over the slab. Rogers and Olsen (1983), while correctly pointing out that the forward scattering theorem takes into account multiple forward scattering, state that the forward multiple-scattering interactions for which the paths are closely aligned 'contribute predominantly to the coherent transmitted wave'.

Ishimaru (1978) discusses the coherence time of a signal transmitted through rain on a line-of-sight link between two narrow-beam antennas. He concludes that the coherence time for the scattered field from a plane wave, received by a point receiver, is in the order of the time needed for a particle to move over a distance equal to the wavelength. For a signal transmitted between two narrow-beam antennas, however, the coherence time is on the order of the time required for a particle to move over the distance of the aperture size. His analysis is based on a small-angle approximation of the scattering function, while assuming that the scattering contributions from the particles are statistically independent and result in a zero mean total field. It is not clear at this moment whether or not the apparent contradiction is inherent to the model, i.e. a fundamental limitation of the transport theory model, with respect to a more exact wave theory solution.

It is concluded that the formulation by Ishimaru should be the basis for consistent modelling of thermal emission, attenuation and scattering by hydrometeors. The basic limitations and the apparent contradictions indicated above should be clarified. Also, since the theory for scintillation is closely related to the theory for incoherent scatter by hydrometeors, the treatment of scintillation should use a compatible formulation.

11.2.2 Bistatic scattering

Bistatic scatter from rain follows the same theoretical treatment as in the previous section and Chapter 5. Assumptions of statistical independence in this case are quite plausible and the treatment does not appear to have basic limitations, provided that the right models of particle properties are available. The total signal resulting from bistatic scatter by rain cells is the result of multiple scattering of incident radiation. Scattering by clouds and precipitation creates in all practical cases an unwanted interfering signal. However, its potentially harmful effect is mitigated by the attenuation associated with the scattering process.

An important parameter in the modelling of the process of bistatic scattering by rain is the size of the common volume, i.e. the volume obtained by intersecting the antenna beams of the transmitting and receiving stations. Theoretical treatment of the problem assumes in most cases a homogeneous medium, the scattering properties of which can be modelled by an integral (over the common volume) of the scattering functions of individual particles. The resulting radiative transfer equation results in a treatment which is highly dependent on the assumptions regarding the simultaneous probability distribution of the scattering and attenuation process. The most probable scenario is that of scattering and attenuation occurring simultaneously, but in an inhomogeneous medium. The result may be strongly variable. For example, assuming a homogeneous rain medium with substantial scattering generally results in a model which entails excessive attenuations. This makes any discussion of interference by rain scattering useless, since the wanted signal suffers too much from attenuation for any practical system to be viable. In contrast, another configuration which is possible (but much less probable) is that of the common volume containing mixed clouds with no attenuating rain in between. This configuration could result in enhanced signal reception from an interfering system. Modelling of bistatic scattering from rain, therefore, is a modelling of the statistical dependence of meteorological processes and geometrical path configurations, rather than an EM propagation modelling problem.

11.3 PREDICTION MODELS

11.3.1 The CCIR model

The CCIR (now ITU-R) model for interference (CCIR, 1990, reports 882-2 and 569-4) is a combination of a simplified scatter model and an adaptation of an attenuation prediction model, both taken from Crane. The scatter model is based on Crane (1974). It is an analytical model, valid basically for cylindrical rain cells. For the asymptotic cases of forward scattering and backscattering, it reduces to expressions valid for first-order scattering by a homogeneous slab.

In the case of backscattering with no attenuation, the bistatic radar equation reduced to

$$\frac{P_r}{P_t} = CM\sigma_s V = C\frac{\Phi_0}{2}\Sigma(1 - e^{-2\alpha l}) \tag{11.3.1}$$

where C (m^{-2}) is a constant dependent on system parameters, $M\sigma_s$ ($m^2\,m^{-3}$) is the average scattering cross-section per unit volume, V (m^3) is the scattering (common) volume, Φ_0 is the scattering albedo (section 5.2.1), Σ (m^2) is the cross-section of the receiving antenna beam, at the location of the slab, α (km^{-1}) is the specific attenuation and l (km) is the thickness of the slab.

For forward scattering:

$$\frac{P_r}{P_t} = CM\sigma_s\Sigma l\, e^{-\alpha l}. \tag{11.3.2}$$

The complete model introduces a path geometry, distinguishing between ice scatter and rain scatter, and introducing a model for total attenuation outside the common volume, based on the 'double-exponential' attenuation model by Crane. This model in particular and its integration into the total interference model have been the subject of much debate.

11.3.2 COST 210 model

In the framework of the European project COST 210 a series of bistatic scatter measurements was carried out to investigate possible improvements in the prediction of interference. The project report (COST 210, 1991) includes an extension of the CCIR procedure mentioned above. The extension includes a more general path configuration and a variable raincell height. This model was tested against available data and appears to agree reasonably well with observations. The extended model will be the basis of a new version of the prediction procedure in CCIR Recommendation 452.

11.3.3 Excell model

A rain scatter model based on a radar database of rain cell observations by CSTS, Milan, was developed under an ESA contract (Capsoni *et al.*, 1992). An exponential rain cell model, the 'Excell model', derived earlier from the same database (Capsoni *et al.*, 1987) is discussed in Chapter 8.

The study indicated that, for many practical cases, simplified assumptions may be used for reasonably accurate predictions.

11.3.4 Conclusions

The CCIR (now ITU-R) model is an accepted algorithm for the evaluation of the question of potential interference by radio relay stations in a satellite–earth link. It is certainly not an accurate prediction model for interference levels in all circumstances, as it has been adapted, where necessary, to provide an upper limit to the estimated levels. The new developments mentioned in sections 11.3.2 and 11.3.3 provide models which have much better 'growth potential' for use in new applications. In particular, the Excell model, being based on radiometeorological data, seems promising in this respect.

The basic questions regarding the distinction between incoherent and coherent fields in transmissions through rain should be addressed in more detail before conclusions can be drawn regarding the importance of the incoherent field. However, it can be expected that for small earth-terminal antennas it could be a potential source of problems with respect to scintillations, owing to the relatively high incoherent forward scattering even at 40–50 GHz (section 4.6.4.)

Formulation of thermal emission, scintillation and rain scatter, all based on transport theory models, should be unified. In microwave propagation studies the formulation by Ishimaru, as followed in this book, should be used as the reference.

REFERENCES

Capsoni, C. and Paraboni, A. (1978) Properties of the forward-scattered incoherent radiation through intense precipitation. *IEEE Trans. Antennas Propag.*, **26**(6), 804–9.

Capsoni, C., Fedi, F., Magistroni, C., Paraboni, A. and Pawlina, A. (1987) Data and theory for a new model of the horizontal structure of rain cells for propagation applications. *Radio Sci.*, **22**(3), 395–404.

Capsoni, C., Paraboni, A., Ordano, L., Tarducci, D., Barbaliscia, F., Martellucci, A. and Poiares Baptista, J.P.V. (1992) Study of interference by rain scatter, *ESA Journal*, **16**(2), 171–92.

CCIR (1990) *Recommendations and Reports of the CCIR*, Vol. V, Propagation in Non-ionized Media.

COST 210 (1991) Influence of the atmosphere on interference between radio communication systems at frequencies above 1 GHz. *EUR 13407*, Commission of the European Communities.

Crane, R.K. (1974) Bistatic scatter from rain. *IEEE Trans. Antennas Propag.*, **22**(2), 312–20.

Ishimaru, A. (1978) *Wave Propagation and Scattering in Random Media*, Vol. 2, Academic Press, New York.

Ishimaru, A., Woo, R., Armstrong, J.W. and Blackmann, D.C. (1982) Multiple scattering calculations of rain effects. *Radio Sci.*, **17**(6), 1425–33.

Rogers, D.V. and Olsen, R.L. (1983) Multiple scattering in coherent radiowave propagation through rain. *COMSAT Tech. Rev.*, **13**(2), 385–401.

12

Reflection and refraction by atmospheric layers

12.1 INTRODUCTION

The negative vertical refractivity gradient during average or standard atmospheric conditions results in the most common form of refraction. Despite its commonness, the variability is considerable and, for significant periods of time, the refractivity gradient departs significantly from standard conditions. Large negative values of the refractivity gradient in the lower atmosphere give rise to ducting, multipath and defocusing or beam spreading. The large positive values of the gradients, which tend to have smaller spatial extent than the negative values, sometimes give rise to obstructive fading, especially on terrestrial links.

For earth–space links, the important resultant effects are bore-sight error from variations in angle of arrival, beam spreading and interference from ducting and multipath. The effects are most significant at low elevations ($< 5°$). In this chapter, we describe briefly the physical models for refraction and relection and then give some information on mathematical techniques for representing wave propagation.

Refraction from weather systems (as opposed to atmospheric layers) is usually neglected (some brief comments are given in section 12.9). Small-scale refractivity associated with atmospheric turbulence is dealt with in Chapter 3. One of the problems in representing satellite–earth propagation at low elevations is that several mechanisms tend to interact (especially ducting, multipath propagation, refraction and scintillation). This aspect is discussed in section 12.9.

12.2 THE INDEX OF REFRACTION

The radio refractivity N, which is generally used to describe the spatial and temporal variations of the refractive index n, is defined as

$$N\ (\text{ppm}) = (n - 1) \times 10^6. \qquad (12.2.1)$$

For clean dry air in the lower atmosphere, the refractivity is given by

$$N_\text{d}\ (\text{ppm}) = 77.6\ P/T \qquad (12.2.2)$$

where P (mbar) is the atmospheric pressure and T (K) is the temperature. The presence of water vapour, however, will modify the refractive index considerably. The 'wet' refractive index is

$$N_\text{w}\ (\text{ppm}) = 375\,000e/T^2 - 5.6e/T \qquad (12.2.3)$$

where e (mbar) is the water vapour pressure. The total refractivity N is the sum of the 'dry' and 'wet' components:

$$N\ (\text{ppm}) = N_\text{d} + N_\text{w} = \frac{77.6}{T}(P + 4832.5e/T - 0.0721e). \qquad (12.2.4)$$

Usually, the two terms in equation (12.2.3) can be combined to give an approximate expression:

$$N_\text{w}\ (\text{ppm}) = 373e/T^2 \qquad (12.2.5)$$

which can be used to reduce the calculation of refractivity to the formula

$$N\ (\text{ppm}) = \frac{77.6}{T}(P + 4810e/T). \qquad (12.2.6)$$

The refractivity can be calculated by referring to various model atmospheres, using the appropriate temperature and pressure profiles with height. The formula

$$\rho\ (\text{g}\,\text{m}^{-3}) = 216.7e/T \qquad (12.2.7)$$

where ρ is the water vapour density may be used to obtain the proper values for water vapour pressure.

The refractivity at a point in space thus varies primarily because of variations in temperature and water vapour concentration. The variations may occur on short scales (a time scale up to the order of minutes, or over scale sizes of a few kilometres or less) as small-scale irregularities or turbulent fluctuations, or on a longer time scale, e.g. diurnal or seasonal variations. The vertical variation of refractivity for an average standard atmosphere can, as a first approximation, be described by an exponential decrease with height:

$$N(h) = N_0 \, e^{-bh} \tag{12.2.8}$$

where N_0 and b are constants that can be statistically determined for different climates. The CCIR (1990, report 563-4) defines an average exponential atmosphere as one in which $N_0 = 315 \times 10^{-6}$, $b = -0.136$ and h is measured from sea level. It is also assumed that, for the first kilometre, the mean refractivity gradient is -40 ppm km^{-1}.

12.3 RAY BENDING

A radio ray passing through the lower layer of the atmosphere undergoes bending caused by the gradient of the refractive index. The bending of a ray at a point is expressed generally by:

$$\frac{1}{a} = \sin\varphi \left| \frac{\nabla n}{n} \right| \tag{12.3.1}$$

where n is the refractive index, φ is the angle of the ray with the index gradient vector at that point and a is the radius of curvature of the ray. This phenomenon is virtually independent of frequency, if the index gradient does not vary significantly over a distance equal to the wavelength. Since the refractive index varies largely with altitude, only the vertical gradient of the refractive index is considered in most cases. The bending at a point is therefore contained in the vertical plane and is expressed by

$$\frac{1}{a} = \frac{\cos\theta}{n} \frac{\mathrm{d}n}{\mathrm{d}h} \tag{12.3.2}$$

where $\mathrm{d}n/\mathrm{d}h$ is the vertical gradient of the refractive index, h is the height of the point above the earth's surface and θ is the angle of the path with the horizontal at the point considered. If the path is approximately horizontal, then θ is small and $\cos\theta \approx 1$. Furthermore, if $n \approx 1$, equation (12.3.2) is simplified as follows:

$$\frac{1}{a} \approx \frac{\mathrm{d}n}{\mathrm{d}h}. \tag{12.3.3}$$

It is therefore clear that, if the vertical gradient is constant, the trajectories are arcs of a circle.

By assuming that the refractive index gradient is constant for the first few hundred metres above the earth's surface and extends homogeneously in the horizontal direction, the radius of curvature a, relative to the radius a_e of the earth, may be expressed in terms of the gradient $\mathrm{d}n/\mathrm{d}h$ by

$$\frac{a}{a_e} = k \approx \frac{1}{1 + a_e \, dn/dh} \tag{12.3.4}$$

where k is normally referred to as the effective earth radius factor. This term refers to a geometrical model in which the ray is a straight line, and the earth has an 'effective' radius ka_e. The actual path of the ray, relative to the earth, is maintained in this model. With an actual radius of 6370 km, k may be expressed in terms of the refractivity gradient as

$$k \approx \frac{1}{1 + (dN/dh)/157} \tag{12.3.5}$$

where dN/dh is in parts per million per kilometre. Figure 12.1 illustrates ray paths associated with various values of k (or dN/dh). The vertical scale of Fig. 12.1 is expanded with respect to the horizontal, to make the differences in curvature noticeable.

The mean value of the refractivity gradient dN/dh near the earth's surface is about -40 ppm km^{-1}, corresponding to $k = 4/3$. If a ray is bent downward less than normal (or bent away from the earth), i.e. $dN/dh > -40$ ppm km^{-1}, the ray is said to be subrefracted, corresponding to $0 < k < 4/3$. If a ray is bent downward more than normal, i.e $dN/dh < -40$ ppm km^{-1}, it is said to

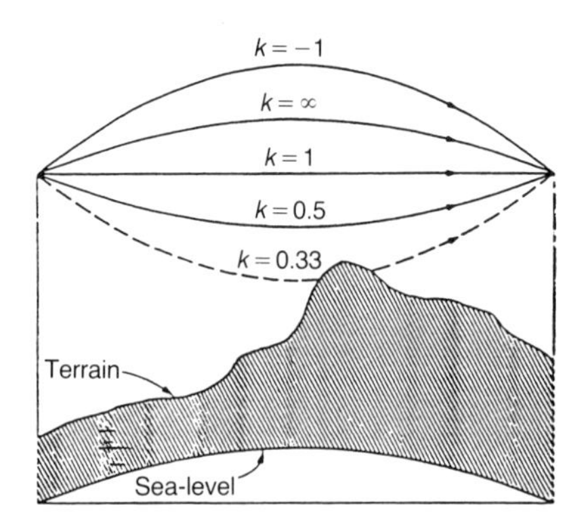

Fig. 12.1 The bending of radiowaves for constant linear refractivity gradients (Dougherty, 1968).

$\Delta N/\Delta h$ (ppm km^{-1})	k
314	0.33
157	0.5
0	1.0
−157	∞
−314	−1.0

be super-refracted, corresponding to $k > 4/3$ or $k < 0$. For negative k values, the ray paths are bent downward so much that trapping or ducting becomes possible. At the critical value $\mathrm{d}N/\mathrm{d}h = -157\ \mathrm{ppm\,km^{-1}}$, i.e. $k = \infty$, the rays are travelling parallel to the curvature of the earth.

12.4 DEFOCUSING

Since the beam of an earth station antenna, usually described by the half-power bandwidth, is not infinitesimally thin, the upper and lower parts of the beam will be bent by slightly different amounts as they pass through the atmosphere. As a result, the beam width is increased, or conversely, the beam becomes defocused. This effect should be negligible for elevation angles above about 3°. Figure 12.2 shows the losses through the complete atmosphere due to atmospheric refraction effects, called 'defocusing loss'.

These results were obtained by ray tracing through numerous refractive index profiles of day and night from Albany, NY, over a period of several years. They may be considered representative for continental conditions. For sites located near coastal areas, on islands, at sea or in tropical regions, the

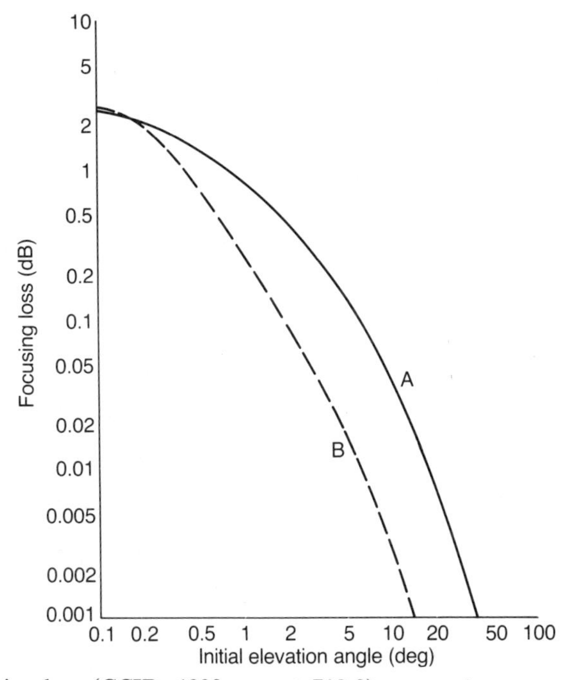

Fig. 12.2 Focusing loss (CCIR, 1990, report 718-3): curve A, average loss; curve B, standard deviation.

focusing losses may increase somewhat, owing to increased surface refractivity and different refractive index gradients near the surface. Focusing losses should be independent of frequency over the range from 1 to 100 GHz, where water vapour is contributing to the refractivity profile. The effects due to dry air alone, and at higher frequencies, have not been estimated, but should be smaller.

12.5 DUCT PROPAGATION

Under certain meteorological conditions, stratification of the lower troposphere occurs in the form of refractivity layering with layers of contrasting refractivity gradients. When the gradient is sufficiently negative, i.e. less than $-157\ \text{ppm km}^{-1}$, and the gradient is maintained over a sufficiently large height interval, tropospheric ducting may occur. The meteorological phenomena involved in the formation of this layering are abnormal vertical structures of temperature and humidity, caused by processes such as subsidence, evaporation, advection (surface heating) and radiation cooling.

In the description of ducting, it is convenient to introduce the concept of a refractive modulus (or 'modified refractivity'):

$$M\ (\text{ppm}) = N + (h/a_e) \times 10^6 \tag{12.5.1}$$

where h is the height above the surface of the earth and a_e is the radius of the earth. Introduction of a refractive modulus profile makes it possible to transform a spherically stratified refractivity structure above a spherical earth into planar layers above a flat earth. The simplified geometry of layers providing ducting conditions is illustrated in Fig. 12.3. From equation (12.5.1) it is seen that ducting conditions correspond to negative gradients of M.

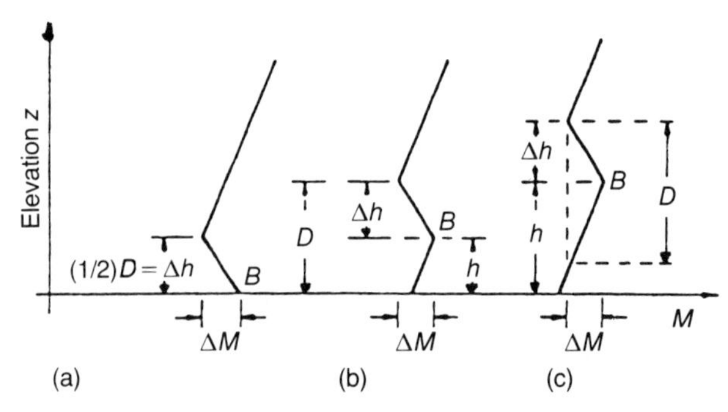

Fig. 12.3 Elevated layer and surface layer parameters for ducting conditions (Dougherty and Hart, 1976): (a) ground-based layer, ground-based duct; (b) elevated layer, ground-based duct; (c) elevated layer, elevated duct.

The effect of ducting conditions on the radio wave propagation depends on the positions of the antennas, relative to the duct geometry. Whether a ducting layer causes, for a radiowave that is incident upon the layer below, scattering, reflection or refraction depends largely on the refractive index gradient, the small-scale fluctuations in refractive index, the angle of incidence and the ratio of the radio wavelength to layer thickness. Ducting occurs for grazing angles of incidence ξ less than a critical value ξ_c:

$$\xi < \xi_c = \arcsin\,(2|\Delta M| \times 10^{-6})^{1/2} \tag{12.5.2}$$

where ΔM is the lapse of refractive modulus across a ducting layer of thickness h_d (Fig. 12.3). For efficient duct propagation, the wavelength λ must be less than a critical value, λ_c:

$$\lambda < \lambda_c = 1.9 \times 10^{-4} {h_d}^{1.8} \tag{12.5.3}$$

where h_d is the duct thickness.

Normally, in free-space propagation, the energy spreads out in the two directions orthogonal to the direction of propagation. This results in the $1/z^2$ dependence of free space transmission. However, in the case of duct propagation, the spread of energy in the vertical direction is eliminated, leaving a $1/z$ dependence. Hence, over a distance z' within the duct, the basic transmission loss L_b is related to that for free space L_{bf} by

$$L_b = L_{bf} - 10 \log z' + A. \tag{12.5.4}$$

It can be seen from equation (12.5.4) that the significant improvement over free-space propagation is offset by the term A for various attenuation mechanisms, including for example leakage losses due to duct irregularities or losses due to ground reflection, etc.

Ducting propagation occurs only in those cases where the elevation angle between the two antennas is about or below 3°. So it is clear that the ducting propagation mechanism does not affect the earth–satellite link, unless very low elevation angles have to be considered.

12.6 ANGLE OF ARRIVAL AND MULTIPATH EFFECTS

The large-scale refractive index variations in the atmosphere are rarely stationary processes, and large-scale fluctuations will cause variations in the amount of ray bending. At the receiving antenna, these will appear as apparent changes in the angle of arrival of the signal and, in extreme cases, in the multipath phenomenon.

Angle of arrival fluctuations can be considered as a single ray that is being deviated from its normal path. In some situations, however, several possible paths can exist simultaneously through the atmosphere between the transmitter and the receiver. The rays travelling the various paths arrive at the receiver with different amplitudes and phases, and hence interference results. This phenomenon is called multipath. On terrestrial paths, multipath is the most common propagation outage in the frequency range 1–10 GHz, because a reflecting surface, the ground in most cases, is present close to the path. On satellite–ground paths above an elevation angle of 10° multipath virtually does not exist. If the elevation angle is low enough, or the beamwidth of the earth station antenna is wide enough, destructive interference due to reflections from the ground can occur.

Multipath, especially that due to a tilted or elevated duct or to a smooth sea surface, generally causes relatively long periods of signal fading ranging from seconds to several minutes. This is because of the large-scale, stable nature of the atmosphere or sea, producing the conditions conducive to multipath.

If the atmosphere becomes mixed, owing to moderate wind or rainfall present in the ray path, the occurrence of multipath will be substantially reduced. Turbulence, however, will cause a degree of phase incoherence across the aperture of the receiving antenna, thereby producing an apparent gain reduction.

12.7 PHASE ADVANCE

In radar and remote sensing systems, the fact that the refractive index of the atmosphere is not unity results in a delay in the received signal. This causes an apparent phase advance or, conversely, an overestimate of the range (the distance of the observed object). If the range delay Δz is the distance by which the range is overestimated by assuming that the velocity of the radiowave in the atmosphere is the same as in vacuum, then

$$\Delta z = \Delta z_{\mathrm{d}} + \Delta z_{\mathrm{w}} \tag{12.7.1}$$

where Δz_{d} (m) is the range delay due to dry air and Δz_{w} (m) is the range delay due to moisture in the air. The variations in Δz due to the moist component of the atmosphere are larger, in general, than those due to the dry component. Crane (1976) has calculated Δz for elevation angles of 0°, 5° and 50° for a standard atmosphere. These are shown in Table 12.1; the ray paths extend to the heights shown. For geostationary communications satellites, the height is about 36 000 km, but the additional range error beyond a height of 80 km is only about 2% of the 80 km value. The values in Table 12.1 are

Table 12.1 Ray parameters for a standard atmosphere (Crane, 1976)

ϵ (deg)	h (km)	z (km)	Bending (mdeg)	$\Delta\epsilon$ (mdeg)	Δz (m)
0.0	0.1	41.2	97.2	48.5	12.63
	1.0	131.1	297.9	152.8	38.79
	5.0	289.3	551.2	310.1	74.17
	25.0	623.2	719.5	498.4	101.1
	80.0	1081.1	725.4	594.2	103.8
5.0	0.1	1.1	2.6	1.3	0.34
	1.0	11.4	25.1	12.9	3.28
	5.0	55.2	91.7	52.4	12.51
	25.0	241.1	176.7	126.3	24.41
	80.0	609.0	181.0	159.0	24.96
50.0	0.1	0.1	0.2	0.1	0.04
	1.0	1.3	1.9	1.0	0.38
	5.0	6.5	7.0	4.0	1.47
	25.0	32.6	14.3	10.3	3.05
	80.0	104.0	14.8	13.4	3.13

representative of a standard atmosphere. To extrapolate to angles above 5°, an equation of the form

$$\Delta z(\epsilon) = \Delta z / \sin \epsilon \tag{12.7.2}$$

can be used, where $\Delta z(\epsilon)$ is the range error at an elevation ϵ.

12.8 MATHEMATICAL TECHNIQUES FOR REPRESENTING TROPOSPHERIC RADIOWAVE PROPAGATION

Until recently, the principal tool for the evaluation of tropospheric propagation has been geometrical optics. The most sophisticated ray tracing programs can show the qualitative effects of super-refraction and ducting, but quantitative calculations are generally not possible.

Quantitative results require the direct solution of Maxwell's equations for the electromagnetic field, rather than ray trajectories. Until recently, the only method available for the solution of this problem in the troposphere has been mode theory. The transmission medium is modelled by analogy with a dielectric-filled waveguide and, by imposing suitable boundary conditions, the equations can be solved for the propagation modes present. However, a parabolic equation method (PEM) has been developed recently (Dockery, 1988) to deal with tropospheric propagation. The PEM overcomes several limitations inherent in ray and mode methods and allows the rapid computation of the coverage of an antenna without the need of unrealistic assumptions concerning the environment. A brief description of these techniques is presented below.

12.8.1 Ray tracing

This method is very well known and has been extensively used as a prediction method for short (line-of-sight) ranges, where it is considered to be a sufficiently good model and can be reasonably efficient at predicting field strengths. At longer ranges, however, ray theory assumptions are invalid, and recourse to a full wave method is required.

12.8.2 Mode theory

The most common full wave technique employed has been mode theory, in which the propagating field is represented as the superposition of normal modes. In the usual mode theory approach, simple multilinear vertical refractive index profile models are used, since these simplify the nature of the solutions. The incorporation of arbitrary profiles is desirable in order to model better the measured refractivity data, but is generally impractical, owing to the numerical difficulties involved. The analysis of tropospheric radiowave propagation as it is derived by mode theory involves the analytical solution of the wave equation, in terms of series of Airy functions in each region where the vertical profile of the refractive index has a linear dependence on height. Matching conditions between the various regions constrain the solutions to lie in a discrete set of modes whose properties (attenuation factors, propagation constants, etc.) are easily calculated in terms of the geometry of the problem (frequency, layer height and thickness, etc.).

One problem with the above approach is the difficulty of attaining an *a priori* cutoff for the number of modes required in the mode scan. At the higher microwave frequencies, several hundreds of modes may be required to obtain a stable solution for a given system configuration.

12.8.3 The parabolic wave equation method

The parabolic equation is a simplification of the wave equation that allows a full wave solution to be obtained for two-dimensional refractive index structures. The derivation begins with the scalar wave equation for an electric or magnetic field component, Ψ, considered in a system with spectral coordinates r, θ and ϕ:

$$\nabla^2 \Psi + k_0^2 n^2 \Psi = 0 \tag{12.8.1}$$

where $k_0 = 2\pi/\lambda$ is the wavenumber in vacuum. The index of refraction n is generally a function of all three variables r, θ and ϕ, but, assuming azimuthal symmetry (which in fact implies that ϕ = constant), n and Ψ can be written as functions of r and θ alone. Then, the hyperbolic equation (12.8.1) reduces to an elliptic partial differential equation containing second derivatives with respect to r and θ.

If one is primarily interested in the variations of the field on scales that are large compared with a wavelength, it is convenient to remove the rapid phase variation by expressing Ψ in terms of an 'attenuation' factor $\mathcal{A}(r, \theta)$:

$$\Psi(r,\theta) = \frac{\mathcal{A}(r,\theta)\, e^{jk_0 a_e \theta}}{r\varepsilon_0 n^2(r,\theta)\, (\sin\theta)^{1/2}} \tag{12.8.2}$$

where a_e is the earth's radius and ϵ_0 the permittivity of free space. Applying the usual earth flattening approximation by transforming to a rectangular coordinate system (z, h) with the origin on the earth's surface beneath the source, i.e.

$$\begin{aligned} z &\approx a\theta \\ h &= r - a \\ n^2(r, \theta) &\rightarrow n_m^2(z, h) \end{aligned} \tag{12.8.3}$$

where z is the range along the earth's surface, h is the height above the surface and $n_m(z, h)$ is the modified refractive index, defined as

$$n_m^2(z, h) = n^2(z, h) + 2h/a, \tag{12.8.4}$$

we arrive at the parabolic equation

$$\frac{\partial^2 \mathcal{A}}{\partial h^2} + 2jk_0 \frac{\partial \mathcal{A}}{\partial z} + k_0^2(n_m^2 - 1)\mathcal{A} = 0 \tag{12.8.5}$$

when the following assumptions are made:

- the variations of the index of refraction are small at the scale of the wavelength;
- the field point is a large number of wavelengths away from the source;
- the fractional change in $\partial\mathcal{A}/\partial z$ over a wavelength is small.

The first two assumptions are nearly satisfied in practical cases. The third assumption is equivalent to a restriction on the maximum angle of propagation above the horizon (15°–20°). In practical terms this is not a limitation for long-range propagation, as only energy emitted within about 2° from horizontal will be trapped and ducted to long ranges. High altitude radiation will penetrate the ducting structures and be lost.

The parabolic equation retains all the diffraction effects associated with the refractive index structure of the propagation medium and is thus valid in those areas where ray tracing methods break down, i.e. near caustics and focal points. It assumes that the backscattered field due to refractive index variations is negligible, which is true for realistic refractive index profiles, and is in any case also assumed by ray theory and coupled-mode methods of solving two-dimensional problems in mode theory. At the same time, the PEM effectively retains the full coupling between waveguide modes which must be modelled for two-dimensional profiles, but in a much simpler way than in the coupled-mode formulation.

The full power of the parabolic equation becomes apparent when a numerical solution of the wave equation is sought. A solution of an elliptic equation, such as equation (12.8.1), requires the specification of boundary conditions on the closed boundary of the two-dimensional domain of interest. In the usual finite difference scheme, the field at each point of the two-dimensional domain is expressed in terms of the field at neighbouring points, requiring the solution of a large system of simultaneous equations with a large number of unknowns. While this is theoretically feasible, the requirements in terms of memory and computation time make it impractical.

A parabolic equation, on the other hand, can be solved by a 'marching' technique. The initial vertical field distribution is specified at the transmitter, and the solution is marched forward in a range given the boundary conditions at the ground and at great height. The numerical method that makes the solution tractable is the 'split-step' algorithm, which makes use of fast Fourier transform techniques.

The PEM has been applied to the problem of propagation in the presence of elevated ducting layers and, in particular, to the effect of these layers on radar coverage diagrams. Figures 12.4 and 12.5 show the field radiated from a 1 GHz vertically polarized Gaussian beam antenna at 200 m height (Craig and Levy, 1991). In each case the basic refractivity profile is modelled by a 'tanh' profile layer superimposed on a standard exponential atmosphere. The refractivity change across the layer is 100 ppm. In Fig. 12.4, the layer slopes at a rate of 2 m km^{-1}, typical of the structure that might be expected in coastal advection. The high level of leakage of energy from the top of the duct is apparent and is not predicted by one-dimensional layer models such as are used by mode theory. Figure 12.5 shows the effects of 'corrugations' on a horizontal layer. These have been observed in stratocumulus layers at the top of subsidence inversions and have typical wavelengths of 5 km. The leakage is much more complex in this case. Clearly the predictions are very sensitive to the detailed structure of the layer.

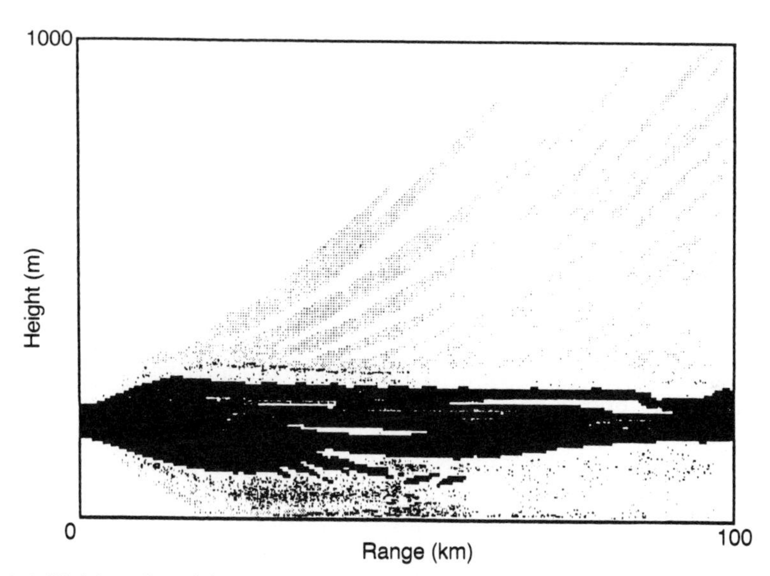

Fig. 12.4 Field radiated by vertically polarized Gaussian beam in 10 ppm duct at 300 m. The duct is 'corrugated', the corrugations having a wavelength of 5 km and an amplitude of 25 m (10 dB contours). (Craig and Levy, 1991.)

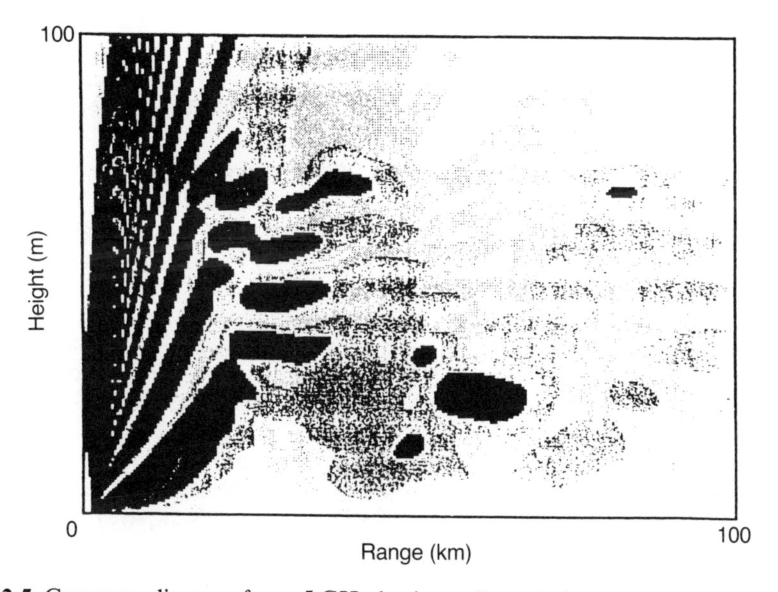

Fig. 12.5 Coverage diagram for a 5 GHz horizontally polarized antenna in the presence of a 10 ppm duct at 70 m (6 dB contours). (Craig and Levy, 1991.)

In conclusion, the PEM is a powerful alternative to ray tracing and mode theory for the prediction of field strength under complex conditions of tropospheric propagation. Coverage diagrams are easier to produce, and the model can handle two-dimensional refractive index structures without the need for *a priori* identification of the structures.

12.9 COMBINATIONS OF TROPOSPHERIC PROPAGATION MECHANISMS

In reality, we have to be able to represent combinations of mechanisms occurring in the troposphere. This is particularly true at low elevations ($< 5°$) when we need to be able to represent refraction, turbulence, ducting and multipath simultaneously. Until now, it has not been possible to combine the representation of these mechanisms in a single technique (especially ducting, multipath and turbulence), but recently Craig and Levy (1991), by using the PEM, have proposed a way to combine the representation of scintillation effects with ducting at low elevation.

The PEM would thus appear to be very useful as a general purpose tool in tropospheric propagation modelling. Also, there is no reason why the small refractive effects from weather systems could not also be included in a total atmospheric model using the parabolic equation method.

REFERENCES

CCIR (1990) *Recommendations and Reports of the CCIR*, Vol. V, Propagation in Non-ionized Media.

Craig, K.H. and Levy, M.F. (1991) Parabolic equation modelling of effects of multipath and ducting on radar systems. *Proc. IEE F*, **138**, 153–62.

Crane, R.K. (1976) Low elevation angle measurement limitations imposed by the troposphere: an analysis of scintillation observations made at Haystack and Millstone. *MIT Lincoln Lab. Rep. 518*, Lexington, MA.

Dockery, G.D. (1988) Modelling electromagnetic wave propagation in the troposphere using the parabolic equation. *IEEE Trans. Antennas Propag.*, **36**(10), 1464–70.

Dougherty, H.T. (1968) A survey of microwave fading mechanism, remedies and applications. *ESSA Tech. Rep. ERL-WPL4*, Environmental Science Administration, Washington, DC.

Dougherty, H.T. and Hart, B.A. (1976) Anomalous propagation and interference fields. *OT Rep. 76-107*, Office of Telecommunications, Boulder, CO (available from National Technical Information Service, Boulder, CO, as PB-262-477).

13 Physical properties of hydrometeors

13.1 INTRODUCTION

In order to characterize the scattering properties of hydrometeors, accurate models representing the physical properties of these particles are required. In addition the variation of these microphysical properties within typical (macrophysical) storm structures must also be known. Therefore, information is required on the following.

- *Composition*: ice (of specified density), water or mixture of ice and water.
- *Temperature*: profile throughout the particle.
- *Shape*: rain, spherical or near spherical (small raindrops) or oblate spheroidal (medium–large raindrops) (the possibility of particle oscillation may also be relevant); ice, needles, plates or conical (graupel) plus a variety of complex or irregular shapes.
- *Size*: distribution of sizes as observed in rainstorms.
- *Orientation*: relation of axis of symmetry to vertical.
- *Macrophysical structure*: typical profile of variation of microphysical properties in specified storm types and climates.

(In addition, if a population of particles and the relationship with ground rainfall intensity is considered, the terminal velocity of particles in still air is also an important quantity.)

Modelling of these properties has been refined gradually with improvements in instrumentation over the past 40 years. Detailed information can be found in Mason (1971), Gossard (1983) and Pruppacher and Klett (1978). In the next sections brief descriptions of the models currently used are given for the most important microphysical properties of hydrometeors. Models for macrophysical structure are described in Chapter 8 in the context of attenuation prediction and in Chapter 10 in the context of depolarization.

Evidently, with such a variety of types, characterization of the distribution of sizes and shapes is no simple matter. Also, knowledge of the average distributions is of little value in radio prediction models, without information on the occurrence probability related to meteorological and geographical factors. For adequate physical modelling it is necessary to understand the origins and development of ice and snow particles, which are exceedingly complex and not fully understood. For details see Mason (1971) or Pruppacher and Klett (1978).

13.2 RAINDROPS

13.2.1 Introduction

The effect of rain (and raindrops) on satellite-to-ground radiowave propagation is considerable, especially at higher frequencies. Rain is frequently encountered in most climates. Therefore it can be considered as the most important atmospheric phenomenon in the design of radio links for frequencies above 10 GHz. Because of its importance, rain has been given a great deal of attention in many papers and publications. In the next sections the most usable existing models for the physical properties of raindrops are summarized.

13.2.2 Dielectric properties of pure water

Raindrops are most commonly assumed to be homogeneous in composition, although in middle and high latitudes this is often not the case, especially in winter. Quite commonly, the larger raindrops in showery rain have an ice core nucleus. Such drops can occur up to 1 km or more below the 0 °C isotherm (Beard, 1976). This phenomenon probably has little effect on the dielectric properties at millimetre wavelengths, except perhaps giving a lower mean drop temperature, since the ice core is embedded completely in liquid water. (It may, however, influence the depolarization at microwave frequencies, since drops can be stabilized to a greater size; section 13.2.3).

For practical purposes at millimetre wavelengths, where in any case scattering from the smaller drops dominates, raindrops may thus be modelled as homogeneous. For homogeneous particles (either water or ice) semi-empirical formulae can be used to model the dielectric properties, as developed by Ray (1972). These models apply from 1 GHz to over 1000 GHz for water and in the temperature range from −20 °C to +50 °C.

The (complex) relative permittivity of pure water, as a function of frequency, has been given by Debye:

$$\varepsilon_r = \varepsilon_r' + j\varepsilon_r'' = \varepsilon_r^{\infty} + \frac{\varepsilon_r^0 - \varepsilon_r^{\infty}}{1 - j\omega\tau} \tag{13.2.1}$$

where ε_r' is the real part of ε_r and ε_r'' is the imaginary part of ε_r, which can be written as

$$\varepsilon_r' = \varepsilon_r^\infty + \frac{\varepsilon_r^0 - \varepsilon_r^\infty}{1 + \omega^2\tau^2} \tag{13.2.2a}$$

$$\varepsilon_r'' = \frac{(\varepsilon_r^0 - \varepsilon_r^\infty)\omega\tau}{1 + \omega^2\tau^2}, \tag{13.2.2b}$$

and τ is the relaxation time of water, ω is the angular frequency, ε_r^0 is the value of ε_r for $\omega \ll 1/\tau$ and ε_r^∞ is the value of ε_r for $\omega \gg 1/\tau$. (Note that a time dependence of $e^{-j\omega t}$ has been assumed in the complex notation.)

In many publications the complex relative permittivity is used to describe the dielectric properties of raindrops. The complex relative permittivity is related to the complex index of refraction by

$$\varepsilon_r = n^2, \tag{13.2.3a}$$

$$\varepsilon_r' = n'^2 - n''^2 \tag{13.2.3b}$$

and

$$\varepsilon_r'' = 2n'n'' \tag{13.2.3c}$$

The relaxation time has been physically explained by Von Hippel (1988) as related to the statistical correlation time between consecutive quantum jumps of a single water molecule. These jumps, induced by thermal phonon excitation in the near-IR, disconnect the water dipoles sufficiently from their surroundings to allow reorientation of these dipoles and a subsequent reorganization of their near surroundings. In an applied electric field, these jumps can be measured as dielectric polarization spike signals.

The values of the parameters of the Debye model have been given by Kaatze (1983) for $T = 25$ °C:

$$\begin{aligned} \varepsilon_r^0 &= 78.36 \\ \varepsilon_r^\infty &= 5.16 \\ \tau &= 8.27 \text{ ps} \quad (10^{-12} \text{ s}). \end{aligned} \tag{13.2.4}$$

Figure 13.1 shows a plot of $\varepsilon_r' + j\varepsilon_r''$ with the frequency as a parameter.

In his model, Ray adopted the equations derived by Cole and Cole (1941).

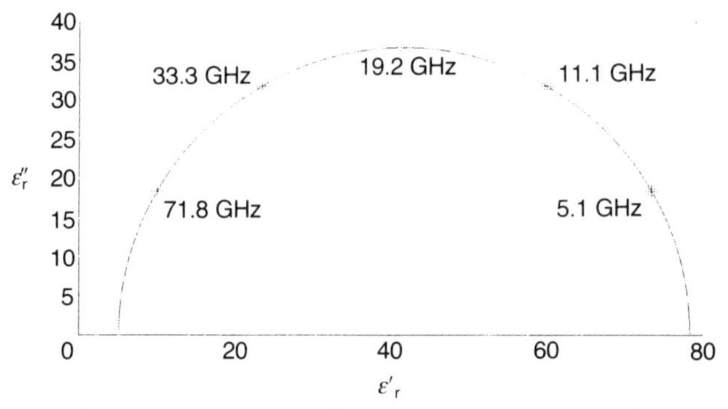

Fig. 13.1 Spectral plot of ε of pure water in the complex plane at $T = 25$ °C.

These equations are extensions of the Debye theory and contain a spread in the relaxation time and an extra frequency-independent conductivity term σ:

$$\varepsilon_r' = \varepsilon_r^\infty + \frac{(\varepsilon_r^0 - \varepsilon_r^\infty)[1 + (\omega\tau)^{1-s}\sin(\frac{1}{2}s\pi)]}{1 + 2(\omega\tau)^{1-s}\sin(\frac{1}{2}s\pi) + (\omega\tau)^{2(1-s)}} \tag{13.2.5}$$

$$\varepsilon_r'' = \frac{(\varepsilon_r^0 - \varepsilon_r^\infty)(\omega\tau)^{1-s}\cos(\frac{1}{2}s\pi)}{1 + 2(\omega\tau)^{1-s}\sin(\frac{1}{2}s\pi) + (\omega\tau)^{2(1-s)}} + \frac{\sigma}{\varepsilon_0\omega} \tag{13.2.6}$$

which reduce to the classical Debye formulae when σ and the spread parameter s are equal to zero. Here $\sigma/\varepsilon_0 = 12.5664 \times 10^8$ (or $\sigma = 0.0111265\ \Omega^{-1}\,\mathrm{m}^{-1}$). The Ray model is illustrated in Fig. 13.2 for several temperatures. The experimental data of the complex index of refraction are represented by the full lines and the broken lines give the result after fitting the parameters.

In principle, all of these parameters are dependent on temperature. Ulaby, Moore and Fung (1986) present expressions for the T dependence of ε_r^0 and τ, obtained by polynomial fit to measurement data and Ray gives the T dependence of s:

$$\varepsilon_r^0 = 88.045 - 0.4147(T - 273) + 6.295 \times 10^{-4}(T - 273)^2$$
$$+ 1.075 \times 10^{-5}(T - 273)^3; \tag{13.2.7a}$$

$$\tau\,(\mathrm{ps}) = \frac{1}{2\pi}[111.09 - 3.824(T - 273) + 0.06938(T - 273)^2]; \tag{13.2.7b}$$

$$s = 16.8123/T + 0.0609265. \tag{13.2.7c}$$

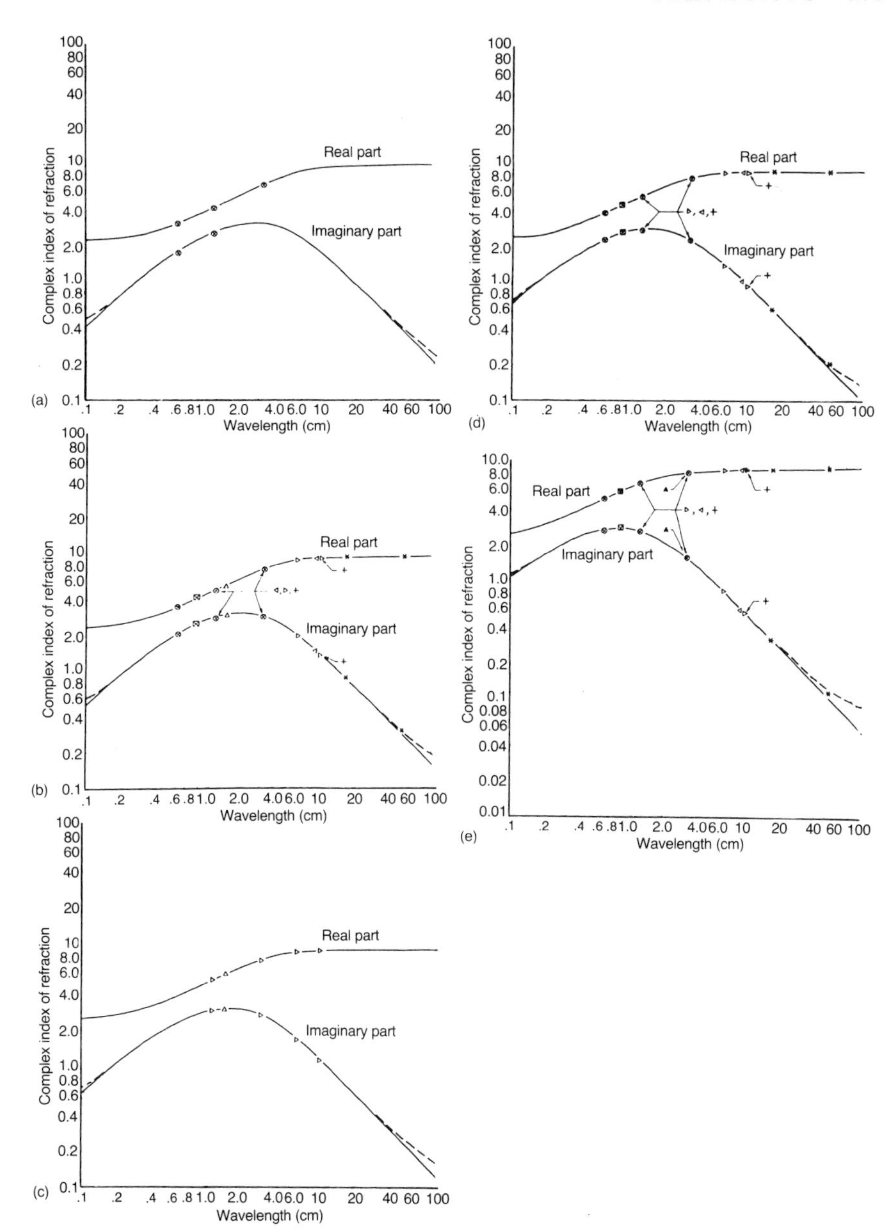

Fig. 13.2 Debye resonance absorption spectra for temperatures of (a) −8 °C, (b) 0 °C, (c) 5 °C, (d) 10 °C and (e) 30 °C (Ray, 1972): ——, fit for each temperature; - - -, fit when the parameters were fitted as a function of temperture; ⊗, experimental results of Lane and Saxton; ∗, experimental results of Grant, Buchanan and Cook; +, experimental results of Collie, Hasted and Ritson; □, experimental results of Grant and Shack; ◁, experimental results of Hasted and Sabeh; ▷, experimental results of Cook; △, experimental results of Saxton; ▲, experimental results of Sandus and Lubitz. For references to experimental data see Ray (1972).

Ulaby, Moore and Fung state that the dependence of ε_r^∞ on T is too weak to be considered.

A different expression for ε_r for 1 GHz $< f <$ 1000 GHz is the new double-Debye formulation (Manabe, Liebe and Hufford, 1987):

$$\varepsilon_r = \frac{\varepsilon_1 - \varepsilon_2}{1 - j\omega\tau_1} + \frac{\varepsilon_2 - \varepsilon_3}{1 - j\omega\tau_2} + \varepsilon_3 \quad (13.2.8a)$$

$$\varepsilon_r' = \frac{\varepsilon_1 - \varepsilon_2}{1 + (\omega\tau_1)^2} + \frac{\varepsilon_2 - \varepsilon_3}{1 + (\omega\tau_2)^2} + \varepsilon_3 \quad (13.2.8b)$$

$$\varepsilon_r'' = \frac{(\varepsilon_1 - \varepsilon_2)\omega\tau_1}{1 + (\omega\tau_1)^2} + \frac{(\varepsilon_2 - \varepsilon_3)\omega\tau_2}{1 + (\omega\tau_2)^2} \quad (13.2.8c)$$

with

$$\varepsilon_1 = 77.66 + 103.3(\theta - 1)$$

$$\varepsilon_2 = 5.48$$

$$\varepsilon_3 = 3.51$$

$$\tau_1 \text{ (ps)} = \{2\pi[0.02009 - 0.1424(\theta - 1) + 0.294(\theta - 1)^2]\}^{-1}$$

$$\tau_2 \text{ (ps)} = \{2\pi[0.59 - 1.50(\theta - 1)]\}^{-1}$$

where $\theta = 300/T$ is the relative inverse temperature.

In the Ray model, the remaining absorption bands are represented by the empirical expression

$$n'' = n''^{D}(\text{Fig. 13.2}) + \sum_i \beta_i \, e^{-|\log(\lambda/\lambda_{0i})/\Delta_i|^{\gamma_i}}, \text{ for } \lambda < 0.3 \text{ cm}, \quad (13.2.9)$$

where β_i, Δ_i and γ_i are empirically determined parameters and λ_{0i} is the centre of the ith band. For $\lambda \geqslant 0.3$ cm, n'' reduces to the Debye contribution:

$$n'' = n''^{D}. \quad (13.2.10)$$

Details are summarized in Table 13.1. The real part is represented by

$$n' = \sum_j (1 + \Theta)\left\{\alpha_j + \sum_i \frac{\beta_{ij}[\omega_{0ij}^2 - (10^4/\lambda)^2]}{[\omega_{0ij}^2 - (10^4/\lambda)^2]^2 + \gamma_{ij}(10^4/\lambda)^2}\right\}^{1/2} \quad (13.2.11)$$

where $\omega_{0ij} = 10^4/\lambda_{0ij}$, λ is in microns, α_j, β_{ij} and γ_{ij} are parameters fitted in

Table 13.1 Parameter values for n'' in the Ray model for water

Spectral interval (μm)	λ_{0i} (μm)	β_i	Δ_i	γ_i
$\lambda \leqslant 2.97$	2.97	0.27	0.025	2.0
	4.95	0.01	0.06	1.0
$2.97 \leqslant \lambda \leqslant 4.95$	2.97	0.27	0.04	2.0
	4.95	0.01	0.06	1.0
$4.95 \leqslant \lambda \leqslant 6.1$	4.95	0.01	0.05	1.0
	6.1	0.12	0.08	2.0
$6.1 \leqslant \lambda \leqslant 17.0$	6.1	0.12	0.042	0.6
	17.0	0.39	0.165	2.4
	62.0	0.41	0.22	1.8
$17.0 \leqslant \lambda \leqslant 62.0$	17.0	0.39	0.45	1.3
	62.0	0.41	0.22	1.8
	300.0	0.25	0.40	2.0
$62.0 \leqslant \lambda \leqslant 300.0$	17.0	0.39	0.45	1.3
	62.0	0.41	0.35	1.7
	300.0	0.25	0.40	2.0
$300.0 \leqslant \lambda \leqslant 3000.0$	17.0	0.39	0.45	1.3
	62.0	0.41	0.35	1.7
	300.0	0.25	0.47	3.0

the jth spectral interval and Θ is empirically determined by comparison with observations to be

$$\Theta = 10^{-4}(T - 298)\,\mathrm{e}^{(\lambda/4.0)^{1/4}}. \tag{13.2.12}$$

Further details are given in Table 13.2. The Ray model is plotted in Fig. 13.3 with a sample of experimental data that illustrates the error envelope.

Table 13.2 Parameter values for n' in the Ray model for water

Spectral interval (μm)	α_j	ω_{0i}	β_i	γ_i
(a) $\lambda \leqslant 6.0$	1.799 07	3352.27	999 140	151 963
		1639.0	50 483.5	9246.27
		588.24	844 697	1076 150
(b) $7.0 \leqslant \lambda \leqslant 340.0$	1.838 99	1639.0	52 340.4	10 399.2
		688.24	345 005.0	259 913.0
		161.29	43 319.7	27 661.2
$6.0 \leqslant \lambda \leqslant 7.0$	$n' = n'^{(a)}(7.0 - \lambda) + n'^{(b)}(\lambda - 6.0)$			
$340.0 \leqslant \lambda \leqslant 1000.0$	$n' = n'^{(b)}\dfrac{1000.0 - \lambda}{660.0} + n'^{D}\dfrac{\lambda - 340.0}{660.0}$			
$1000.0 \leqslant \lambda$	$n' = n'^{D}$			

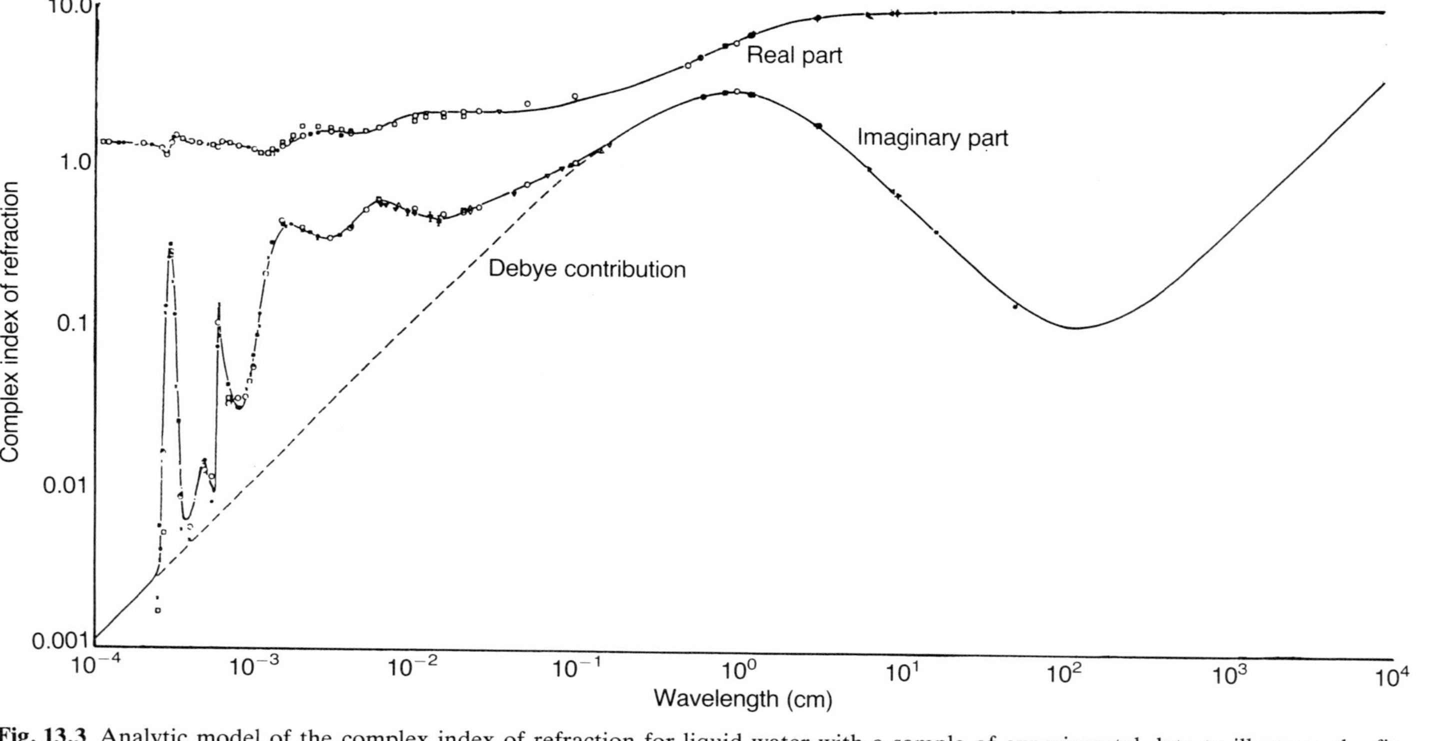

Fig. 13.3 Analytic model of the complex index of refraction for liquid water with a sample of experimental data to illustrate the fit ($T = 25$ °C) (Ray, 1972): ○, Zolotarev *et al.*; △, Stanevich and Yaroslavskii; □, Irvine and Pollack; ▽, Chamberlain; •, Rusk, Williams and Querry; X, Robertson and Williams; ▲, Sanous and Lubitz; ●, Lane and Saxton, temperature adjusted; ◀, Hasted and Sabeh, temperature adjusted; ▪, Pontier and Dechambenoy, temperature adjusted; ▷, Cook, temperature adjusted; ■, Grant and Shack, temperature adjusted; ∗, Grant, Buchanan and Cook, temperature adjusted; +, Collie, Hasted and Ritson, temperature adjusted; ▼, Dragert, temperature adjusted. For references to experimental data see Ray (1972).

13.2.3 The dielectric constant of acid rain

(a) General

During recent decades, air pollution by nitric oxide and sulphur dioxide (NO_x and SO_2 respectively) has caused *inter alia* an increased rate of acidity in rain water. The acids present in the water are sulphuric and nitric acid (H_2SO_4 and HNO_3), which have a serious impact on the environment. This phenomenon is subject to many discussions nowadays. Also other ions such as $NH_4{}^+$, Na^+, Ca^{2+} and Cl^- have been found in rain water.

Another consequence of this phenomenon, not as much discussed and not as alarming either, is the possible impact on the dielectric properties of rain water. It may be useful to know how strongly it affects the scattering mechanism of radiowaves by raindrops and other hydrometeors. It has long been known (Hasted, Ritson and Collie, 1948) that the complex dielectric constant of water depends on the concentration of dissolved salts. In this section, the effect of the currently measured concentration of dissolved materials in rain water is investigated for frequencies above 1 GHz. If the currently observed acidity of rain can be noticed in the scattering properties of the drops, this effect should be taken into account in a propagation model. Conversely, in that case it must be possible to measure the acidity of rain using a radiometer.

(b) Saline water

Saline water is water containing dissolved salts. The presence of dissolved ions causes a deviation in the permittivity of the solution. This difference can be expressed in a difference in the parameter values of equations (13.2.1) and (13.2.2), together with the addition of the ohmic conductance loss, since the solution is an electrolyte. Obviously, the permittivity depends on salinity (concentration of solutes). The static permittivity ε^0 decreases with salinity. The explanation of this can be found in two phenomena: dielectric saturation and kinetic depolarization (Kaatze, 1983).

In the dielectric saturation effect, the electric dipole moments of water molecules around small ions are assumed to be more or less strongly aligned in the electrostatic field of these solutes, resulting in a reduction of ε^0.

Kinetic depolarization is a coupling between hydrodynamic and dielectric properties. An ion drifting through a dipolar liquid in an external field sets up a non-uniform solvent flow. The solvent dipoles are turned thereby in other directions than that in which they would orient without being affected by the moving ion. The result is again a reduction of ε^0.

As for the high frequency permittivity ε^∞, Kaatze presents values of ε^∞ for electrolytic solutions that differ from that of pure water, these being higher or lower depending on the solute. Hasted (1973) shows that there even is a second relaxation spectrum in the submillimetre region for electrolytic solutions.

The relaxation time τ also decreases with salinity. This is explained by Hasted (1973) as being due to the ion breaking the structure of the water by a reduction in the hydrogen bonding; this is similar to the breaking of the water structure when the temperature is raised.

(c) *Analysis*

The result of differing parameters and addition of the conductance loss can be expressed as

$$\varepsilon_s = \varepsilon_s^{\infty} + \frac{\varepsilon_s^0 - \varepsilon_s^{\infty}}{1 - (j\omega\tau_s)^{1-h_s}} + j\frac{\sigma}{\varepsilon_0\omega} \tag{13.2.13}$$

where ε_s^{∞}, ε_s^0 and τ_s are the values of ε_r^{∞}, ε_r^0 and τ for the solution, h_s is a measure of the width of the relaxation time spread, σ is the conductivity and $\varepsilon_0 = 8.854\,19 \times 10^{-12}\ \mathrm{F\,m^{-1}}$ is the (absolute) permittivity of vacuum. The deviation in ε_s^0, ε_s^{∞}, τ_s and σ from the values for pure water can approximately be represented as proportional to solute concentration (Hasted, 1973; Kaatze, 1983; Ulaby, Moore and Fung, 1986):

$$\begin{aligned} \varepsilon_s^0 &\approx \varepsilon_r^0 + am \\ \varepsilon_s^{\infty} &\approx \varepsilon_r^{\infty} + bm \\ \tau_s &\approx \tau + cm \\ \sigma &\approx dm \end{aligned} \tag{13.2.14}$$

where m is the molarity of the solution and a, b, c and d are constants, depending on T and on the solute. For an investigation of the significance of the various terms, we make use of some environmental measurements.

ECN (Keuken *et al.*, 1989) performs measurements on the composition of rain water in The Netherlands. Some of the results are presented in Table 13.3. In Table 13.4, the values of the concentration coefficients for ε_r^0 and τ, a and c for most of these ions are presented. Combining Table 13.4 with the data of week 5 (maximum concentration) in Table 13.3, assuming that the values of a and c for the other ions are of the same order of magnitude and adding the effect of all the different ions, the following relations can be derived:

Table 13.3 Measured concentration of dissolved ions in rain water in Den Helder, The Netherlands, for the first five weeks with rain in 1989, and averaged over the first quarter of the year (Keuken *et al.*, 1989)

	m ($\times 10^{-6}$ mol l^{-1})					
Ion	Week1	Week 2	Week 3	Week 5	Week 6	Quarter
H^+	54.16	63.11	58.20	39.05	19.43	32.32
$NH_4{}^+$	41.48	57.86	80.69	70.83	44.36	47.30
Na^+	122.9	113.1	5.99	334.8	332.9	187.8
K^+	3.310	3.520	2.706	7.868	7.180	4.280
Ca^{2+}	8.552	7.724	6.169	13.17	11.87	8.233
Mg^{2+}	16.37	13.34	8.340	38.74	39.19	22.14
Zn^{2+}	0.4909	0.9508	0.4896	0.4170	0.4454	0.3939
$SO_4{}^{2-}$	43.13	43.43	47.88	54.88	40.16	35.02
$NO_3{}^-$	59.26	54.00	53.56	50.36	29.76	37.41
Cl^-	153.4	134.9	88.57	405.1	390.2	221.4
F^-	1.090	0.9508	1.831	0.5818	0.3784	0.7058

Table 13.4 The coefficients a and c for some ions (from Hasted, 1973)

Ion	a ((mol l^{-1})$^{-1}$)	c (ps (mol l^{-1})$^{-1}$)
H^+	−17	+0.21
Na^+	−8	−0.21
K^+	−8	−0.21
Mg^{2+}	−24	−0.21
$SO_4{}^{2-}$	−7	−0.58
Cl^-	−3	−0.21
F^-	−5	−0.21

$$\varepsilon_s^0 = \varepsilon_r^0 + \sum_{\text{ions}} am \approx \varepsilon_r^0(1 - 6 \times 10^{-5})$$

$$\tau_s = \tau + \sum_{\text{ions}} cm \approx \tau(1 - 1.5 \times 10^{-5}). \tag{13.2.15}$$

From Kaatze (1983), it can be seen that b (the variation in ε_r^∞) will not be of another order of magnitude. It is now useful to estimate the sensitivity of equations (13.2.3) to small variations in ε_s^0, ε_s^∞ and τ_s. In order to do this, the following function is considered:

$$E(x) = \varepsilon_r^\infty x + \frac{\varepsilon_r^0 x - \varepsilon_r^\infty / x}{1 - j\omega\tau/x}. \tag{13.2.16}$$

In this expression, x is close to unity. It has been placed in numerators and denominators, depending on the sign of the dependence of $E(x)$, in order to maximize the effect. h_s has been assumed zero for simplicity, and the effect of σ has not (yet) been taken into account. The sensitivity of $E(x)$ to fluctuations in x around a certain value can be expressed by the condition number $c(x)$:

$$c(x) = \frac{|E'(x)|\,|x|}{|E(x)|}. \tag{13.2.17}$$

The condition number is the factor by which a relative deviation in x may be multiplied in order to obtain the relative deviation in $E(x)$. This will now be calculated for $x = 1$.

$$E'(x) = \frac{d}{dx} E(x) = \varepsilon_r^\infty + \frac{(\varepsilon_r^0 + \varepsilon_r^\infty/x^2)(1 - j\omega\tau/x) - (\varepsilon_r^0 x - \varepsilon_r^\infty/x) j\omega\tau/x^2}{(1 - j\omega\tau/x)^2}$$

$$E'(1) = \varepsilon_r^\infty + \frac{\varepsilon_r(1 - 2j\omega\tau) + \varepsilon_r^\infty}{(1 - j\omega\tau)^2}$$

and

$$E(1) = \varepsilon_r^\infty + \frac{\varepsilon_r^0 - \varepsilon_r^\infty}{1 - j\omega\tau}$$

so

$$\begin{aligned} c(1) &= \frac{|\varepsilon_r^\infty(1 - j\omega\tau)^2 + \varepsilon_r^0(1 - 2j\omega\tau) + \varepsilon_r^\infty|}{|\varepsilon_r^0 - \varepsilon_r^\infty j\omega\tau|\,|1 - j\omega\tau|} \\ &= \left\{\frac{[\varepsilon_r^\infty(2 - \omega^2\tau^2) + \varepsilon_r^0]^2 + [2\omega\tau(\varepsilon_r^0 + \varepsilon_r^\infty)]^2}{[\varepsilon_r^{0\,2} + (\varepsilon_r^\infty \omega\tau)^2](1 + \omega^2\tau^2)}\right\}^{1/2} \qquad (13.2.18) \\ &= \begin{cases} 1.13 & \text{for } f = 1 \text{ GHz} \\ 1.68 & \text{for } f = 1/2\pi\tau \\ 1.98 & \text{for } f = 100 \text{ GHz} \\ 1.32 & \text{for } f = 200 \text{ GHz}. \end{cases} \end{aligned}$$

So, the relative deviation of ε_r as a consequence of dissolved salts is never larger than about $6 \times 10^{-5} \times 1.98 \approx 1.2 \times 10^{-4}$. Since this deviation is smaller than the measurement inaccuracy (Kaatze, 1983), it may be neglected. Moreover, it will be seen that σ has a greater influence on ε_r than all the other parameters. The imaginary part of equation (13.2.13) (with $h_s = 0$) is

$$\varepsilon_s'' = \frac{(\varepsilon_s^0 - \varepsilon_s^\infty)\omega\tau_s}{1 + \omega^2\tau_s^2} + \frac{\sigma}{\varepsilon_0\omega}. \tag{13.2.19}$$

In Table 13.5, values of the ionic conductivity coefficient d for the solutes of Table 13.4 are tabulated. Using these, the conductivity can be calculated of every sample of rain water for which ECN has measured the concentration of different solutes. This way, it can be found that the average value of σ in rain water in 1989 in The Netherlands is 5.30×10^{-3} S m^{-1} and the maximum value is 8.52×10^{-3} S m^{-1}. This results for the last term of equation (13.2.19), say $\Delta\varepsilon''$, in

$$\Delta\varepsilon'' = \frac{\sigma}{\varepsilon_0\omega} = \begin{cases} 0.095/f & \text{average} \\ 0.153/f & \text{maximum} \end{cases} \tag{13.2.20}$$

with f in gigahertz. This decreases with increasing frequency, just as the other term in equation (13.2.19), for higher frequencies. It would be interesting to know whether it decreases faster or slower than the other term. In Fig. 13.4, both terms of this equation are shown, using the average and maximum values of conductivity.

Table 13.5 Ionic conductivity coefficients at $T = 25$ °C (Weast, Astle and Beyer, 1987)

Ion	d (S m^{-1} (mol l^{-1})$^{-1}$)
H^+	34.965
NH_4^+	7.35
Na^+	5.008
K^+	7.348
Ca^{2+}	11.894
Mg^{2+}	10.6
Zn^{2+}	10.56
SO_4^{2-}	16.0
NO_3^-	7.142
Cl^-	7.631
F^-	5.54

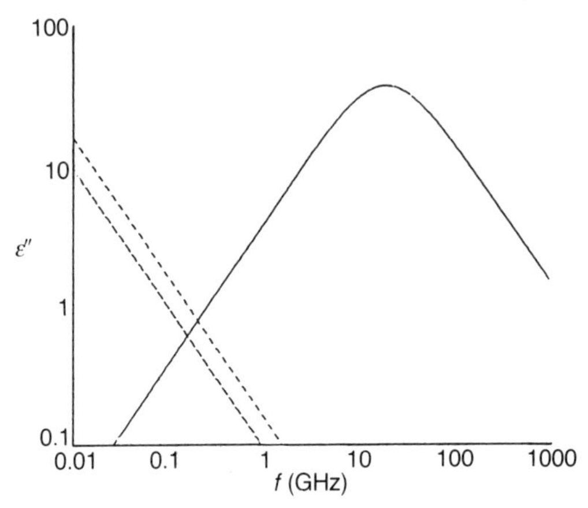

Fig. 13.4 Both terms of the expression for ε'' in equation (13.2.19) ($T = 25$ °C): ——, dielectric loss; – – –, average conductance loss; - - - - -, maximum conductance loss.

In order to know the consequence of this result in terms of attenuation, the following very coarse approximation of the attenuation coefficient α, as a function of frequency and permittivity, can be considered:

$$\alpha \approx \varepsilon'' f. \tag{13.2.21}$$

This implies that the attenuation will become approximately constant for higher frequencies (since $\varepsilon'' \propto 1/f$) and that the attenuation due to conductance loss will be approximately constant for all frequencies. This fraction of attenuation due to conductance loss, say ΔA, can be calculated approximately as (since $\alpha \propto \varepsilon''$)

$$\Delta A = \frac{\Delta A}{A} A = \frac{\Delta \alpha}{\alpha} A = \frac{\Delta \varepsilon''}{\varepsilon''} A \tag{13.2.22}$$

where A, α and ε'' are the values purely due to dielectric loss. With maximum values of $\Delta\varepsilon''/\varepsilon''$ following from equation (13.2.20) and values of A taken from the CCIR (1986), this is calculated in Table 13.6 for various frequencies. It is seen that the currently observed concentration of solutes causes a maximum rise in attenuation of about 0.01 dB. Since other stochastic parameters, such as the geometry of the rainstorm, may cause larger differences in attenuation, the conductance loss of rain water may be neglected.

It would, of course, be interesting to know how strong the concentration of solutes would have to be before the effect can be measured. This is most easily expressed in terms of σ, since $\Delta\varepsilon'' \propto \sigma$, and thus $\Delta A \propto \sigma$. If we consider the effect noticeable when $\Delta A \geqslant 1$ dB, this means a raise by a factor of 100 in ΔA and thus in σ. So, σ should be about 0.8 S m^{-1}. (As a reference,

Table 13.6 The fraction of attenuation due to conductance loss

f (GHz)	$\Delta\varepsilon''/\varepsilon''$	A (dB)	ΔA (dB)
1	4.037×10^{-2}	0.01	4.0×10^{-4}
3	4.582×10^{-3}	0.08	3.7×10^{-4}
10	5.113×10^{-4}	6	3.1×10^{-3}
30	1.534×10^{-4}	50	7.7×10^{-3}
100	1.127×10^{-4}	100	1.1×10^{-2}
300	1.091×10^{-4}	100	1.1×10^{-2}
1000	1.087×10^{-4}	70	7.6×10^{-3}

Assumptions: rain rate, 50 mm h^{-1}; path length, 5 km.

σ of germanium is 1.67 S m^{-1}.) To convert this result to a more popular value, pH can be used. Estimating the pH of rain water currently at about 4.5, it can be concluded that a pH value of 2.5 would cause 1 dB extra attenuation. This is about the value of vinegar and lemon juice.

(d) Conclusion

The difference in dielectric behaviour of saline water with respect to pure water is mainly caused by conductance loss. For $f > 1$ GHz, and the currently measured concentrations of solutes in acid rain, this loss is too low to be measured. Therefore, the conductance loss of acid rain does not have to be taken into account in the modelling.

13.2.4 Drop shape and size

Photographs in still air and wind tunnels (e.g. Magono, 1954) showed that raindrops could be approximated to oblate spheroids. The average size-dependent shapes of raindrops are believed to be well represented by those given by Pruppacher and Beard (1970), which have a flattened base for large diameters. Nevertheless, for many calculations at millimetre wave frequencies or higher, where the smaller drops dominate, approximations to oblate spheroids with size-dependent eccentricity are adequate.

An important phenomenon that has been observed for water drops supported in wind tunnels is that they vibrate around a mean shape. Similar effects are believed to occur in the atmosphere, especially for diameters near 1 mm or less.

If the terminal fall velocity of the drop is known, the shape of a water drop can be calculated. For this calculation, the equation describing the balance of the internal and external pressures at the surface of the drop must be solved. This is currently done using numerical techniques.

Pruppacher and Pitter (1971) derived the aerodynamic pressure around the drop surface from the measured aerodynamic pressure around the surface of a rigid sphere and compensating for the lack of internal circulation. They were able to solve the pressure balance numerically, yielding the shape of water drops falling through the atmosphere at rest. The agreement between the calculated and the measured drop shapes appeared to be satisfactory. The Pruppacher and Pitter drop shape is used in most papers on microwave propagation; Fig. 13.5 shows these shapes for various drop sizes.

Raindrop radius varies between $\approx 100\ \mu m$ and 3.5 mm, although some attention must be paid to semantics. Non-precipitating water particles in rain clouds are composed of minute droplets of up to a few tens of microns in diameter and possess velocities of only a few centimetres per second. Con-

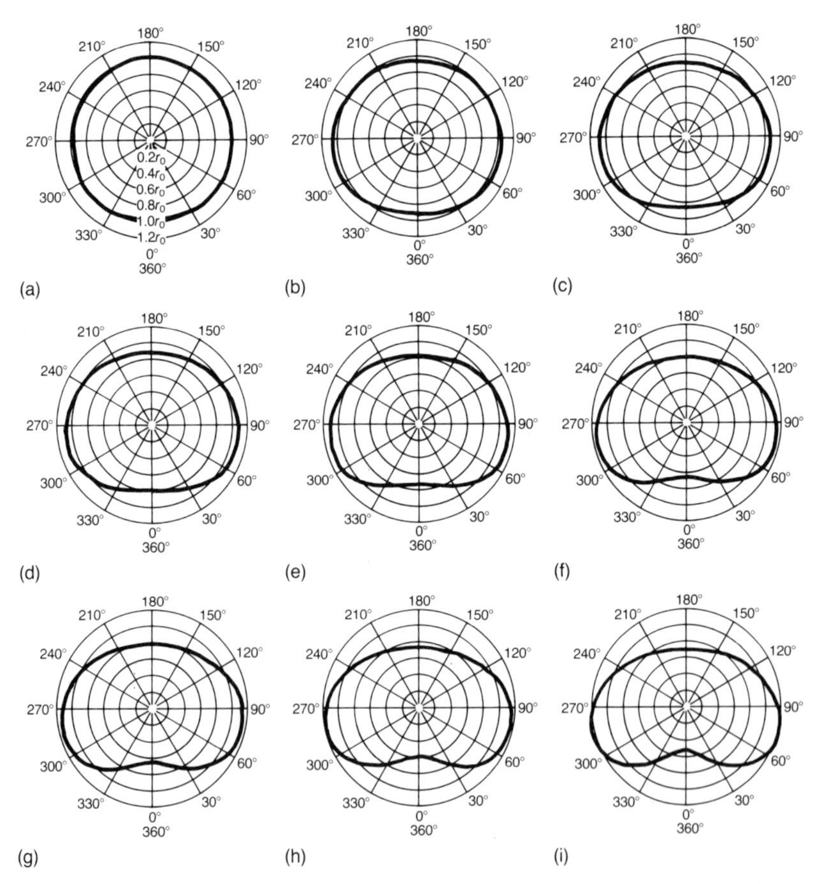

Fig. 13.5 Computed shape of selected drops of various radii (Pruppacher and Pitter, 1971): (a) 0.11 cm; (b) 0.14 cm; (c) 0.18 cm; (d) 0.20 cm; (e) 0.25 cm; (f) 0.29 cm; (g) 0.30 cm; (h) 0.35 cm; (i) 0.40 cm.

siderable growth of particle size must thus be attained before a drop can gain a falling speed sufficient for it to fall out of a cloud and to be called a raindrop.

Widespread layer clouds are associated with upcurrents of speed usually less than 0.5 $m\,s^{-1}$, so that droplets with a diameter of 100 μm can fall out of them. If it is assumed that the distance over which a drop can fall through unsaturated air before evaporating completely is proportional to the fourth power of the initial diameter (Findeisen, 1939), then in an atmosphere of 90% humidity this distance is 3.3 cm for a diameter of 20 μm, 150 m for a diameter of 100 μm and 42 km for a diameter of 1 mm. Since, except in special circumstances, the base of clouds lies a few hundred metres above the ground, 100 μm may be regarded as a lower limit on raindrop size.

Precipitation composed entirely of drops a little larger than 100 μm commonly falls in damp weather from larger clouds which are not far above the ground. This type of rain is known as drizzle, and will not persist downwards more than 100 or 200 m. Thus when talking about drop size distributions care must be taken, since the terms or semantics are traditionally based on ground observation. If we talk about drizzle (seen at the ground) the effective height (on a satellite path) will be small. If convective rain or stratiform rain is considered, each probably has a lower limit on raindrop size at the ground, while, at higher altitudes, smaller drop sizes may occur.

The upper limit on drop size is determined by stability. Pure raindrops with radii above ≈ 3.5 mm break up during their fall to the ground, although drops stabilized by an ice core can persist with a slightly larger radius.

13.2.5 Orientation

Aerodynamic, gravitational and electrostatic forces affect particle alignment. For rain the combination of aerodynamic and gravitational forces is by far the most important. Rain particles falling under terminal velocity in the atmosphere will, generally, have their axis of symmetry distributed around the vertical direction. Steady wind shear forces may modify this alignment, but for slant paths such effects are small since they decrease rapidly with increasing height above ground. A physical model for the aerodynamic–gravitational forces on a raindrop as they affect drop canting has been given by Brussaard (1976). He assumed that the symmetry axis of the drop is parallel to the air flow around the drop. The air flow around the drop can be split into two components: (1) a vertical component caused by the fall of the raindrops and (2) a horizontal component due to the wind shear that is caused by the friction between the ground and the wind. Hence the canting angle decreases as the height to the ground increases. Figure 13.6 shows the results obtained by the model of Brussaard. The model shows that the canting angle

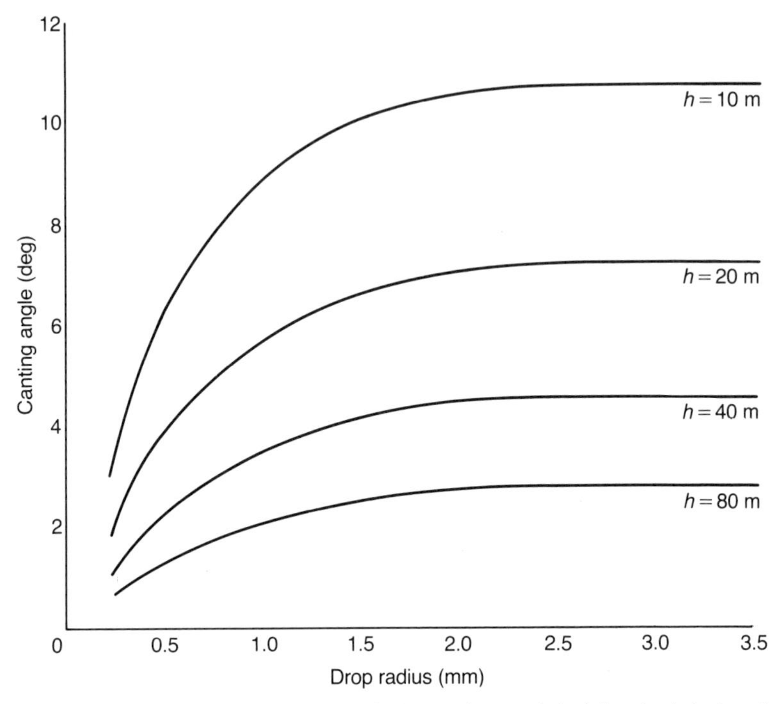

Fig. 13.6 Canting angle as a function of drop size and height h (wind velocity, $15\ m\,s^{-1}$).

is a function of the drop size; it increases as the drop radius increases until a radius of 2 mm is reached, while for larger drops the canting angle is nearly constant, since the fall speed of larger drops is nearly constant (shown in Fig. 13.7).

The model of Brussaard does not account for the angle distribution of the drops; it only predicts the mean canting angle of the drops. To account for the angle distribution, Maher, Murphy and Sexton (1977) developed a model based on the effect of wind gusting. They also assumed that the drops are axisymmetric with their axes in the direction of the air flow, but in their model the wind velocity is composed of a steady state component plus a sinusoidally oscillating component with a small amplitude. They calculated the angle distribution from measurements of the amplitude and frequency of the horizontal wind speed fluctuations.

Howard and Gerogiokas (1982) solved the differential equation for the horizontal drop movement also with the introduction of a sinusoidal variation, but they used the energy spectrum of wind to derive this variation.

Photographic measurements of drop canting angle have been reported by Saunders (1971), who analysed the photographs of 463 raindrops, taking a two-dimensional projection of the three-dimensional shape. These measurements are unsatisfactory in many ways:

- the sample size is very small in space, time and number of drops;
- the distribution is irregular (it can hardly be described as Gaussian);
- ambiguity exists between shape and orientation, owing to the two-dimensional approach.

Probably the best ways to measure raindrop orientation effects are radio methods (dual linear polarization radar and dual polarization link measurements). Direct sampling of rain using photographic techniques, if done properly, would generate an enormous quantity of data. To take photographs of raindrops at one location for a limited period gives little information of value for radio path prediction where the spatiotemporal mean must be considered.

However, the single-location photographic information does illustrate that there is probably a temporal distribution of canting angles with a small sample mean and modest sample deviation. If the single-location photographs were taken over a long period of time in typical city locations, a mean of $\approx 0°$ would be expected to result. For open sites with a prevalent wind direction, a small temporal mean canting angle might be expected, which would decrease with height above ground (provided that the camera orientation was correct or a three-dimensional approach to photography was taken). It is nevertheless the temporal fluctuations of the spatial mean orientation which matters for radio path purposes.

Temporal distributions of spatial mean canting angles as observed on radio links give small mean values ($< 2°$) both on terrestrial links (Watson and Arbabi, 1975) and on satellite links (Arnold *et al.*, 1980). Maggiori, Migliori and Paraboni (1983) found, in measurements on a terrestrial link, that the temporal variance of spatial mean canting angle decreases with increasing rain intensity, from $\approx 3°$ for light rain to $< 1°$ for very heavy rain. This behaviour points to an increasing tendency to stability for larger drop sizes, owing to their larger inertia.

13.2.6 Fall velocity

In order to determine ground rainfall intensity from the spatial distribution of drops aloft, it is necessary to model the fall velocities of rain particles. Invariably the fall velocities determined by Gunn and Kinzer (1949) are used, giving a relationship between terminal velocity in still air and raindrop size. Figure 13.7 shows the measured terminal velocities of drops from Gunn and Kinzer as a function of their drop diameter. It can be seen from this figure that the terminal velocity increases with increasing drop radius until about 2.5 mm, where a maximum of the terminal velocity is reached. Drops with

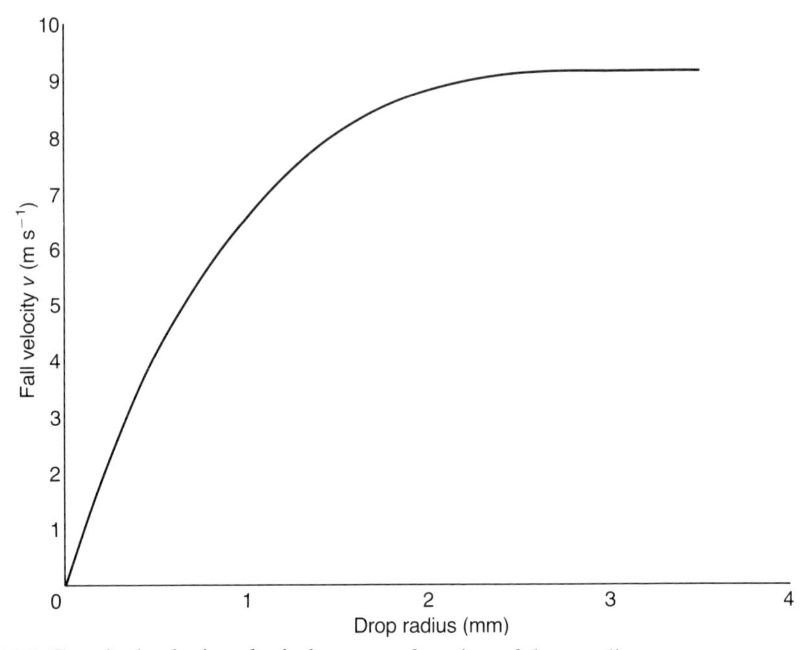

Fig. 13.7 Terminal velocity of raindrops as a function of drop radius.

radii greater than approximately 3.5 mm are not stable and break up into smaller drops during their fall.

Gunn and Kinzer measured the terminal fall velocity of water drops by giving them an electric charge and letting them fall past two inductor electrodes. They presented the result in the form of tables and plots of velocity versus drop mass. Very small droplets appeared to obey Stokes law:

$$\gamma = 3\pi D\eta \tag{13.2.23}$$

where γ is the friction coefficient and $\eta = 17.1 \times 10^{-6}$ Pa s is the viscosity of air. Combining this with the force equilibrium

$$mg = \gamma v \tag{13.2.24}$$

where $m = \rho_w \pi D^3/6$ is the mass, $\rho_w = 10^3$ kg m^{-3} is the density of water and $g = 10$ m s^{-2} is the gravitational acceleration yields

$$v = \rho_w g D^2/18\eta^2 = 32.5 D^2 \tag{13.2.25}$$

where D is in millimetres. In the data of Gunn and Kinzer, the coefficient seems to be a bit lower, about 28. The rest of the data can also be fitted in linear and parabolic curves, yielding

$$v(D) = \begin{cases} 28D^2 & \text{for } D \leqslant 0.075 \text{ mm} \\ 4.5D - 0.18 & \text{for } 0.075 \text{ mm} < D \leqslant 0.5 \text{ mm} \\ 4.0D + 0.07 & \text{for } 0.5 \text{ mm} < D \leqslant 1.0 \text{ mm} \\ -0.425D^2 + 3.695D + 0.8 & \text{for } 1.0 \text{ mm} < D \leqslant 3.6 \text{ mm.} \end{cases} \tag{13.2.26}$$

13.2.7 Size distribution

(a) Introduction

The drop size distribution most commonly used in propagation studies is the one of Laws and Parsons (1943). They measured the size distribution by the flour method: a pan containing fine flour was exposed to the rain and the size of the pellets produced by the raindrop was measured. The distributions for each rain rate were averaged, because the distribution varied for a fixed rain rate. The flour method yields the percentage of the total volume reaching ground contributed by drops of different size ranges. To convert this quantity to the number of drops per unit volume in space the distribution of the fall velocity of drops is used. Figure 13.8 shows the raindrop size distribution in space.

Marshall and Palmer (1948) proposed the use of a negative exponential relation to express the number of drops per unit volume in space as a function of the rain rate (discussed below). The resulting curves are also plotted in Fig. 13.8.

Today, better measurements are possible by using new instrumentation: electromechanical sensors or disdrometers, electrostatic sensors and optical detectors. Measurements of the average distribution of drop sizes above 0.5 mm diameter, however, have confirmed many times the accuracy of the Laws and Parsons distribution. More recent measurements have resulted in quantified drop size distributions for different types of rain: Joss, Thams and Waldvogel (1968) derived three types of distributions for three types of rain: drizzle, widespread rain and thunderstorm rain, as listed in Table 13.7.

Drops with a small radius ($<$ 0.1 mm) are difficult to measure, and therefore the above-mentioned distributions are not valid for this range. This is a serious shortcoming of the distributions, because these small drops considerably influence the propagation of radio waves at higher frequencies (above 10 GHz). Ugai *et al.* (1977) measured the distribution for small drops by measuring the size of the raindrops fallen in a pan filled with oil. Figure 13.9

Table 13.7 The parameters of exponential size distributions of Joss, Thams and Waldvogel (1968)

Type of rain	M_0 ($\mathrm{m^3\,mm^{-1}}$)	Λ ($\mathrm{mm^{-1}}$)
Drizzle	6×10^4	$11.4R^{-0.21}$
Widespread	1.4×10^4	$8.2R^{-0.21}$
Thunderstorm	2.8×10^4	$6R^{-0.21}$

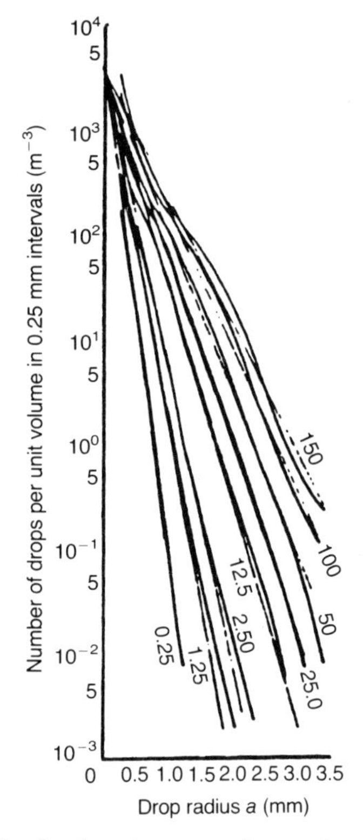

Fig. 13.8 Raindrop size distribution in space for various rain rates from 0.25 to 150 mm h^{-1}: ——, from measurements by Laws and Parsons (1943); ——, negative exponential relation of Marshall and Palmer (1948).

shows their results. The number of drops with a small radius is very large, yielding a considerable influence on the attenuation at high frequencies.

In all of the measurements discussed so far, the terminal velocity distribution of raindrops in stagnant air is used to convert ground drop size distributions into distributions in space. Because of updrafts and downdrafts, the

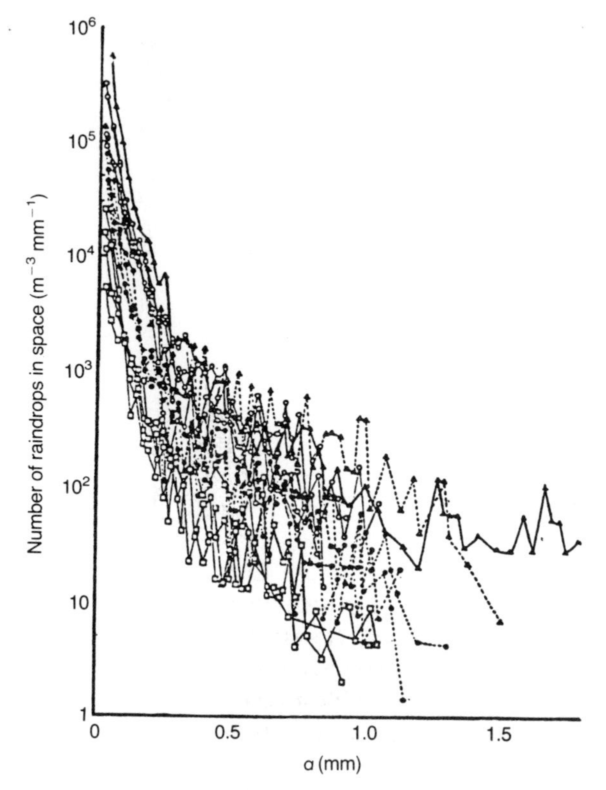

Fig. 13.9 Examples of raindrop size distributions for rainfall intensities R of $46.4\ \mathrm{mm\,h^{-1}}$ (▲), $32.2\ \mathrm{mm\,h^{-1}}$ (△), $5.6\text{–}7.1\ \mathrm{mm\,h^{-1}}$ (○), $3.6\ \mathrm{mm\,h^{-1}}$ (●) and $0.3\text{–}1.1\ \mathrm{mm\,h^{-1}}$ (□). (Ugai *et al.*, 1977.)

rainfall velocity can differ from the ideal fall velocity, yielding erroneous results. By measuring the velocity and size of every drop, this problem can be alleviated. This can be done by using an optical detector.

A further problem relates also to the fact that the size distribution is measured near the ground and not at higher altitudes. Because of the growth mechanisms of drops on their way to the ground, the size distribution changes with height. These growth processes are called 'warm rain processes'; they are (1) condensation of water vapour, (2) coalescence between drops, and (3) drop breakup. These processes can be incorporated into a transport equation, which describes the variation of the size distribution with time. Figure 13.10 shows the results of a calculation done by Young (1975) using the transport equation. This model is very promising, because it also accounts for the relatively large number of small drops, which is in agreement with the measurements shown in Fig. 13.10.

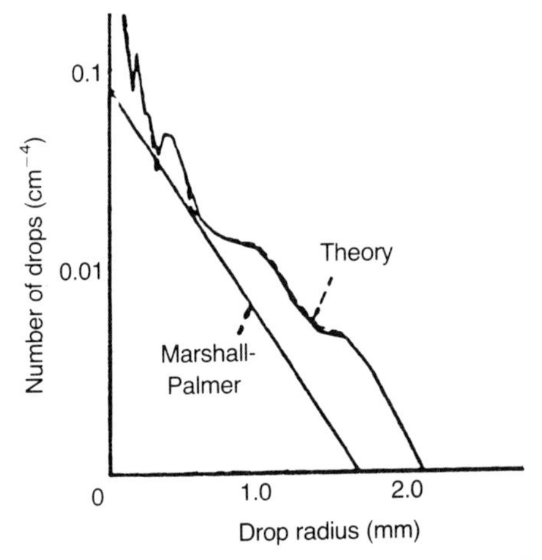

Fig. 13.10 Calculated drop size distribution and the Marshall–Palmer distribution, both for $R = 225.55$ mm h^{-1}.

(b) *Physical factors*

The exact interactive effects of the various physical factors that shape the drop size distribution (DSD) are not fully understood, because of their random and complex nature. However, in order to make use of a theoretical model, and to know about the range of validity, it takes at least a qualitative description of these effects. As most important effects the following can be mentioned.

- *Drop generation in clouds*. Non-precipitating water particles in rain clouds are composed of minute droplets of up to a few tens of microns in diameter and possess velocities of only a few centimetres per second. Considerable growth of particle size must thus be attained before a drop can gain a falling speed sufficient for it to fall out of a parent cloud and to be called a raindrop.
- *Evaporation*. While the generation mechanism may imply a certain lower limit on the drop size measured just below the cloud, this is eliminated by the effect of evaporation for larger fall distances. This causes shrinking of all drops during their fall, and complete evaporation for the smallest drops. The evaporation velocity is dependent on drop size, relative humidity, temperature and relative wind speed, although not all of these parameters are mutually independent.
- *Coalescence of drops*. The result of drop collision can be coalescence or breakup. Especially for small drops, coalescence is more probable. It seems that the relative importance of the coalescence versus the breakup process is dependent on rain rate.

- *Drop breakup*. Drops can break up into smaller drops not only as a result of collision, but also spontaneously. This is especially valid for drops larger than about 7 mm. Therefore, this can be regarded as an upper limit on drop size, although drops with an ice core can persist with a slightly larger size.
- *Fall velocity*. Fall velocity is more qualitatively known than the other effects. It has to be taken into account when the DSD is measured at the ground (per square metre per hour) and it is converted to the DSD in space (per cubic metre). Gunn and Kinzer (1949) have measured terminal fall velocity for water drops as a function of size. It can be assumed that the fall velocity, influenced by drag force and gravity, is not at all heights equal to the terminal velocity. However, Park, Mitchell and Bubenzer (1983) found that raindrops obtain 99% of their terminal velocity after a few metres of fall.

For a more complete review of the work done on DSD the reader is referred to Assouline and Mualem (1989). The most useful mathematical models for dropsize distributions are reviewed below.

(c) Exponential distribution

Marshall and Palmer (1948) proposed an exponential distribution:

$$M_a(D)\,(\mathrm{m^{-3}\,mm^{-1}}) = M_0\,\mathrm{e}^{-\Lambda D} \tag{13.2.27a}$$

$$M_0 = 8000\,\mathrm{m^{-3}\,mm^{-1}} \tag{13.2.27b}$$

$$\Lambda\,(\mathrm{mm^{-1}}) = 4.1 R^{-0.21} \tag{13.2.27c}$$

where $M_a(D)$ is the number of drops aloft, D (mm) is the diameter and R ($\mathrm{mm\,h^{-1}}$) is the rain rate. This result has been obtained from measurements of raindrops on dyed filter paper, performed in Montreal, Canada. The DSD appeared to be in agreement with the result of Laws and Parsons, who used the flour pellet method in Washington, DC. The range of validity is stated to be up from 1.5 mm. Below this value, the exponential distribution overestimates the number of drops.

(d) Shifted lognormal distribution

Park, Mitchell and Bubenzer (1983) presented a 'shifted lognormal' distribution:

$$M_g(D)\,(\mathrm{m^{-2}\,s^{-1}\,mm^{-1}}) = M_{gt}\,p(D) \tag{13.2.28a}$$

$$p(D)\ (\mathrm{mm}^{-1}) = \frac{1}{\sigma(2\pi)^{1/2}} \frac{1}{(D+s)\sigma} \mathrm{e}^{-[\log(D+s)-\mu]^2/2\sigma^2} \qquad (13.2.28b)$$

$$M_{\mathrm{gt}}\ (\mathrm{m}^{-2}\,\mathrm{s}^{-1}) = 154R^{1/2} \qquad (13.2.28c)$$

where

$$\mu = aR^{\mathrm{b}} \qquad (13.2.28d)$$

is the mean of $\log(D+s)$ and

$$\sigma = cR^{d} \qquad (13.2.28e)$$

is the standard deviation of $\log(D+s)$ and $p(D)$ is the probability density function, $M_{\mathrm{g}}(D)$ is the number of drops on the ground and $M_{\mathrm{gt}} = \int M_{\mathrm{g}}(D)\,\mathrm{d}D$ is the total number of drops on the ground. Proposed values are as follows:

$$s\ (\text{shifting value}) = 1\ \mathrm{mm}$$

$$a = 0.33$$

$$b = 0.12$$

$$c = 0.09$$

$$d = 0.08.$$

Park, Mitchell and Bubenzer made use of data from different experiments, which are summarized in Table 13.8.

The number of drops in the air $M_{\mathrm{a}}(D)$ can be obtained by dividing $M_{\mathrm{g}}(D)$ by the fall velocity v:

$$M_{\mathrm{a}}(D) = M_{\mathrm{g}}(D)/v(D). \qquad (13.2.29)$$

Table 13.8 Experiments used for the model of Park, Mitchell and Bubenzer (1983)

Source	Method	Location
Laws and Parsons (1943)	Flour pellet	Washington, DC
Chapman (1948)	Flour pellet	New Haven, CT
Rogers *et al*. (1967)	Photographs	Urbana, IL
Carter *et al*. (1974)	Flour pellet	Baton Rouge, LA
Bubenzer (1979)	Flour pellet	Pullman, WA

This does, however, have the following consequence. Because of the shifting value, the distribution for $M_g(D)$ does not become zero for $D \to 0$. When dividing this by v, which does become zero, the result becomes infinity for $D \to 0$. Since this is not desirable, this model can only be used from a lower limit of about 0.1 mm upwards.

(e) Γ distribution

Ulbrich (1983) proposed a Γ distribution:

$$M_a(D)\ (\mathrm{mm^{-1}\,m^{-3}}) = M_0 D^m\, \mathrm{e}^{-(3.67+m)D/D_0} \tag{13.2.30}$$

where M_0 ($\mathrm{mm^{-1}\,m^{-3}}$) is a proportionality factor, D_0 is the median diameter, ranging from 0.5 to 3 mm and m is the dispersion factor, ranging from -2 to 6.

(f) Modified Γ distribution

Ugai *et al.* (1977) present a 'modified Γ' distribution function of drop volume:

$$M_g(V)\ (\mathrm{mm^{-3}\,m^{-2}\,s^{-1}}) = M_0 V^{\mu-1}\, \mathrm{e}^{-\alpha V-\beta/V} \tag{13.2.31a}$$

$$M_0 = M_{gt}\frac{(\alpha/\beta)^{\mu/2}}{2K_\mu(2(\alpha\beta)^{1/2})}\ (\mathrm{mm^{-3}\,m^{-2}\,s^{-1}}) \tag{13.2.31b}$$

where M_0 is the proportionality factor, $V = \pi D^3/6$ ($\mathrm{mm^3}$) is the drop volume, α is a parameter ranging from 0.1 to 1.5, $\beta = 10^{-5}$ is a parameter, μ is a parameter ranging from -1.4 to 0.5 and $K_\mu(x)$ is the μth-order modified Bessel function of argument x. The mode of the distribution function is found to depend mainly on the value of μ, the largest diameter on the value of α and the smallest on the value of β. μ is dependent on rain rate, but not on this alone. Various distribution functions are found to exist under a certain fixed rainfall intensity, by variation of μ and α.

The distribution function was fitted to results of measurements of drop size. The experiment made use of a pan filled with castor oil to catch the raindrops and was performed in Tokyo, Japan.

(g) Weibull distribution

Assouline and Mualem (1989) present the Weibull distribution:

$$M_g(x)\ (\mathrm{mm^{-1}\,m^{-2}\,s^{-1}}) = \frac{M_{gt}}{D_0}\varphi n x^{n-1}\, \mathrm{e}^{-\varphi x^n} \tag{13.2.32a}$$

where

$$x = D/D_0 \tag{13.2.32b}$$

$$D_0 = aR^b \, e^{-cR} \tag{13.2.32c}$$

$$\varphi = \Gamma^n(1 + 1/n) \tag{13.2.32d}$$

and D_0 is the mean drop size, $\Gamma(\ldots)$ is the standard Γ function and a, b, c and n are parameters. Assouline and Mualem derived this expression theoretically, starting from the mechanism of drop coalescence and breakup. The expression is adjustable to climatic variations. It was compared with measurement results from different experiments in different locations, which resulted in values of the parameters as summarized in Table 13.9. A good correlation was found between the model and the measurement results.

The model can be simplified and presented as a 'universal distribution function', when n is taken to be 3 ($\varphi = 0.71$):

$$M_g(x) = \frac{M_{gt}}{D_0} 2.13x^2 \, e^{-0.71x^3}. \tag{13.2.33}$$

(h) Comparison of the DSDs

The drop size distributions of the previous section are plotted simultaneously for different rain rates in Fig. 13.11. They are all taken as the number of drops in space $M_a(D)$.

The expression for the modified Γ distribution of Ugai *et al.* is not adjustable to rainfall intensity using merely the information in the paper. In

Table 13.9 Model parameters derived for data from different sites in the world

Source	Method	Location	a	b	c	n
Laws and Parsons (1943)	Flour	Washington, DC	1.314	0.181	0.297×10^{-3}	2.94
Hudson (1965)	Flour	Zimbabwe	0.858	0.313	0.337×10^{-2}	2.32
Cateneo and Stout (1968)	Camera	Coweeta, NC	0.807	0.173	0.198×10^{-2}	3.49
Carter *et al.* (1974)	Flour	Baton Rouge, LA Holly Springs, MS	0.941	0.336	0.471×10^{-2}	2.39
Feingold and Levin (1986)	Distro-meter	Hadera, Israel	0.343	0.512	0.891×10^{-2}	2.68

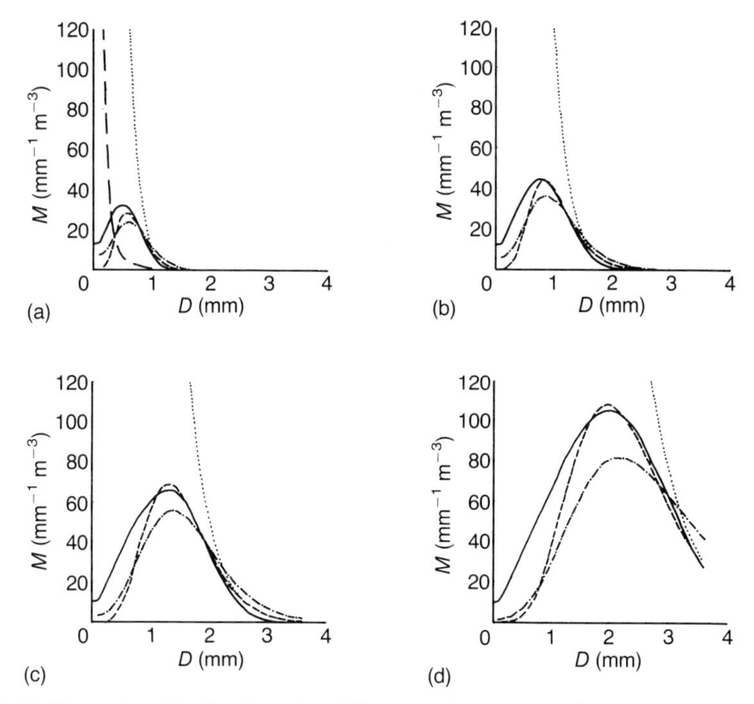

Fig. 13.11 Drop size distributions for different rain rates R: (a) $R = 0.1$ mm h^{-1}; (b) $R = 1$ mm h^{-1}; (c) $R = 10$ mm h^{-1}; (d) $R = 100$ mm h^{-1}; · · · · ·, exponential distribution; - · - · -, shifted lognormal distribution; - - - - -, Γ distribution; — —, modified Γ distribution; ——, Weibull distribution.

Fig. 13.11(a), it is plotted for the parameter values which made it lie closest to the others, within the range specified before. Still, this curve deviates strongly from the others.

It is evident from Fig. 13.11 that the exponential distribution of Marshall and Palmer overestimates the number of drops with radius below a certain limit. This agrees with the conclusions of Marshall and Palmer when they compared it with the data of Laws and Parsons. For the small drop range which is of interest for millimetre wave propagation, consequently, this model can not be used.

For the Γ, shifted lognormal and Weibull distributions, it is found that they do not differ essentially from each other. Each of these has been compared with measurement results by the corresponding researchers, who all say that good agreement is found. It can be concluded that each of these could be used for the calculations of scattering by small raindrops.

The Weibull distribution has the advantage of simplicity and ability to be easily adjusted to different climatic variations. With the simplification of equation (13.2.33), it is even more simple, and still adjustable to rain rate variations, so in this form it is applicable to any region in the world where the parameter values are not known.

13.3 ICE

13.3.1 Introduction

The most important effect of ice particles on radiowave propagation is depolarization. Ice particles occur in ice clouds when the particles are more or less aligned. This alignment causes depolarization of the radiowaves. It is mainly for this reason that they have been studied. The attenuation of ice particles is much less than the attenuation of raindrops.

13.3.2 Dielectric properties

(a) General picture

Surveys of current knowledge on the complex permittivity of ice, covering the entire electromagnetic spectrum, have been given by Ray (1972), Warren (1984) and Hufford (1991). Although the general characteristics of the dielectric properties of ice are believed to be well understood, there are important gaps of detail where existing data are sparse and/or inconsistent (Hufford, 1991).

The general picture is one with dipole relaxation at low frequencies and a set of absorption lines in the IR region. In between there are no new phenomena. Hufford has pointed to the strikingly low value of frequency for the dipole relaxation process (7.3 kHz at 0 °C), coupled with the fact that the first absorption line is at ≈ 4.8 THz. In between ice should be a near-perfect dielectric; it exhibits absorption only in so far as the processes on either side of the radio spectrum have far wings that are still active.

(b) Models for ice permittivity

Wörz and Cole (1969) measured ε_r' and ε_r'' near the low-frequency end and produced a semi-empirical model for frequency dependence including a frequency-independent conductivity term.

Ray (1972) then developed the model of Wörz and Cole to include the spread parameter s given in equations (13.2.5), (13.2.6) and (13.2.7). Then s and ε_r^0 were adjusted so that $n' = 1.78$ over the entire microwave spectrum and so that n'' agreed with the values found by Cumming (1952) at 3.2 cm as a function of temperature. The results are

$$\varepsilon_r^\infty = 3.168 \tag{13.3.1a}$$

$$s = 16.01 - 0.12038T + 0.00023T^2 \tag{13.3.1b}$$

$$\sigma\,(\Omega^{-1}\,\text{m}^{-1}) = 1.26\,\text{e}^{-6291.2/T} \times \varepsilon_0 \tag{13.3.1c}$$

$$2\pi c\tau = \lambda_s\,(\text{cm}) = 9.990288 \times 10^{-4}\,\text{e}^{6643.5/T} \tag{13.3.1d}$$

$$\varepsilon_r^0 = 10700 - 79.4T + 0.15T^2. \tag{13.3.1e}$$

Substitution of equations (13.3.1) into equations (13.2.2), (13.2.5) and (13.2.6) gives the Debye absorption for ice as a function of temperature. This has been plotted for −20 °C and 0 °C in Fig. 13.12(a) together with the results of Wörz and Cole. The remaining absorption bands were represented using equation (13.2.9) with the values of n'' taken from Irvine and Pollack (1968). The parameter values are listed in Table 13.10; n' in the IR region was obtained using equation (13.2.10) with $T = 273$ K and the data from Irvine and Pollack. Details are given in Tables 13.11 and 13.12 (Ray, 1972).

Hufford (1991) has re-examined the fit of Ray's model (and the original Wörz and Cole model) in the frequency range above 10 GHz for ε'' (the loss factor). He points out that three sets of experimental data are normally quoted in survey papers, and refers to a more recent set of data (Mätzler and Wegmüller, 1987). All of these are seen to be quite inconsistent, to the extent that they seem to describe different phenomena, although they all seem to contain careful measurements using good equipment. Hufford, in the light of a careful interpretation of the experimental data available (including some recent IR data), has thus given an alternative semi-empirical fit to the experimental data for ε'', which is believed to be more accurate at frequencies above 10 GHz than that of Ray. This is illustrated in Fig. 13.2(b), and Hufford's equations are given below:

$$\varepsilon_i' = 3.15 \tag{13.3.1f}$$

and

$$\varepsilon_i''(f) = A_i/f + B_i f \tag{13.3.1g}$$

where

$$A_i = [50.4 + 62(\theta - 1)]10^{-4} \exp[-22.1(\theta - 1)] \tag{13.3.1h}$$

and

$$B_i = (0.633/\theta - 0.131)10^{-4} + [7.36 \times 10^{-4}\theta/(\theta - 0.9927)]^2 \tag{13.3.1i}$$

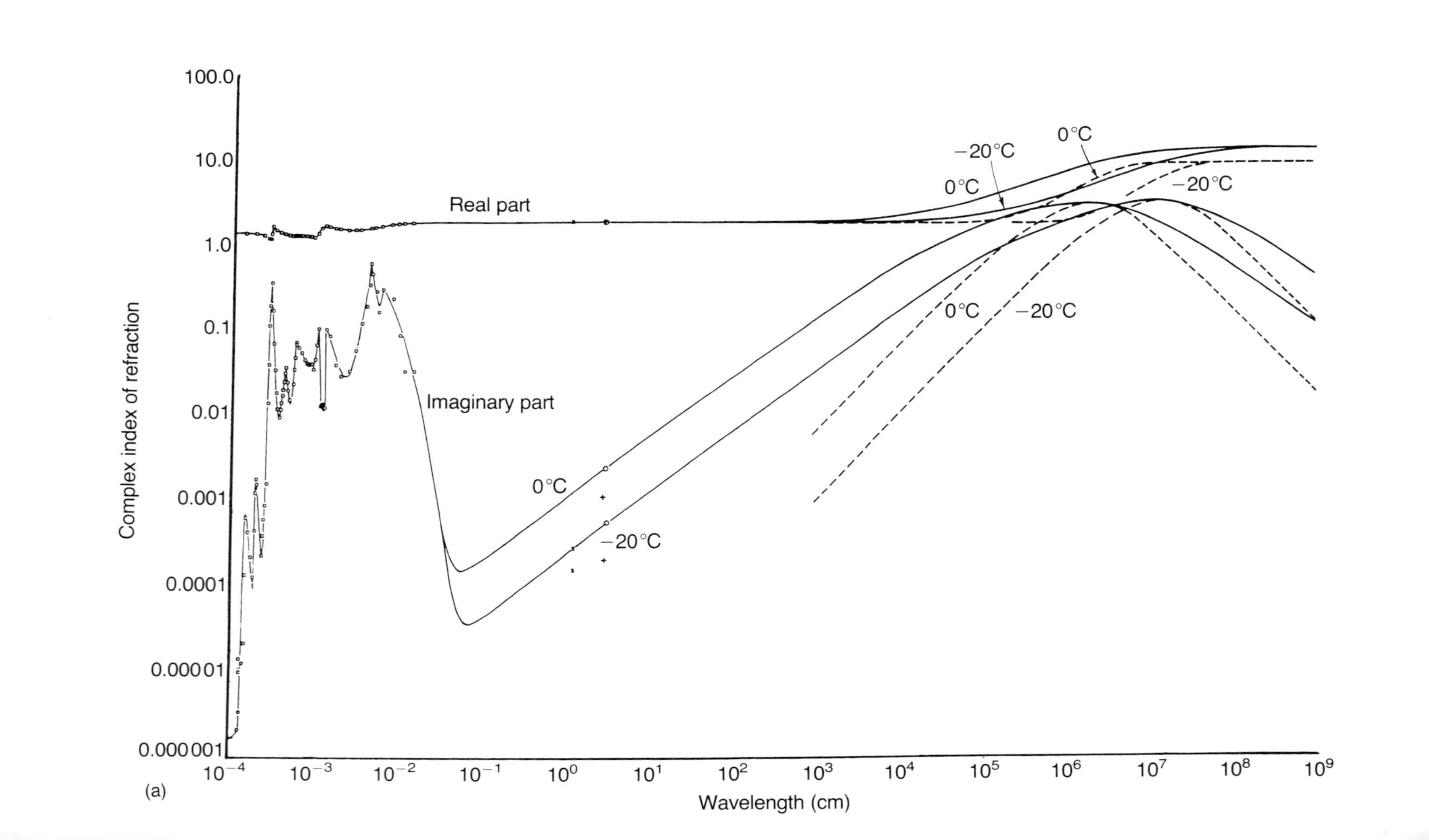

Complex index of refraction
100.0
10.0
1.0
0.1
0.01
0.001
0.0001
0.00001
0.000001
Real part
Imaginary part
0°C
−20°C
0°C
−20°C
0°C
−20°C
0°C
−20°C
0°C
−20°C
10^{-4}
10^{-3}
10^{-2}
10^{-1}
10^{0}
10^{1}
10^{2}
10^{3}
10^{4}
10^{5}
10^{6}
10^{7}
10^{8}
10^{9}
Wavelength (cm)
(a)

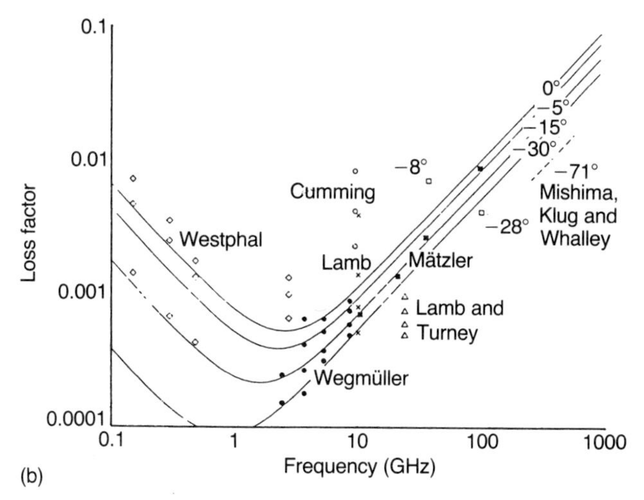

Fig. 13.12 (a) Ray's (1972) semi-empirical model of the complex index of refraction for ice with a sample of experimental data as used by Ray to develop the model: – – –, Wörz and Cole; □, Irvine and Pollack; ○, Cumming; ×, Lamb and Turney; +, Lamb. For references to experimental data see Ray (1972). (b) The loss factor ε'' of ice as a function of frequency, as presented by Hufford (1991): ——, Hufford's model; ◊, ●, ×, △, □, ■, - - - - -, experimental data.

with

$$\theta = 300/T \tag{13.3.1j}$$

and the temperature T is in kelvin. It should be noted that the loss factor at ≈ 30 GHz in Hufford's model is an order of magnitude greater than that in the model of Ray (Fig. 13.12(a)).

(c) Conclusions on ice permittivity

Evidently the order of magnitude uncertainty in the experimental data for the loss factor of ice requires explanation and removal. The possibility of the presence of impurities has been raised by Hufford, who has also suggested an appropriate programme of measurements.

However, in terms of extinction, where for ice the greater contribution is made by scattering rather than absorption, uncertainty in the loss factor in the lower millimetre range (say 30–100 GHz) has a negligible effect on attenuation. For example at 100 GHz the specific attenuation for a monodispersed medium, with equivalent ice–water content of 1 g m^{-3} consisting of 1 mm radius spheres, is 12.46 dB km^{-1} and 12.35 dB km^{-1}, predicted using respectively the models of Ray and Hufford.

Table 13.10 Parameter values for n_i in the model for ice

Spectral interval (μm)	λ_{0i} (μm)	β_i	Δ_i	γ_i	Spectral interval (μm)	λ_{i0} (μm)	β_i	Δ_i	γ_i
$\lambda \leqslant 1.5$	1.5	0.00063	0.027	1.4	$9.0 \leqslant \lambda \leqslant 11.0$	6.05	0.064	0.13	1.0
$1.5 \leqslant \lambda \leqslant 2.0$	1.5	0.00063	0.043	1.5		6.65	0.011	0.075	0.85
	2.0	0.00158	0.017	1.6		9.0	0.008	0.01	2.0
	4.5	0.0325	0.052	1.0		11.0	0.0855	0.038	0.91
$2.0 \leqslant \lambda \leqslant 3.075$	2.0	0.00158	0.03	1.35		44.8	0.581	0.06	0.90
	3.075	0.3428	0.022	1.2	$11.0 < \lambda < 13.5$	6.05	0.064	0.13	1.0
	4.5	0.0325	0.052	1.0		6.65	0.011	0.075	0.85
$3.075 \leqslant \lambda \leqslant 4.5$	3.075	0.3428	0.019	1.14		11.0	0.0855	0.0008	0.3
	4.5	0.0325	0.052	1.0		13.5	0.087	0.0025	0.6
$4.5 \leqslant \lambda \leqslant 6.05$	4.5	0.0325	0.043	1.05		44.8	0.58	0.06	0.9
	6.05	0.064	0.032	1.32	$13.5 < \lambda < 44.8$	6.05	0.064	0.13	1.0
	6.65	0.011	0.009	0.7		6.65	0.011	0.075	0.85
	9.0	0.008	0.06	1.3		11.0	0.0855	0.0008	0.3
	11.0	0.0855	0.038	0.91		13.5	0.087	0.14	1.1
$6.05 \leqslant \lambda \leqslant 6.65$	4.5	0.0325	0.043	1.05		44.8	0.581	0.06	0.9
	6.05	0.064	0.13	1.0		62.0	0.242	0.04	1.0
	6.65	0.011	0.009	0.7	$44.8 < \lambda < 62.0$	13.5	0.087	0.14	1.1
	9.0	0.008	0.06	1.3		44.8	0.581	0.055	1.0
	11.0	0.0855	0.038	0.91		62.0	0.242	0.04	1.0
$6.65 \leqslant \lambda \leqslant 9.0$	4.5	0.0325	0.043	1.05	$62.0 < \lambda$	44.8	0.581	0.055	1.0
	6.05	0.064	0.13	1.0		62.0	0.242	0.23	1.6
	6.65	0.011	0.075	0.85					
	9.0	0.008	0.06	1.3					
	11.0	0.0855	0.038	0.91					

Table 13.11 Parameter values for n_r in the model for ice

Spectral interval (μm)	α_a	ω_{0i}	β_i	γ_i
(a) $\lambda \leqslant 3.5$	1.75	3252.27	10 18140.0	76 183.4
		1652.9	588 684.0	73 801.1
		909.09	46×10^{-11}	46×10^{-11}
(b) $4.0 \leqslant \lambda \leqslant 7.5$	2.156 22	3252.27	29.8023×10^{-9}	10.7118×10^{18}
		1652.9	744 332.0	1 100 350.0
		909.09	1 645 160.0	434 408.0
(c) $9.5 \leqslant \lambda \leqslant 200.0$	1.2225	1652.9	1 120 820.0	46×10^{-11}
		909.09	416 441.0	118 852.0
		223.2	47 031.8	126 834.0
$3.5 \leqslant \lambda \leqslant 4.0$	$n_r = n_r^{(b)} \dfrac{\lambda - 3.5}{0.5} + n_r^{(a)} \dfrac{4.0 - \lambda}{0.5}$			
$7.5 \leqslant \lambda \leqslant 9.5$	$n_r = n_r^{(c)} \dfrac{\lambda - 7.5}{2.0} + n_r^{(b)} \dfrac{9.5 - \lambda}{2.0}$			
$200.0 \leqslant \lambda \leqslant 800.0$	$n_r = n_r^{(D)} \dfrac{\lambda - 200.0}{600.0} + n_r^{(c)} \dfrac{800.0 - \lambda}{600.0}$			
$800.0 \leqslant \lambda$	$n_r = n_r^{D}$			

Table 13.12 Characteristics of some of the data used in the model for ice

Reference	Spectral interval (μm)	Temperature (°C)	Percent relative error: real part, imaginary part
Irvine and Pollack (1968)	1–152	−50	1.64, 22.95
Cumming (1952)	3.2×10^4	−18 to 0	–, 1.70
Lamb (1946)	1.25×10^4	−50 to 0	–, 22.19
Lamb and Turney (1949)	3.0×10^4	−180 to 0	–, 84.81

13.3.3 Shape and size

(a) Classification by shape

The ice elements formed in natural clouds are of four main types; ice crystals, snowflakes, ice pellets and hailstones. Snowflakes are the result of coagulation of several ice crystals. Ice pellets and hailstones may originate from ice crystals or frozen drops. The classification of snow and ice agreed by an international commission in 1949 follows that given in Table 13.13. The terms

Table 13.13 Classification of solid precipitation (Mason, 1971)

Code	Graphic symbol	Typical forms	Term	Remarks
Type of particle (F)				
1			Plates	Also combinations of plates with or without very short connecting columns
2			Stellar crystals	Also parallel stars with very short connecting columns
3			Columns	Also combinations of columns
4			Needles	Also combinations of needles
5			Spatial dendrites	Spatial combinations of feathery crystals
6			Capped columns	Columns with plates on either (or one) side
7			Irregular particles	Irregular aggregates of microscopic crystals
8			Graupel (soft hail)	Isometric shape, central crystal cannot be recognized

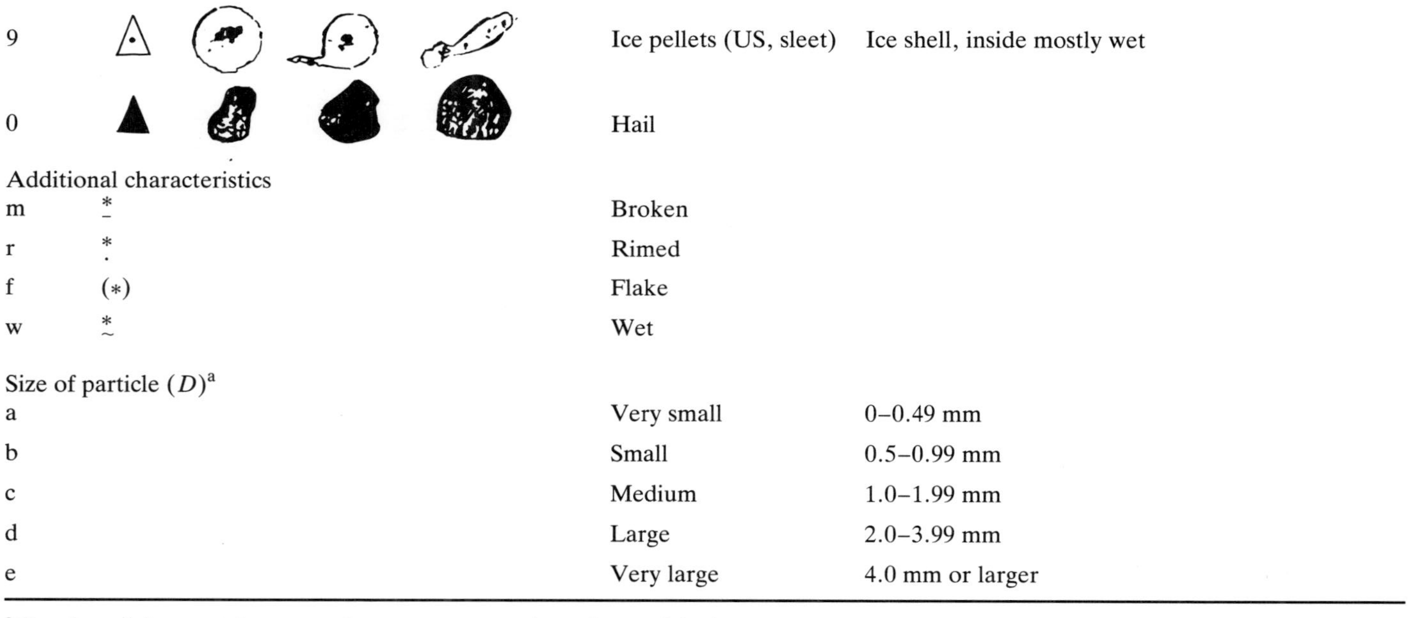

9		Ice pellets (US, sleet)	Ice shell, inside mostly wet
0		Hail	
Additional characteristics			
m	$\underset{-}{*}$	Broken	
r	$\underset{\cdot}{*}$	Rimed	
f	(*)	Flake	
w	$\underset{\sim}{*}$	Wet	
Size of particle (D)[a]			
a		Very small	0–0.49 mm
b		Small	0.5–0.99 mm
c		Medium	1.0–1.99 mm
d		Large	2.0–3.99 mm
e		Very large	4.0 mm or larger

[a]The size of the particle means the greatest extension of a particle (or average when many are considered). For a cluster of crystals it refers to the average size of the crystals composing the flake.

used for the basic structures are as follows: plates, stellar crystals, columns, needles, spatial dendrites, capped columns, irregular particles, graupel (or soft hail), ice pellets and hail.

(b) Shape and size

Each crystal shape develops under certain special conditions that are present while it is growing. Temperature controls the direction of growth, and the effect of water vapour supersaturation with respect to ice controls the branching and detail. In the ranges from 0 °C to −3 °C and from −8 °C to −25 °C plates are formed; in the range from −3 °C to −8 °C and below −25 °C needles and prisms. Growth at about ice saturation is slow and gives unbranched crystals, whereas growth at high ice supersaturation is rapid and conducive to forming branches. Growth under changing conditions results in crystals composed of several of the basic types.

Crystal sizes range from a few microns to about 2 mm. Auer and Veal (1970) have made an extensive study on the dimensions of ice crystals collected at Elk mountain. Individual ice crystals occurring in orographic and cumuliform clouds have been collected, replicated and studied under a microscope. The empirical relationships for the ice crystal data were derived by fitting polynomial or power functions to the diameter and thickness data for plate crystals and length, width and thickness data for columnar crystals. The best-fit equations relating dimensions for various types of crystals are given in Table 13.14.

In view of the complexity of ice particle modelling, only limited further information will be given on the properties of particles with significant radiowave interaction, i.e. melting ice and snow (especially as found in the melting zone of stratiform rain; section 13.5) and particles with high orientation capability (needles and plates) which can cause depolarization. Dry ice particles, being almost pure dielectric in nature, cause little radio path attenuation. Simple calculations, assuming spherical dry ice particles of relatively high density and high number concentration, show little contribution to attenuation up to 100 GHz.

13.3.4 Orientation

Ice particles show both aerodynamic and electric field orientation mechanisms. Ice needles and plates have a preferred horizontal orientation for their longer axes under aerodynamic and gravitational forces.

As a result of radio measurements with ATS-6, it has been deduced that ice needles can also be oriented in a preferred direction in the horizontal plane by the electric fields associated with thunderstorms. Haworth, McEwan and Watson (1977) gave models for this effect. McEwan *et al.* (1981) have also

Table 13.14 A summary of the empirical dimensional relationships for a variety of crystal types identified as to type and temperature

Crystal type	Temperature regime (°C)	Dimensional relation (μm)
Hexagonal plate	−10 to −13; −17 to −20	
Plate with sector-like branches	−10 to −13; −17 to −20	$h = 2.020D^{0.449}$
Plate with simple extensions	−13 to −20	
Plate with sector-like extensions	−13 to −20	
Crystal with broad branches	−13 to −17	
Stellar	−13 to −17	$h = 2.028D^{0.431}$
Stellar with end plates	−13 to −17	
Stellar with sector ends	−10 to −17	
Ordinary dendritic	−13 to −17	
Fernlike crystal	−13 to −17	
Dendritic crystal with end plates	−10 to −17	$h = 2.801D^{0.377}$
Plate with dendritic extensions	−13 to −20	
Four-branched crystal	−13 to −17	
Dendritic crystal with 12 branches	−13 to −17	
Solid thick plate	−9.5 to −11; −18.5 to −20	$h = 0.402D^{1.018}$
Thick plate of skeleton form	−9.5 to −11; −18.5 to −20	
Solid column	−8 to −10; <−20	$w = -8.479 + 1.002l - 0.00234l^2$; $l \leq 200$ μm
Hollow column	−8 to −10; <−20	$w = 11.3l^{0.413}$; $l > 200$ μm; $h = 0.866D$
Elementary needle	−4 to −6	$w = 1.099l^{0.61078}$
Long solid column	−6 to −8	$h = w$

Dimensions: h, height; D, diameter; w, width, l, length.

argued, on the basis of radio measurements with OTS, that alignment out of the horizontal plane also takes place for ice needles and plates, because of electric fields. Subjective reports from aircraft of rapidly changing optical glint over a wide area from the ice particles in thunderstorms also supports the models for electric field alignment of ice particles.

The model for electric field alignment of ice particles is based on the following observation and deduction: large and very rapid changes of depolarization are observed on satellite–earth paths with an ice particle population along the line of sight. The rate of change of depolarization is too large to be explained by aerodynamic effects, even if these effects propagate at the speed of sound. Correlations between depolarization and electric field at the ground have also been observed.

13.3.5 Fall velocity

Very small ice particles may remain suspended or slowly falling in ice clouds and, subject to convective updraught or downdraught, may be aggregated through the process of riming to form hail or graupel (a more spongy form of hail). Ice and hail thus have very variable particle velocities which may, especially in the case of hail, be upward as well as downward. Because of this, the terminal velocity of a hail-like particle in still air is not a particularly meaningful or relevant parameter.

13.4 ICE, WATER AND AIR MIXTURES

13.4.1 Introduction

There is a large variety of forms of hydrometeors other than raindrops, consisting of varying proportions of ice, air and water, with length varying from fractions of a millimetre (e.g. ice needles) to several centimetres (hail) and with an enormous variety of shapes. Dry snow and hail are properly regarded as ice–air mixtures, whereas melting snow and melting hail are ice–air–water mixtures. Mixtures of ice and water with little air content should also be considered. Because these types of hydrometeors occur less frequently in European climates, they have been given less attention than raindrops. However, the impact on radiowave propagation can be as severe as for raindrops and ice particles. In the melting layer these mixtures occur during the melting process of ice particles. The melting layer and its particles are treated separately in Chapter 7 and section 13.5. Most models presented in this section describe the properties of snowflakes, for which many measurements and experiments have been carried out.

13.4.2 Dielectric properties

Finding the 'effective' dielectric properties of a mixture is a quite difficult problem, since a large number of interactions can occur among the component materials. The solutions can only be obtained by various approximations. Bohren and Battan (1980) examined the four dielectric functions that are frequently used, and showed similarities and differences of the functions in the light of recently developed theory. Their work suggests that the agreement between theory and measurements is not always satisfactory.

One method of finding the effective dielectric functions of mixtures is based on a simple and intuitive consideration by Wiener and is applied for evaluating the dielectric constants of snow. Consider a two-component mixture made up of materials 1 and 2, the fractions of the total volume occupied by them being q and $1 - q$, respectively; further, let the dielectric constants of materials 1 and 2 be ε_1 and ε_2, respectively, and the mean electric fields in materials 1 and 2 be E_1 and E_2 respectively; then

$$E = qE_1 + (1 - q)E_2 \tag{13.4.1}$$

$$\varepsilon E = \varepsilon_1 qE_1 + \varepsilon_2(1 - q)E_2 \tag{13.4.2}$$

where ε and E are the dielectric constant and the mean electric field of the effective medium. The ratio of equations (13.4.1) and (13.4.2) defines ε. Also the ratio

$$\frac{E_1}{E_2} = \frac{\varepsilon_2 + u}{\varepsilon_1 + u} \tag{13.4.3}$$

is considered, where the quantity u may depend on the shapes of the component materials and is termed the 'form number'. When the materials are arranged in flat layers which are parallel to the direction of the electric field, u becomes infinity and $E_1 = E_2$. Conversely, when the layers are perpendicular to the direction of the electric field, $u = 0$ and $\varepsilon_1 E_1 = \varepsilon_2 E_2$ (i.e. $D_1 = D_2$). Further, when spherical particles are sparsely distributed in vacuum, $u = 2$, and this conforms with the result of electrostatic calculations. For shapes other than flat layers or a sphere, we have to obtain u from comparisons between the defining equation (13.4.3) and measurements. Although this method is not purely theoretical, it has the advantage of flexibility for handling material of various shapes. The dielectric constants obtained by the above procedure have been used in the calculation of attenuation due to snow. The expression of the dielectric function for wet snow (ice–air–water mixture) is simply the extension of equations (13.4.1) and (13.4.2) for three components. The final expression is thus

$$\frac{\varepsilon_s - 1}{\varepsilon_s + u} = q_i \frac{\varepsilon_i - 1}{\varepsilon_i + u} + q_a \frac{\varepsilon_a - 1}{\varepsilon_a + u} + q_w \frac{\varepsilon_w - 1}{\varepsilon_w + u} \tag{13.4.4}$$

where ε_s, ε_i, ε_a and ε_w are the dielectric constants of snow, ice, air and water respectively and q_i, q_a and q_w are the fractions of the total volume occupied by ice, air and water respectively ($q_i + q_a + q_w = 1$).

A recent description of a general model for the effective permittivity of ice–air and ice–water mixtures based on the theory by Maxwell Garnett (1904) is given by de Wolf, Russchenberg and Ligthart (1990).

13.4.3 Shape and size of snowflakes

The classification of snow and ice agreed by an international commission in 1949 follows that given in Table 13.13. Snowflakes are aggregates of columnar or needle-like crystals. The maximum diameter of snowflakes is 15 mm; normally they have a diameter between 2 and 5 mm. Snowflakes are mostly classified on basis of their water content, because they have varying and often indefinite shapes.

From a study of mass, dimensions and fall velocity of snow crystals made on Mount Tokai, Japan, at temperatures ranging from −8 °C to −15 °C, Nakaya (1970) has given the following empirical formulae relating mass and maximum dimension:

rimed plates and stellar dendrites $m\ (\mathrm{mg}) = 0.027D^2$

powder snow and spatial dendrites $m\ (\mathrm{mg}) = 0.01D^2$

plane dendrites $m\ (\mathrm{mg}) = 0.0038D^2$

needles $m\ (\mathrm{mg}) = 0.0029l$

where D (mm) is the diameter of the sphere which just contains the crystal and l (mm) is the length of the needle. From the above relations and those given in Table 13.14, the density of snow crystals is found to vary between $0.15\ \mathrm{g\,cm^{-3}}$ and $0.45\ \mathrm{g\,cm^{-3}}$ (disregarding rimed crystals).

Gunn and Marshall (1958) derived from measurements a formula for the size distribution of aggregate snowflakes:

$$M(a)\,\mathrm{d}a\ (\mathrm{m^{-3}}) = M_0\,\mathrm{e}^{-\Lambda a}\,\mathrm{d}a \tag{13.4.5a}$$

with

$$M_0\ (\mathrm{m^{-3}\,mm^{-1}}) = 7.6 \times 10^3 R^{-0.87} \tag{13.4.5b}$$

$$\Lambda\,(\text{mm}^{-1}) = 5.1R^{-0.48} \tag{13.4.5c}$$

where $M(a)\,\mathrm{d}a$ is the number of snowflakes per unit volume in space with melted radius between a and $a + \mathrm{d}a$ and R is the rate of snowfall, in millimetres of water per hour.

Other investigators have proposed different parameters based on other measurements, but the Gunn and Marshall distribution is often used.

13.4.4 Orientation of snowflakes

Snow particles fall with their longer axis near horizontal, analogous to a falling leaf.

13.4.5 Fall velocities of snowflakes

Magono and Nakamura (1965) derived a semi-empirical formula for the fall velocity of snowflakes with a wide range of densities:

$$v = 3.14[a(\rho_s - \rho_a)]^{1/2} \tag{13.4.6}$$

where v ($\mathrm{m\,s^{-1}}$) is the terminal fall velocity, ρ_s and ρ_a ($\mathrm{g\,cm^{-3}}$) are the densities of snowflakes and air and a (mm) is the mean radius of snowflakes.

Pruppacher and Klett (1978) summarized the results of the measurements done by several investigators:

- most snowflakes fall with speeds between 1.0 and 1.5 $\mathrm{m\,s^{-1}}$;
- the terminal fall velocity increases with snow particle size;
- the terminal fall velocity increases with the mass of the snow particle.

13.5 NATURE OF PARTICLES IN THE MELTING ZONE

13.5.1 Introduction

The melting zone, observed as a radar bright band mainly in stratified precipitation, is of interest, simply because of its high radar reflectivity and potentially high specific attenuation (Jain and Watson, 1985). Unfortunately the characteristics of the particles and melting processes necessary to explain such a bright band are not simple (Klaassen, 1989). In this section an attempt will be made to summarize the assumptions on the nature of the particles made in the most recent models proposed for the European region.

Figure 13.13 shows some properties of the melting layer, or bright band, as described by Klaassen (1988). The importance of the bright band is related to the density of the particles before melting. According to the model of Klaassen, the reflectivity is very sensitive to the dielectric properties and density of the melting particles, but the influence of aggregation is restricted. The bright band is mainly observed in stratiform precipitation, but even there it may suddenly disappear. Hence its transient nature causes problems in choosing a realistic density of ice particles.

The melting layer is studied because of

- the attenuation of radio waves,
- the withdrawal of melting heat from the atmosphere, which may result in downdrafts, and
- the increased reflectivity in the layer, which may cause errors in radar-derived rain intensity predictions.

The explanations for the high reflectivity in the bright band are reviewed by Battan (1973). A reflectivity excess of about 5 dB, compared with the underlying rain, can be explained by the increase of the dielectric constant, followed by an increase of the fall velocity of the melting snowflakes. However, even higher reflectivities are common in the bright band. Two types of explanations are found in the literature to explain these high reflectivities.

- The first is the influence of aggregation and breakup: recent studies show

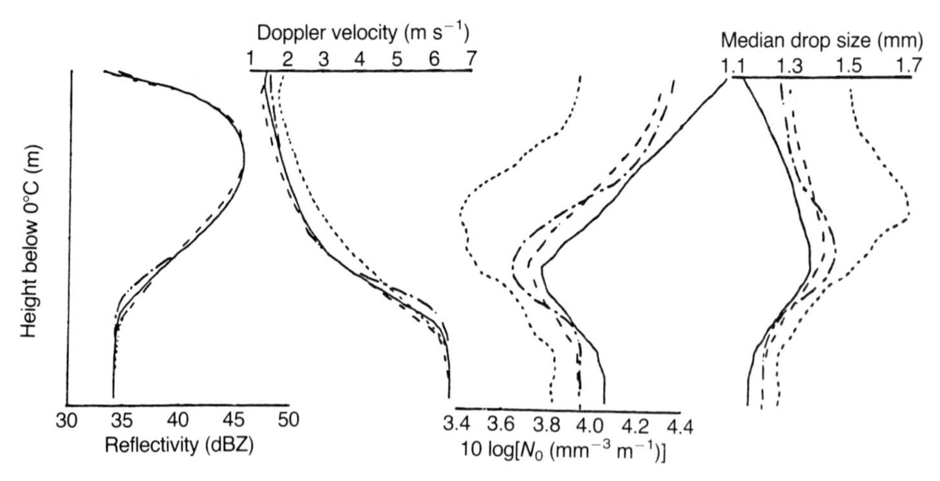

Fig. 13.13 Simulated profiles of reflectivity, mean Doppler velocity and the particle number density and size distribution in the melting layer, using different versions of the model (Klaassen, 1988); - · - · -, model without variations in ice density; – – –, basic model; ——, model with collisional aggregation–breakup; - - - - -, model with spontaneous breakup.

that this influence cannot be neglected. However, it is insufficient to explain the high reflectivity.
- The second is the assumption that melted water forms a shell around the particle. This may hold for hail, but for snow the melted water lodges inside, owing to surface tension considerations.

For estimating the upper boundary of the bright band, two methods are available. The first fixes the boundary at the altitude where the reflectivity Z drops to the value of the underlying rain; the second uses the first height above the height of maximum reflectivity where $\mathrm{d}^2Z/\mathrm{d}h^2 > 0$. Generally, with smooth profiles, both methods agree within 100 m. The lower boundary is fixed at the height where the Doppler velocity reaches its maximum value minus $0.5\ \mathrm{m\,s^{-1}}$.

13.5.2 Dissanayake and McEwan

Following Aden and Kerker (1951), Dissanayake and McEwan (1978) took a model for melting particles, in which melted water is distributed in a shell around the particle. The following additional assumptions are made:

- one-to-one correspondence between snowflakes and melted drops (substantiated by Ohtake (1969));
- spherical snowflakes and raindrops;
- latent heat of fusion provided by condensation in a modest updraught;
- particle breakup and aggregation negligible;
- density and size relation for snowflakes as given by Magono and Nakamura (1965).

This was the first model used to produce practical estimates of attenuation in the melting zone. Its major weaknesses are two-fold: (a) snowflakes, dry or partially melted, cannot be regarded as spherical, and (b) while for hail or graupel it may be acceptable to model the particles by a solid core surrounded by a water shell, for snowflakes the situation is much more complicated.

13.5.3 Dissanayake, Chandra and Watson

Dual-polarization radar measurements (Hall, Goddard and Cherry, 1982) showed a peak in differential reflectivity in the melting zone, at a height slightly lower than the peak in reflectivity. This could not, of course, be explained in terms of spherical scatterers. Dissanayake, Chandra and Watson (1983a) thus modelled the melting particles as oblate two-layer spheroids and calculated the scattering properties in two polarizations using the Waterman (1965) T-matrix method. A reasonable explanation for the separate horizontal and differential reflectivity peaks was thus given (Fig. 13.14).

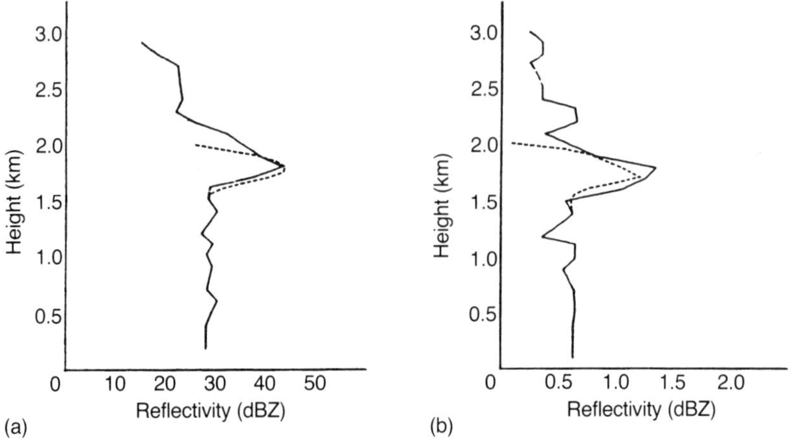

Fig. 13.14 Comparison of observed (——) and predicted (exponential DSD) (- - - - -) reflectivity profiles in the melting layer (event recorded by the Chibolton radar on May 20, 1980): (a) absolute reflectivity (horizontal polarization); (b) differential reflectivity.

13.5.4 Jain

Jain (1984) extended Dissanayake's approach yet further by including coalescence and breakup in the melting zone model. In order to maintain agreement with Ohtake (1969) the two processes were taken to counterbalance approximately, overall maintaining a one-to-one correspondence between snowflakes and raindrops. Jain includes a coalescence factor, indicating the intensity of coalescence, considering up to five-fold coalescence (a coalescence factor of 1 means a one-to-one correspondence, i.e. one snowflake results in one raindrop, on average).

Coalescence is assumed to take place down to a certain height below which the falling hydrometeors maintain a one-to-one correspondence, except in the final descent of 100 m where they break up as they melt. For the model of coalescence, better agreement with observed horizontal and differential reflectivity in the melting zone is claimed. The consequences of the Jain model in terms of specific attenuation in melting snow are described by Jain and Watson (1985).

13.5.5 Klaassen

(a) Introduction

Klaassen (1989) recognized the weakness in the underlying assumption of the previous models, in that the melting snowflake is not properly represented as an ice core surrounded by a water shell. This representation may hold for hail and graupel, but for snow the melted water lodges inside (Knight, 1979; Matsuo and Sasyo, 1981; Fujiyoshi, 1986).

Klaassen calculated the reflectivity using a Mie scattering matrix for spherical particles, with an average dielectric constant for ice–water mixtures. Hence although the approach by Klaassen is better at representing the dielectric properties of the particle, it does not and cannot represent differential reflectivity and depolarization in the melting zone.

(b) Properties of the model

The model of Klaassen

- includes a new formulation of the dielectric properties of the particles,
- can handle all ice particles with densities ranging from pure snow to hail,
- calculates air temperature from the vertical air velocity (it does not assume a particular value),
- can simulate the aggregation and breakup of the melting particles,
- gives results that are in good agreement with Doppler radar observations, and
- assumes stationarity, and therefore the best results are found in stratiform precipitation.

The following variables have to be determined in the model:

- the air pressure at the onset of melting;
- the precipitation intensity at the onset of melting;
- the median melted particle size;
- the empirical ice density factor;
- the vertical air velocity.

Figure 13.15 shows the observations compared with the simulations of the model of Klaassen.

(c) Assumptions

- The thickness of the melting layer is calculated from the energy balance of the rising air with a simple expression for the melting heat released.
- A one-dimensional model is chosen. With this assumption, it is possible to include many microphysical phenomena in the model, while keeping the computation time acceptable.
- The reflectivity is calculated assuming spherical particles.
- The vertical air velocity is assumed constant in the melting layer.
- The air is assumed to be saturated with water. In convective clouds this is a coarse approximation, but Matsuo and Sasyo (1981) found that the melting rate is hardly affected by the humidity of the air when it is calculated from the wet bulb temperature.

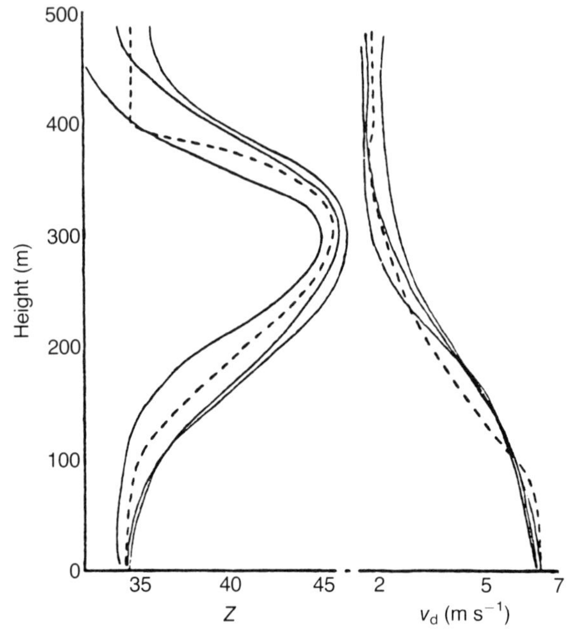

Fig. 13.15 Vertical profiles of reflectivity Z and mean Doppler velocity v_d in the melting layer during stratiform precipitation (Klaassen, 1988): ——, observed; – – –, simulated.

- The model assumes stationarity, i.e. equal inflow and outflow of particles in a certain air volume.
- The assumptions on particle breakup and aggregation, and the densities of the particles are described in section 13.5.5(d).
- The average dielectric constant of the ice–water–air mixtures is calculated using the Maxwell Garnett (1904) theory.

(d) Melted fraction and air temperature

The melted fraction of an ice particle is obtained by integration of the melting rate, which follows from simple physical expressions.

The vertical air temperature gradient is calculated from the wet adiabatic lapse rate, with a correction for the released melting heat. As melting proceeds and a larger number of particles have melted, the lapse rate approaches the wet adiabatic lapse rate. The highest part of the melting layer is almost isothermal. The height of this layer is proportional to the precipitation intensity and inversely proportional to the vertical air velocity. The dependence on vertical air velocity can be understood intuitively: without supply of new air, the air would continuously cool down, because of the release of melting heat. So, with increasing vertical air flow, the minimum temperature of the bright band increases and the height of the isothermal part decreases.

The expressions for the melted fraction and the air temperature form a couple of differential equations with a negative feedback: a high air temperature increases the melting rate, resulting in a decrease of air temperature. The thickness of the bright band increases with rain intensity: according to this model, this is mainly caused by a wider (almost) isothermal layer at the upper part of the melting layer.

(e) Particle size distribution

The model starts with a Γ distribution of particle sizes that may change during melting because of three effects:

- variations in fall velocity;
- condensation of water vapour on the melting particles;
- aggregation and breakup.

Variations in fall velocity In the stationary situation, the requirement of mass conservation results in an equal inflow and outflow of mass in a certain air volume, giving a number density inversely proportional to the fall velocity:

$$M(a)v(a) = \text{constant}. \tag{13.5.1}$$

The reflectivity is proportional to the particle number density, and thus inversely proportional to the fall velocity. For example, because of melting, the fall velocity in stratiform precipitation increases from 2 to 6 $\mathrm{m\,s^{-1}}$, resulting in a decrease in reflectivity of about 5 dB.

Condensation of water vapour on the melting particles The melting heat is received from sensible heat (the particle is colder than the surrounding air) and latent heat (the saturated water vapour pressure on the cold particle is smaller than the water vapour pressure of the surrounding air). Because of the latent heat flux, the melted diameter of the particle grows during melting. As the model does not simulate the transport of water vapour, the air is assumed to be saturated with water. This results in an increase of about 10% in water mass during melting.

Aggregation and breakup Aggregation and breakup are caused by the collision of particles with different fall speeds. The collision results in one particle of the combined mass (aggregation) or a large number of smaller particles (breakup). This model uses the criterion of List and Gillespie (1976), who find breakup when the smallest drop exceeds 1 mm. For melting particles, the model assumes breakup when the amount of water in the smallest particle exceeds the amount of water in a 1 mm drop. According to Brazier-Smith, Jennings and Latham (1973), an average of three satellite drops is formed per breakup. The mass distribution per fragment is given as 1.6%, 11.4% and 87%.

Spontaneous breakup may occur when the particle becomes unstable, for instance because of melting. This effect is negligible for hail with a diameter below 9 mm. According to Fujiyoshi (1986), spontaneous breakup is observed frequently during the melting of snow, and the largest breakup fragments are found in the final stage of melting. From these qualitative findings, the model optionally includes spontaneous breakup for all particles with initial density below 0.2 $\mathrm{kg\,m^{-3}}$ when 80% of the mass is melted, with a fragment distribution equal to that given for collisional breakup.

In order to conserve mass, the number density of the particles that result from aggregation and breakup is corrected for deviations in fall velocity between the initial and resulting particles.

(f) Density of the particles

The density of the melting particles affects the chance of collision between the particles (through their cross-section), the fall velocity and the reflectivity. For the density of dry snowflakes, the model uses a new relation as a function of size, derived from measurements of Magono and Nakamura (1965). This relation contains stochastic variables to model the scatter of the density of snowflakes. For melting particles, the model assumes that the volume of the particle is proportional to the melted fraction until the particle is saturated with water. After saturation, the density is linearly interpolated between the densities of ice and water.

(g) Fall velocity

The terminal fall velocity of the particles is calculated from the air friction. This friction depends on

- a friction coefficient, which depends on the shape of the particle,
- the horizontal cross-section,
- the (constant) gravity acceleration,
- the density of the surrounding air, and
- the diameter of the melting particle.

Klaassen uses an expression of the friction coefficient that is derived from the results of Langleben (1954) and Locatelli and Hobbs (1974). This expression is used until the particle is saturated with water; after saturation, the fall velocity is linearly interpolated to the value for raindrops given by Atlas,

Srivasta and Sekhon (1973). The fall velocity is corrected for air density according to Beard (1985).

In the model, the friction coefficient is related to the heat exchange coefficient, which is important for the determination of the melted fraction of the particles. Too few measurements of the heat exchange coefficient are available to check this relationship on accuracy.

13.5.6 Conclusions

The existence of a bright band is found to be related to the ice particle density, rather than to the rain intensity. However, the thickness increases with rain intensity. According to the model, this is mainly caused by a thicker (almost) isothermal layer at the upper part of the melting layer.

The radar reflectivity and mean Doppler velocity can be simulated accurately with the melting layer model in stratiform and slightly convective precipitation. In strong convective situations, the observations show too many meteorological fluctuations, which cannot be simulated by a stationary model.

The dielectric constant of the melting particles was calculated successfully by taking the air as inclusions within the melting ice frame. The ice density was fitted between the snow and hail densities, resulting in a realistic simulation of melting layer reflectivity. Including variations around the mean ice density and spontaneous breakup near the end of the melting improved the simulation, but it appeared not useful to include collisional aggregation or breakup.

Calculation of the temperature profile in the melting layer is essential in order to simulate the observed thickness of the bright band, although the vertical air velocity resulting in the model is an averaged value that does not account for actual fluctuations.

The relation between the reflectivity of the melting layer and the attenuation of radio waves passing through the melting layer has not been investigated.

The model of Klaassen perhaps gives the most accurate description of the melting process, but it has not been applied to give accurate scattering predictions. Indeed unless simplifications are made on the shape of the scatterers, the scattered field is very difficult to evaluate. Klaassen used Mie scattering and spheres. In their first model, Dissanayake and McEwan (1978) took two-layer spheres, but in their later work Dissanayake, Chandra and Watson (1983b) took two-layer spheroids, which is slightly more realistic.

There still exists a fair degree of uncertainty regarding the significance and proper modelling of the effects of the melting layer on millimetre wave propagation. Because of the correlation in the occurrence of rain and melting ice, it is very difficult to distinguish their effect in propagation experiments.

REFERENCES

Aden, A.L. and Kerker, M. (1951) Scattering of electromagnetic waves by two concentric spheres. *J. Appl. Phys.*, **22**, 1242–6.

Arnold, H.W., Cox, D.C., Hoffman, H.H. and Leck, R.P. (1980) Characteristics of rain and ice depolarization for a 19- and 28-GHz propagation path from a Comstar satellite. *IEEE Trans. Antennas Propag.*, **28**(1), 22–8.

Assouline, S. and Mualem, Y. (1989) The similarity of regional rainfall: a dimensionless model of drop size distribution. *Trans. ASAE*, **32**(4), 1216–22.

Atlas, D., Srivasta, R.C. and Sekhon, R.S. (1973) Doppler radar characteristics of precipitation at vertical incidence. *Rev. Geophys. Space Phys.*, **11**(1), 1–35.

Auer, A.H. and Veal, D.L. (1970) The dimensions of ice crystals in natural clouds. *J. Atmos. Sci.*, **27**, 919–26.

Battan, L.J. (1973) *Radar Observation of the Atmosphere*, University of Chicago Press.

Beard, K.V. (1976) Terminal velocity and shape of cloud and precipitation drops aloft. *J. Atmos. Sci.*, **33**, 851–64.

Beard, K.V. (1985) Simple altitude adjustments to raindrop velocities for Doppler radar analysis. *J. Atmos. Oceanic Technol.*, **2**, 468–71.

Bohren, C.F. and Battan, L.J. (1980) Radar backscattering by inhomogeneous precipitation particles. *J. Atmos. Sci.*, **37**, 1821–7.

Brazier-Smith, P.R., Jennings, S.G. and Latham, J. (1973) Raindrop interactions and rainfall rates within clouds. *Q. J. R. Meteorol. Soc.*, **99**, 260–72.

Brussaard, G. (1976) A meteorological model for rain induced cross-polarisation. *IEEE Trans. Antennas Propag.*, **24**, 5–11.

Bubenzer, G.D. (1979) Rainfall characteristics important for simulation, in *Proc. Rainfall Simulation Workshop*, Tucson, AZ, USDA-SEA, ARM-W-10.

Carter, C.E., Green, J.D., Brand, H.J. and Floyd, J.M. (1974) Raindrop characteristics in south central United States. *Trans. Am. Soc. Agric. Eng.*, **17**(6), 1033–7.

Cateneo, R. and Stout, G.E. (1968) Raindrop-size distribution in humid continental climates and associated rainfall rate–radar reflectivity relationships. *J. Appl. Meteorol.*, **7**, 901–7.

CCIR (1986) *Recommendations and Reports of the CCIR*, Vol. V, Propagation in Non-ionised Media, p. 200.

Chapman, G. (1948) Size of raindrops and their striking force at the soil surface in a Red Pine plantation. *Trans. Am. Geophys. Union*, **29**, 664–70.

Cole, K.S. and Cole, R.H. (1941) Dispersion and absorption in dielectrics. *J. Chem. Phys.*, **9**, 341–51.

Cumming, W.A. (1952) The dielectric properties of ice and snow at 3.2 centimeters. *J. Appl. Phys.*, **23**, 768–73.

Dissanayake, A.W.N.J. and McEwan, N.J. (1978) Radar and attenuation properties of rain and bright band. Int. Conf. on Antennas and Propagation, November 1978. *IEE Conf. Publ.*, **169**, 125–9.

Dissanayake, A.W., Chandra, M. and Watson, P.A. (1983a) Backscattering and differential backscattering characteristics of the melting layer. Proc. URSI Comm. F. Symposium, Louvain, pp. 363–70.

Dissanayake, A.W., Chandra, M. and Watson, P.A. (1983b) Prediction of differential reflectivity due to various type of ice particles and ice–water mixtures. *IEE Conf.*

Publ. ICAP, **219** (Part 2), 56–9.

Feingold, G. and Levin, Z. (1986) The lognormal fit to raindrop spectra from frontal convective clouds in Israel. *Am. Meteorol Soc.*, 1–19.

Findeisen, W. (1939) The evaporation of cloud and raindrops. *Meterol. Zeitschrift*, **56**(453), 282.

Fujiyoshi, Y. (1986) Melting snowflakes. *J. Atmos. Sci.*, **43**, 307–11.

Gossard, E. (1983) *Radar Observations of Clear Air and Clouds*, Elsevier, New York.

Gunn, K.L.S. and Marshall, J.S. (1958) The distribution with size of aggregate snowflakes. *J. Meteorol.*, **15**, 452–61.

Gunn, R. and Kinzer, G.D. (1949) The terminal velocity of fall for water droplets in stagnant air. *J. Meteorol.*, **6**, 243–8.

Hall, M.P.M., Goddard, J.W.F. and Cherry, S.M. (1982) Identification of hydrometeors and other targets by dual polarisation radar. URSI Commission F, Open Symp., Bournemouth.

Hasted, J.B. (1973) *Aqueous Dielectrics*, Chapman & Hall, London, pp. 136–65.

Hasted, J., Ritson, D. and Collie, C. (1948) Dielectric properties of aqueous ionic solutions, parts I and II. *J. Chem. Phys.*, **16**, 1–11.

Haworth, D.P., McEwan, N.J. and Watson, P.A. (1977) Crosspolarisation for linearly and circularly polarised waves propagating through a population of ice particles on satellite–earth paths. *Electron. Lett.*, **13**, 703–4.

Howard, J. and Gerogiokas, M. (1982) A statistical raindrop canting angle model. *IEEE Trans. Antennas Propag.*, **30**, 141–7.

Hudson, N.W. (1965) The influence of rainfall mechanics on soil erosion. MSc Thesis, University of Cape Town.

Hufford, G. (1991) A model for the complex permittivity of ice. *Int. J. Infrared Millimeter Waves*, **12**(7), 677–81.

Irvine, W.M. and Pollack, J.B. (1968) Infrared optical properties of water and ice spheres. *Icarus*, **8**, 324.

Jain, Y.M. (1984) Microwave scattering from melting zone particles and oscillating raindrops. PhD Thesis, University of Bradford.

Jain, Y.M. and Watson, P.A. (1985) Attenuation in melting snow on microwave and millimetre wave terrestrial radio links. *Electron. Lett.*, **21**(2), 68–9.

Joss, J., Thams, J.C. and Waldvogel, A. (1968) The variation of raindrop size distributions at Locarno. *Proc. Int. Conf. on Cloud Physics*, Toronto, pp. 369–73.

Kaatze, U. (1983) Dielectric effects in aqueous solutions of 1:1, 2:1 and 3:1 valent electrolytes: kinetic depolarization, saturation and solvent relaxation. *Z. Phys. Chem. NF*, **135**, 51–75.

Keuken, M.P., Baard, J.H., Möls, J.J. and Slanina, J. (1989) Landelijk Meetnet regenwatersamenstelling in Nederland, Meetresultaten: Regenwater: Den Helder en Bilthoven, I kwartaal 1989, *ECN Memo.*, ECN, Petten.

Klaassen, W. (1988) Radar observations and simulation of the melting layer of precipitation. *J. Atmos. Sci.*, **45**(24), 3741–53.

Klaassen, W. (1989) From snowflake to raindrop: Doppler radar observations and simulations of precipitation. Doctoral Thesis, University of Utrecht.

Knight, C.A. (1979) Observations of the morphology of melting snow. *J. Atmos. Sci.*, **36**, 1123–9.

Lamb, J. (1946) Measurements of the dielectric properties of ice. *Trans. Faraday Soc. A*, **42**, 238–53.

Lamb, J. and Turney, A. (1949) The dielectric properties of ice at 1.25 cm wavelength. *Proc. Phys. Soc. London B*, **62**, 272–3.

Langleben, M.P. (1954) The terminal velocity of snowflakes. *Q. J. R. Meteorol. Soc.*, **80**, 174–81.

Laws, J.O. and Parsons, D.A. (1943) The relation of raindrop size to intensity. *Trans. Am. Geophys. Union*, **24**, 452–60.

List, R.J.R. and Gillespie, C.A. (1976) Evolution of raindrops spectra with collision induced breakup. *J. Atmos. Sci.*, **33**, 2007–13.

Locatelli, J.D. and Hobbs, P.V. (1974) Fall speeds and masses of solid precipitation particles. *J. Geophys. Res.*, **79**, 2185–98.

Maggiori, D., Migliori, P. and Paraboni, A. (1983) Assessment of principal planes orientation in a terrestrial link at 18 GHz during intense rainfall. *Electron. Lett.*, **19**(16), 617–19.

Magono, C. (1954) On the shape of water drops falling in stagnant air. *J. Meteorol.*, **11**, 77–9.

Magono, C. and Nakamura, T. (1965) Aerodynamic studies of falling snowflakes. *J. Meteorol. Soc. Jpn.*, **43**, 139–47.

Maher, B.O., Murphy, P.J. and Sexton, M.C. (1977) A theoretical model of the effect of wind-gusting on rain induced cross-polarisation. *Ann. Telecommun.*, **32**, 404–8.

Manabe, T., Liebe, H.J. and Hufford, G.A. (1987) Complex permittivity of water between 0 and 30 THz. Conf. Dig. 12th Int. Conf. on Infrared and Millimeter Waves, Lake Buena Vista, December 14–18.

Marshall, J.S. and Palmer, W.M. (1948) The distribution of raindrops with size. *J. Meteorol.*, **5**, 165–6.

Mason, D.J. (1971) *The Physics of Clouds*, Clarendon, Oxford.

Matsuo, T. and Sasyo, Y. (1981) Empirical formula for the melting rate of snowflakes. *J. Meteorol. Soc. Jpn.*, **159**, 1–8.

Mätzler, C. and Wegmüller, U. (1987) Dielectric properties of fresh-water ice at microwave frequencies. *J. Phys. D*, **20**(12), 1623–30.

Maxwell Garnett, J.C. (1904) Colours in metal glasses and in metallic films. *Philos. Trans. R. Soc. London, Ser. A*, **203**, 385–420.

McEwan, N.J., Alves, A.P., Poon, H.W. and Dissanayake, A.W. (1981) OTS propagation measurements during thunderstorms. *Ann. Telecommun.*, **36**, 102–10.

Nakaya, U. (1970) The formation of ice crystals, in *Compendium of Meteorology*, American Meteorological Society, pp. 207–20.

Ohtake, T. (1969) Observations of size distributions of hydrometeors through the melting layer. *J. Atmos. Sci.*, **26**, 545–57.

Park, S.W., Mitchell, J.K. and Bubenzer, G.D. (1983) Rainfall characteristics and their relation to splash erosion. *Trans. ASAE*, **26**(3), 795–804.

Pruppacher, H.R. and Beard, K.V. (1970) A windtunnel investigation of the internal circulation and shape of water drops falling at terminal velocity. *Q. J. R. Meteorol. Soc.*, **96**, 247–56.

Pruppacher, H.R. and Klett, J.D. (1978) *Microphysics of Clouds and Precipitation*, Reidel, Boston, MA.

Pruppacher, H.R. and Pitter, R.L. (1971) A semi-empirical determination of the shape of cloud and raindrops. *J. Atmos. Sci.*, **28**, 86–94.

Ray, P.S. (1972) Broadband complex refractive indices of ice and water. *Appl. Opt.*, **11**, 1836–44.

Rogers, J.S., Johnson, D.C., Jones, D.M.A. and Jones, B.A. (1967) Sources of error in calculating the kinetic energy of rainfall. *J. Soil Water Conserv.*, **22**(4), 140–2.

Saunders, M.J. (1971) Cross-polarisation at 18 and 30 GHz due to rain. *IEEE Trans. Antennas Propag.*, **19**, 273–7.

Ugai, S., Kato, K., Nishijima, M., Kan, T. and Tazaki, K. (1977) Fine structure of rainfall. *Ann. Telecommun.*, **32**(11–12), 422–9.

Ulaby, F.T., Moore, R.K. and Fung, A.K. (1986) *Microwave Remote Sensing Active and Passive*, Vol. III, From Theory to Applications, Artech, Norwood, MA, pp. 2022–5.

Ulbrich, C.W. (1983) Natural variations in the analytical form of the raindrop size distribution. *J. Climatol. Appl. Meteorol.*, **22**, 1764–75.

Von Hippel, A. (1988) The dielectric relaxation spectra of water, ice and aqueous solutions, and their interpretation, parts 1 and 2. *IEEE Trans. Electr. Insul.*, **23**(5), 801–23.

Warren, S.G. (1984) Optical constants of ice from the ultraviolet to the microwave. *Appl. Opt.*, **23**(8), 1206–25.

Waterman, P.C. (1965) Matrix formulation of electromagnetic scattering. *Proc. IEEE*, **53**, 805–12.

Watson, P.A. and Arbabi, M. (1975) Semi-empirical law relating cross-polarisation discrimination to fade depth for rainfall. *Electron. Lett.*, **11**(2), 42–4.

Weast, R.C., Astle, M.J. and Beyer, W.H. (eds) (1987) *CRC Handbook of Chemistry and Physics*, CRC Press, Boca Raton, FL, p. D-167.

de Wolf, D.A., Russchenberg, H.W.J. and Ligthart, L.P. (1990) Effective permittivity of and scattering from wet snow and ice droplets at weather radar wavelengths. *IEEE Trans. Antennas Propag.*, **38**(9), 1317–25.

Wörz, O. and Cole, R.H. (1969) Dielectric properties of ice. *J. Chem. Phys.*, **51**, 1546–51.

Young, K.C. (1975) The evolution of drop spectra due to condensation, coalescence and breakup. *J. Atmos. Sci.*, **32**, 965–73.

Index